Evolutionary Origin of Sensory and Neurosecretory Cell Types

T0188157

Series Preface (Evolutionary Cell Biology)

In recent decades, evolutionary principles have been integrated into biological disciplines such as developmental biology, ecology and genetics. As a result, major new fields emerged, chief among which are Evolutionary Developmental Biology (or Evo-Devo) and Ecological Developmental Biology (or Eco-Devo). Inspired by the integration of knowledge of change over single life spans (ontogenetic history) and change over evolutionary time (phylogenetic history), Evo-Devo produced a unification of developmental and evolutionary biology that generated unanticipated synergies: Molecular biologists employ computational and conceptual tools generated by developmental biologists and by systematists, while evolutionary biologists use detailed analysis of molecules in their studies. These integrations have shifted paradigms and enabled us to answer questions once thought intractable.

Major highlights in the development of modern Evo-Devo are a comparison of the evolutionary behavior of cells, evidenced in Stephen J. Gould's 1979 proposal of changes in the timing of the activity of cells during development — heterochrony — as a major force in evolutionary change, and numerous studies demonstrating how conserved gene families across numerous cell types 'explain" development and evolution. Advances in technology and in instrumentation now allow cell biologists to make ever more detailed observations of the structure of cells and the processes by which cells arise, divide, differentiate and die. In recent years, cell biologists have increasingly asked questions whose answers require insights from evolutionary history. As just one example: How many cell types are there and how are they related? Given this conceptual basis, cell biology — a rich field in biology with history going back centuries — is poised to be reintegrated with evolution to provide a means of organizing and explaining diverse empirical observations and testing fundamental hypotheses about the cellular basis of life. Integrating evolutionary and cellular biology has the potential to generate new theories of cellular function and to create a new field, "*Evolutionary Cell Biology.*"

Mechanistically, cells provide the link between the genotype and the phenotype, both during development and in evolution. Hence the proposal for a series of books under the general theme of "*Evolutionary Cell Biology: Translating Genotypes into Phenotypes*", to document, demonstrate and establish the central role played by cellular mechanisms in in the evolution of all forms of life.

Brian K. Hall and Sally A. Moody

Evolutionary Origin of Sensory and Neurosecretory Cell Types

Vertebrate Cranial Placodes, Vol. 2

Gerhard Schlosser

CRC Press
Taylor & Francis Group
Boca Raton London New York

CRC Press is an imprint of the
Taylor & Francis Group, an **informa** business

First edition published 2021
by CRC Press
6000 Broken Sound Parkway NW, Suite 300, Boca Raton, FL 33487-2742

and by CRC Press

2 Park Square, Milton Park, Abingdon, Oxon, OX14 4RN

© 2021 Taylor & Francis Group, LLC

Reasonable efforts have been made to publish reliable data and information, but the author and publisher cannot assume responsibility for the validity of all materials or the consequences of their use. The authors and publishers have attempted to trace the copyright holders of all material reproduced in this publication and apologize to copyright holders if permission to publish in this form has not been obtained. If any copyright material has not been acknowledged please write and let us know so we may rectify in any future reprint.

Except as permitted under U.S. Copyright Law, no part of this book may be reprinted, reproduced, transmitted, or utilized in any form by any electronic, mechanical, or other means, now known or hereafter invented, including photocopying, microfilming, and recording, or in any information storage or retrieval system, without written permission from the publishers.

For permission to photocopy or use material electronically from this work, access www.copyright.com or contact the Copyright Clearance Center, Inc. (CCC), 222 Rosewood Drive, Danvers, MA 01923, 978-750-8400. For works that are not available on CCC please contact mpkbookspermissions@tandf.co.uk

Trademark Notice: Product or corporate names may be trademarks or registered trademarks and are used only for identification and explanation without intent to infringe.

ISBN: 9780367748524 (hbk)
ISBN: 9780367749804 (pbk)
ISBN: 9781003160625 (ebk)

Typeset in Times
by KnowledgeWorks Global Ltd.

Contents

Preface

Our senses provide us with a richness of experiences that we all too often take for granted. Without our sense organs, our world would be a silent and dark place (not a place at all, really). Yet, we share our sophisticated ears, eyes and noses only with our fellow vertebrates. The senses of their closest living relatives, the tunicates (including sea squirts) and lancelets (or amphioxus), are much simpler. Accordingly, these animals live in a much simpler sensory world. Many other animals get around with an even more basic outfit of sensory cells – isolated cells scattered through their outer tissues – which do not form complex sense organs at all. So how did the vertebrate head become equipped with these complicated new sensory organs? How do they develop in the vertebrate embryo? And how did they evolve from the simpler sensory systems of our ancestors? These are the core questions that this book tries to answer.

Most of the cranial sense organs of vertebrates arise from embryonic structures known as cranial placodes, which in turn arise from a common embryonic precursor in development. In addition to sense organs, cranial placodes also give rise to sensory neurons that transmit the sensory information to the brain as well as to many neurosecretory cells – neuron-like cells that produce hormones (including the cells of the anterior pituitary, the major hormonal control organ of the vertebrate body). Due to the close developmental and evolutionary links between sensory, neuronal and neurosecretory cells derived from cranial placodes, they will be considered jointly in this book. While the photoreceptors of the vertebrate retina and pineal organ are not derived from cranial placodes, they will nevertheless be briefly discussed here. The main reason is the close evolutionary relationship between photoreceptors and other sensory cell types (e.g. mechano- and chemoreceptors) that develop from placodes.

The first volume of this book will focus on the development of sensory and neurosecretory cell types from the cranial placodes in vertebrates. Chapter 1 introduces the vertebrate head with its sense organs and neurosecretory organs. Chapter 2 then provides an overview of the various cranial placodes and their derivatives. Chapter 3 presents evidence that all cranial placodes develop from a common embryonic primordium and discusses how the latter is established in the early embryo. How individual placodes originate from the common primordium is addressed in Chapter 4. Chapters 5 to 8 then review, how the various placodally derived sensory and neurosecretory cell types as well as photoreceptors develop.

The second volume of the book will then consider the evolutionary origin of the sensory and neurosecretory cell types developing from cranial placodes and of photoreceptors. Chapter 1 summarizes our current understanding of vertebrate evolution and presents a brief survey of body plans and embryonic development of their closest relatives (tunicates, amphioxus, hemichordates and echinoderms). Chapter 2 clarifies conceptual issues relating to homology and evolutionary innovation and introduces an evolutionary concept of cell types. Chapters 3 to 5 then compare the sensory and neurosecretory cell types of the vertebrate head with similar cell types in other animals to get insights into their evolutionary origins. While Chapter 3 reviews the evolution of mechano- and chemosensory cells, Chapter 4 considers photosensory cells and Chapter 5 neurosecretory cells. These comparative chapters will show that many sensory cell types have a long evolutionary history. The final chapter, Chapter 6, then addresses the question of how cranial placodes evolved as novel structures in vertebrates by redeploying pre-existing and sometimes evolutionarily ancient cell types.

The book is aimed at a wide audience ranging from graduate students to fellow scientists in the fields of developmental and evolutionary biology. To make it accessible also to newcomers to these fields, I attempted to provide relatively broad background information in the introductory chapters of each volume (Chapter 1 of Volume 1; Chapters 1 and 2 of Volume 2). Because these introductory chapters cover a lot of ground, it was impossible to always provide original references and I have mostly cited reviews there. In the other chapters, I tried to cite relevant original papers in addition to reviews, wherever possible. Nevertheless, I will almost certainly have missed some relevant sources or may have had to limit the number of papers cited due to space constraints. Therefore, I ask all my colleagues, whose relevant papers I left out, to please accept my apologies. Some of the chapters on development in the first volume may be too detailed for those readers with more evolutionary interests. However, I provide summaries of the main points at the end of each major section of each chapter, and it should be possible for those readers to get the main message by reading the summaries and skipping the rest.

I wish to express my gratitude to all my colleagues who have inspired me over the years and without whose ideas and research advances this book could never have been written. I am also very grateful to the following colleagues for providing valuable comments on parts of the book: Eric Bellefroid, Bernd Fritzsch, Benjamin Grothe, Volker Hartenstein, Nick Holland, Anne-Helene Monsoro-Burq, Sally Moody, Seb Shimeld, Andrea Streit, and Günter Wagner. I specifically want to thank my wife, Elke Rink, who read and commented on the entire manuscript, apart from having to bear recurrent spells of absentmindedness over the greater part of two years. Furthermore, I acknowledge my editor, Chuck Crumly, and his team at CRC Press for their support during the publication process. Finally, it must be noted that parts of Chapter 6 in Volume 1 on hair cells and of Chapter 3 in Volume 2 on the evolution of mechanosensory cells have been previously published in a chapter entitled "Evolution of hair cells" in: Fritzsch, B. (Ed.) and Grothe, B. (Volume Editor), *The Senses: A Comprehensive Reference*, Vol. 2. Elsevier, Academic Press 2020, pp. 302–336.

1 The Evolutionary Origin of Vertebrates

Within the large branch of animals known as deuterostomes, only vertebrates possess a complex head equipped with complex paired sense organs such as the eyes, the ears (visible from the outside only in some species by the presence of openings, eardrums, or pinnae), the lateral line system (in fishes and amphibians), and the nose. This is immediately obvious in a comparison of vertebrates (Fig. 1.1) with other deuterostomes (Fig. 1.2). Beneath the surface, these profound differences continue with only vertebrates possessing a cartilaginous or bony skull protecting an enlarged brain and a series of cranial nerves which connect the sensory organs to the brain and the brain to the cranial muscles and glands. Some of these muscles help to draw water and food particles through the mouth into the pharynx and then expel it through perforations in the wall of the pharynx, the pharyngeal slits, to the outside. Comparable complex sense organs, a skull and the muscular ventilation of the pharynx are absent from other deuterostomes indicating that they are evolutionary novelties that arose in the vertebrate lineage. In that sense, vertebrates were proposed to have a "New Head" by Northcutt and Gans (Northcutt and Gans 1983; Gans and Northcutt 1983).

In their "New Head" hypothesis, Northcutt and Gans suggested that the New Head evolved in a series of steps during which filter-feeding ancestors adopted a more active and predatory life-style. Filter-feeding is indeed the predominant feeding mode of other chordates and deuterostomes suggesting that it was the feeding mode of the ancestral chordates (Fig. 1.3). Most deuterostomes use mucus to capture small food particles which are then transported by cilia into the pharynx and onwards into the digestive tract. In chordates, most of this mucus is produced in a special ciliated groove in the ventral pharynx termed the endostyle. The pharynx of tunicates, amphioxus as well as hemichordates, and some extinct groups of echinoderms, is perforated by pharyngeal slits to provide an exit route for the water drawn in during filter feeding. The invention of muscular ventilation of the pharynx allowed to generate larger forces to suck in larger food particles precipitating the transition to a more active and exploratory behavior which in turn favored the evolution of increasing sensory capacities and skeletal protection of sense organs and brain.

Northcutt and Gans further pointed out that many of the evolutionary novelties of the vertebrate head develop from only two embryonic tissues, the neural crest and the cranial placodes (Fig. 1.4) which also originated first in the vertebrate lineage (Northcutt and Gans 1983). The neural crest contributes to the development of the skull and to glial cells and sensory neurons of the cranial nerves, while cranial placodes also contribute sensory neurons to the cranial nerves and in addition give rise to many cranial sense organs and to the neurosecretory cells of the anterior pituitary gland.

FIGURE 1.1 (**A**) Hagfish (*Myxine* sp.). (**B**) Sea lamprey (*Petromyzon marinus*). (**C**) Blacktip reef shark (*Carcharhinus melanopterus*). (**D**) Short-nose Sturgeon (*Acipenser brevirostrum*). (**E**) Arowana (*Scleropages legendrei*). (**F**) White margined unicornfish (*Naso annulatus*). (**G**) Broadbarred firefish (*Pterois antennata*). (**H**) Fire salamander (*Salamandra salamandra*). (**I**) Frog (*Rana* sp.). (**J**) Large Scaled Forest Lizard (*Calotes grandisqamis*). (**K**) Blue-legged Chameleon (*Calumma crypticum*). (**L**) Albino Burmese python (*Python bivittatus*). (**M**) Red-tailed green rat snake (Gonyosoma oxycephalum). (**N**) Greylag goose (*Ansera anser*). (**O**) Bald eagle (*Haliaeetus leucocephalus*). (**P**) Male mongoose lemur (*Eulemur mongoz*). (**Q**) *Dermanura* sp. (**R**) Jaguar (*Panthera onca*). (Photos by [A] NOAA Okeanos Explorer Program; [B] P. Lameiro.; [C] Luc Viatour; [D] US Fish and Wildlife service; [E] Marcel Burkhard (CC BY-SA 3.0); [F] Bernard Spragg; [G] H. Zell; [H] Didier Descouens; [I] Luk; [J] Babujayan; [K] Frank Vassen (cc-by-2.0); [L] Guy Lejeune; [M] Bernard Dupont; [N] Charlesjsharp; [O] US Fish and Wildlife Service/Mike Lockhart; [P] IParjan; [Q] Guilherme Garbino; [R] Cburnett.)

FIGURE 1.2 A collection of non-vertebrate deuterostomes. A head with complex sense organs is lacking. (**A**) Sea star (*Asterias rubens*). (**B**) *Saccoglossus kowalevskii*. (**C**) *Ciona savignyi*. (**D**) Amphioxus (*Branchiostoma floridae*). ([A] Photo by Hans Hillewaert; [B] reprinted with permission from Gerhart, Lowe, and Kirschner 2005; [C] photo by Steve Lonhart/NOAA MBNMS; [D] Photo kindly provided by David Jandzik.)

In the first volume (Schlosser 2021), I introduced the specialized sensory (organs olfactory, ear, lateral line), neurosecretory organs (anterior pituitary) and cranial ganglia of the vertebrate head that arise from cranial placodes. I then reviewed how sensory and neurosecretory cell types arise from cranial placodes during vertebrate embryonic development. I also briefly sketched the development of photoreceptors, even though these do not arise from cranial placodes, because I hope to show that

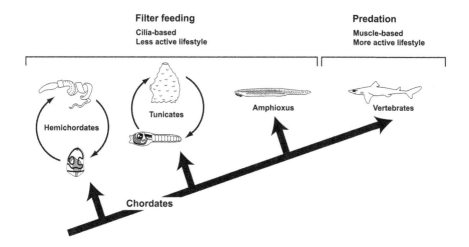

FIGURE 1.3 Origin of vertebrates from suspension feeding chordates. Larval and adult stages are shown for hemichordates and tunicates. See text for detailed explanation. (Silhouettes reprinted with permission from Northcutt 2005.)

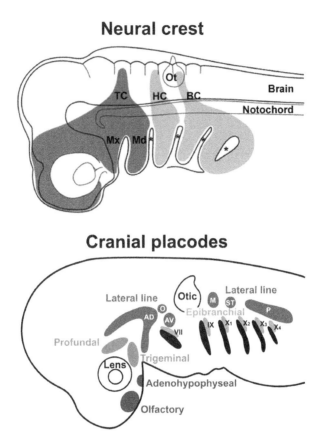

FIGURE 1.4 Neural crest and cranial placodes. Upper panel shows the cranial neural crest streams in a generalized vertebrate embryo. Asterisks mark ectodermal pharyngeal grooves, which will fuse with the underlying pharyngeal pouches to form the pharyngeal slits. BC, branchial crest; HC, hyoid crest; Md, mandible (lower jaw); Mx, maxilla (upper jaw); Ot, otic vesicle; TC, trigeminal crest. Lower panel shows cranial placodes in a generalized vertebrate embryo. These include the adenohypophyseal, olfactory, lens and otic placode, the profundal and trigeminal placodes, and several epibranchial and lateral line placodes. Epibranchial placodes are closely associated with the pharyngeal pouches (shown in black). AD, anterodorsal lateral line placode; AV, anteroventral lateral line placode; O, otic lateral line placode; M, middle lateral line placode; P, posterior lateral line placode; ST, supratemporal lateral line placode; VII, epibranchial placode of facial nerve; IX, epibranchial placode of glossopharyngeal nerve; X_{1-4}, epibranchial placodes of vagal nerve. (Upper panel: Reprinted with permission from Kuratani et al. 2012; Lower panel: Redrawn and modified from Northcutt 1997.)

they are evolutionarily related to other, placode-derived sensory cells. In the second volume, I will now attempt to trace the evolutionary history of cranial placodes and their derivative cell types. In this first chapter, I will place vertebrates into their proper phylogenetic context. In Chapter 2, I will then make some general remarks on cell types, and discuss how we can recognize homology and novelty in cell type evolution. In Chapters 3–5, I compare the sensory and neurosecretory cell types of the vertebrate head with similar cell types in other animals to get insights into their evolutionary origins. In the final chapter (Chapter 6), I will then discuss how cranial placodes evolved as novel structures in vertebrates by redeploying pre-existing and sometimes evolutionarily ancient cell types.

To understand how the vertebrate head originated during evolution and how its complex sense organs evolved, we first need to know more about the pedigree of the vertebrates. What did their ancestors look like? How did they develop and live and how did their development and lifestyle change over time? Answering these questions is by no means trivial. Obviously, we have no time machine that would allow us to travel back in time to study the extinct ancestors of the vertebrates living today. And even if this were possible, we still wouldn't know where to position them on the tree of life. We, thus, must try to reconstruct the evolutionary history and infer the bauplan and life history of ancestors by comparisons with other animals living today or with the fossil remnants of animals of the past. In the remainder of this chapter, I will first summarize our current understanding of the phylogenetic relationships of vertebrates with other deuterostomes and discuss how deuterostomes are phylogenetically related to other metazoans. I will then introduce the vertebrates and their fossil relatives and present a brief survey of other deuterostome groups as the closest living relatives of vertebrates. Finally, I will review different scenarios on how vertebrates and other deuterostomes evolved from their common ancestors.

1.1 VERTEBRATES AND THE TREE OF LIFE

The phylogenetic relationships between vertebrates and other chordates and between chordates and other animals has been controversial for a long time. While not all of these controversies are resolved yet, modern phylogenetic methods provide an increasingly well supported picture of chordate phylogenetic relationships.

1.1.1 A BRIEF PRIMER ON PHYLOGENETIC SYSTEMATICS

The methodological principles underlying these phylogenetic methods were first formulated by Willi Hennig in his influential book on phylogenetic systematics (Hennig 1966). A basic assumption of this methodology is that the diversification of taxa (species or other units of classification; singular: taxon) in evolution occurs typically by dichotomous branching, each branching point giving rise to two side branches termed sister groups. This is represented in a branching diagram called a cladogram. Phylogenetic systematics aims to classify taxa in a way that reflects this branching pattern by constructing a hierarchy of monophyletic groups. A monophyletic group (or clade) contains all the side branches emanating from a branching point, i.e. all descendants of a common ancestor (Fig. 1.5A). Birds, for example, are a monophyletic group. All birds arose from a last common ancestor that only gave rise to birds. In contrast, a polyphyletic group would unite two branches arising from different ancestors (such as birds and bats). Finally, a paraphyletic group contains some but not all descendants of a common ancestor. "Reptiles", for example, are a paraphyletic group because not only "reptiles" but also birds and mammals descended from the last common ancestor of "reptiles". Only monophyletic groups but not poly- or

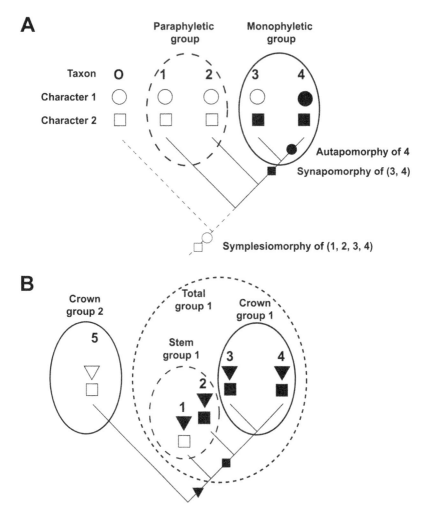

FIGURE 1.5 Terminology of phylogenetic systematic. (**A**) Cladogram depicting the phylogenetic relationships for four taxa (1–4) and an outgroup (O). Different character states for two characters (circle, square) are indicated by empty or filled symbols. (**B**) Illustration of the concepts of crown group, stem group, and total group. Taxa 1 and 2 are extinct; taxa 3–5 are extant. Distribution and inferred origin of character states for the various taxa are indicated. See text for detailed explanation.

paraphyletic groups are acceptable in phylogenetic systematics; therefore, quotation marks are often used when referring to groups that are paraphyletic such as "reptiles".

Monophyletic groups can be recognized because characters that originated in the branch leading to their last common ancestor are inherited by all these taxa and only by those (e.g. the black square for taxa 3 and 4 in Fig. 1.5A). Such shared-derived characters or synapomorphies are the only informative characters for establishing phylogenetic relationships. In contrast, there are shared primitive characters or symplesiomorphies. These are shared not only between the taxa under consideration (e.g. taxa 1 and 2 in Fig. 1.5A), but also with more distantly related taxa (e.g. taxon O), indicating that they did not originate in the last common ancestor of the taxa considered but in more distant ancestors. Therefore, symplesiomorphies do not allow to identify monophyletic groups. For example, taxa 1 and 2 in Fig. 1.5A are not a monophyletic group since they only share characters with each other that are also present in O and, thus, likely originated already in distant ancestors preceding the last common ancestor of 1 and 2. Characters that originated in only one of the taxa to be classified, so-called autapomorphies, likewise do not provide any useful information for recognizing relationships between taxa (e.g. the presence of the black circle in taxon 4 in Fig. 1.5A does not provide any information about its relation to taxon 3).

It is not possible to decide simply by inspection of character states (such as black or white square in Fig. 1.5A) which one of those are primitive and which ones derived within a group of taxa to be classified (such as taxa 1–4 in Fig. 1.5A). Information on character polarity has to come from comparison with another group, the "outgroup", independently known to be a more distant relative (e.g. O in Fig. 1.5A). For example, if we want to establish phylogenetic relationships between vertebrates, tunicates, and amphioxus, our outgroup will have to be chosen from taxa clearly not belonging to the chordates such as hemichordates.

While paraphyletic groups are not accepted as units of classification in phylogenetic systematics, it can be useful to refer to such groups when considering the fossil relatives of living taxa (Fig. 1.5B). A monophyletic group defined by the relationship among extant taxa is known as a crown group. It includes all descendants of the last common ancestor of the living members of a group (note that some of these descendants may be now extinct). Between a crown group (e.g. the birds) and the last common ancestor with its sister crown group (e.g. the crocodiles) there is often a lineage of extinct sister taxa, known as the stem group, that is a paraphyletic collection of taxa (e.g. various lineages of dinosaurs and pterosaurs). Crown and stem group together are known as the total group. Fossils of stem group taxa are particularly interesting because they allow insights into the sequence, in which the shared-derived characters defining the crown group were acquired in evolution (Fig. 1.5B).

Given the principles of phylogenetic systematics outlined in the preceding paragraphs, how can unknown phylogenetic relationships be inferred by comparing character distributions between taxa? Let us consider a simple situation with three taxa and one outgroup (1–3 and O, respectively in Fig. 1.6). There are only two possible phylogenetic relationships in this case: (1) taxon 2 and 3 are most closely related and form the sister group to taxon 1 (tree 1 in Fig. 1.6) or (2) taxon 1 and 2 are most closely related and form the sister group to taxon 3 (tree 2 in Fig. 1.6).

If taxa 2 and 3 share a character state B (symbolized by the black square) that is not shared by taxon 1 which instead displays a different character state W (symbolized by the white square), comparison to the outgroup will allow us to infer, which of the character states is likely to be primitive and was present in the last common ancestor of taxa 1, 2, 3, and which one is likely to be derived. Assuming that B is thus identified as the derived character state (Fig. 1.6A), the first phylogenetic tree, in

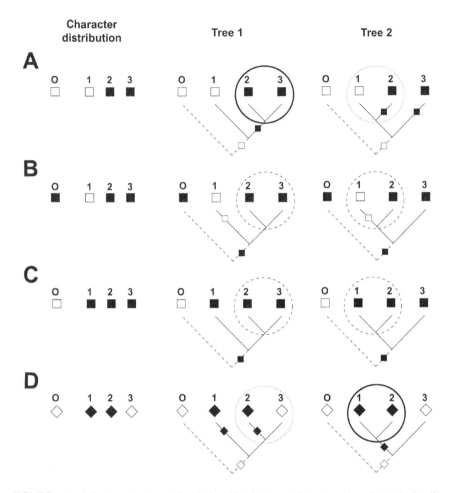

FIGURE 1.6 Inferring phylogenetic relationships between three taxa. In column 1, the distribution of character states (black or white) for one character (square) is shown for three taxa (1–3) and one outgroup (O). Columns 2 and 3 depict the inferred character state changes for the two possible phylogenetic trees, in which either taxon 2 and 3 are more closely related (tree 1) or taxon 1 and 2 are more closely related (tree 2). If one grouping of taxa is more likely than the alternative because it requires fewer changes of character states this is indicated by bold vs stippled outlines. If both alternative groupings of taxa are equally likely this is indicated by broken outlines. (**A**) Taxon 2 and 3 share a derived character state that is absent from the outgroup and is a candidate for a synapomorphy of 2 and 3 (tree 1) but may also have evolved convergently (tree 2). (**B**) Taxon 2 and 3 share a primitive character state that is present in the outgroup. This distribution is equally likely explained by tree 1 and 2. (**C**) All taxa 1, 2, and 3 share the same character state. This distribution is also equally likely explained by tree 1 and 2. (**D**) Taxon 1 and 2 share a derived character state that is absent from the outgroup. The distribution of these character states favors a different tree than the ones for the character in **A**.

which B is interpreted as shared-derived character of the monophyletic group comprising taxon 2 and 3 requires only one change of character state, whereas the second phylogenetic tree, in which B is interpreted as being independently (convergently) acquired in taxon 2 and 3 requires two such changes. Tree 1 therefore is more parsimonious (requires fewer steps) and offers a more likely explanation of the observed character distribution than tree 2.

The situation is different, if B is identified as the primitive character state because it is shared with the outgroup and thus was likely present in the last common ancestor of taxa 1, 2, and 3 (Fig. 1.6B). B in this case is a shared primitive character of taxon 2 and 3 in both trees and in each case one change of character state is required to explain the different character state in taxon 1. Both trees therefore require an equal number of steps and the character distribution does not favor one tree over the other. Similarly, characters that are shared between all three taxa are not informative for inferring phylogenetic relationships (Fig. 1.6C).

When looking at additional characters, the distribution of character states between taxa may either be congruent with the first character and support the same tree or may be at odds with the first character and support a different tree (e.g. the square in Fig. 1.6A versus the diamond in Fig. 1.6D). According to the principle of parsimony, with multiple characters analyzed, the tree that is supported by the majority of characters and, thus, requires overall the smallest number of changes is considered the best supported tree.

While the principle of parsimony works well in inferring phylogenetic relationships when the rates of evolutionary change are identical in all branches of the phylogenetic tree, it can lead to wrong conclusions if rates are significantly different between branches. More sophisticated tools to infer phylogenetic trees such as Maximum likelihood allow taking different evolutionary rates into account. Additional computational methods have been developed to correct for artifacts such as "long branch attraction", the tendency of fast evolving groups to cluster together in phylogenetic trees or for the heterogeneity of nucleotide or amino acid substitution events in different lineages (Telford, Budd, and Philippe 2015). Applicable to both morphological and molecular data, the increasingly sophisticated methods of phylogenetics have led to an increasingly detailed and well supported view of animal phylogeny unbiased by preconceived ideas on the relative importance of different characters. With more and more genomic data becoming available, such phylogenetic trees are now often reconstructed from sequence comparisons between a large number of genes in what is known as "phylogenomics".

1.1.2 The Phylogenetic Relationships of Vertebrates

Because of its overall resemblance to vertebrates, amphioxus was long considered to be the closest living relative of vertebrates or even a degenerate vertebrate. In contrast, tunicates were long thought to be mollusks and only after the discovery of the tunicate larva by Kowalevsky were they recognized to belong to the chordate clade, together with amphioxus and vertebrates (Kowalevsky 1866). Due to their highly divergent adult body plan, tunicates were then considered to be only a distant relative of vertebrates, sister group to the amphioxus-vertebrate clade (Gee 1996, 2018;

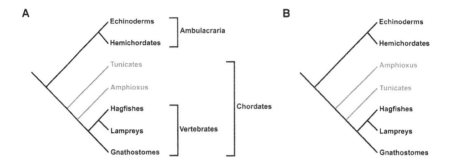

FIGURE 1.7 Two different views of deuterostome phylogeny. The traditional view suggested that amphioxus is the closest living relative of chordates (**A**), but modern phylogenomic analyses indicate that tunicates are more closely related to vertebrates (**B**). (Redrawn from Schubert et al. 2006.)

Swalla and Smith 2008; Holland 2015) (Fig. 1.7A). However, such reasoning based on overall similarities can be misleading because many of the similarities between amphioxus and vertebrates (such as the pharyngeal apparatus, notochord, postanal tail, and segmented muscle blocks) may be shared primitive characters of chordates rather than shared-derived characters of an amphioxus-vertebrate clade (Holland et al. 2008; Putnam et al. 2008; Holland 2015). Conversely, many of the specializations of tunicates were specifically acquired in the fast evolving tunicate lineage while some of the primitive chordate traits and, more importantly, some of their shared-derived characters with vertebrates were modified or lost obscuring the phylogenetic relationships of tunicates (Putnam et al. 2008; Paps, Holland, and Shimeld 2012; Holland 2015).

Recent phylogenomic analyses have confirmed that chordates are a monophyletic group but have overturned our traditional view of chordate relationships in providing strong support for tunicates being the sister group of vertebrates with amphioxus being more distantly related (Bourlat et al. 2006; Delsuc et al. 2006) (Figs. 1.7B, 1.8). The tunicate-vertebrate clade has been described as "Olfactores" (Jefferies 1991), but I will avoid this misleading term here, since no homolog of the vertebrate olfactory organ is present in tunicates (see Chapter 3). Phylogenomic analyses together with morphological evidence from shared-derived larval characters also support the grouping of echinoderms and hemichordates into a monophyletic clade, the ambulacrarians, and confirm that these are the sister group of chordates within the deuterostomes (Fig. 2.3) (Bourlat et al. 2006; Swalla and Smith 2008; Philippe et al. 2011).

Several other phyla that were traditionally often affiliated with the deuterostomes, in particular the lophophorates (ectoprocts, phoronids, and brachiopods) as well as the chaetognaths are now firmly placed in the sister group to deuterostomes, the protostomes (Fig. 1.8) (Halanych et al. 1995; Matus et al. 2006). However, some phylogenomic analyses suggest that a group of worms, the so-called xenacoelomorphs, are the sister group of the ambulacrarian-chordate clade within the deuterostomes (Philippe et al. 2011). Xenacoelomorphs are relatively simple animals comprising the enigmatic *Xenoturbella* and the acoelomorph flatworms and are characterized by the absence of coeloms and a blind ending gut. Their position within the deuterostomes is still contested since other studies position them much more basally in the animal tree as a sister group to the combined protostome and deuterostome clade (Fig. 1.8) (Ruiz-Trillo et al. 1999; Cannon et al. 2016; Rouse et al. 2016).

Since the 1990's molecularly based phylogenies, initially based on the genes encoding ribosomal RNA and later on phylogenomic analyses, have also overturned our view of the phylogenetic relationships among other metazoans (animals) (Fig. 1.8).

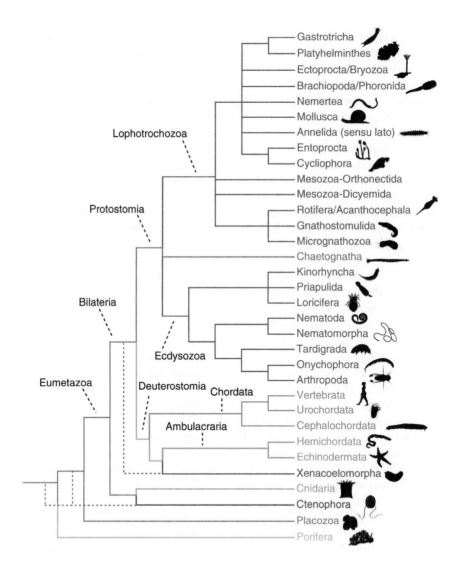

FIGURE 1.8 Consensus tree of metazoan phylogeny based on phylogenomic data. The position of xenacoelomorphs and ctenophores is still contentious (hatched lines). (Reprinted with permission from Telford et al. 2015.)

Some widely accepted groups such as the "Articulata", uniting the segmented annelids and arthropods were shown to be invalid, whereas arthropods and nematodes were found to be surprisingly closely related (Halanych 2004). Although controversies about the placement of some groups remain, the main branches of this new metazoan tree of life appear to be increasingly consolidated. The sister group to the deuterostomes, the protostomes, comprise two major groups called the ecdysozoans and the lophotrochozoans (Fig. 1.8) (Halanych et al. 1995; Aguinaldo et al. 1997; Halanych 2004; Philippe, Lartillot, and Brinkmann 2005; Telford et al. 2015). The ecdysozoans include the arthropods, nematodes, and a number of smaller and less well known phyla. Although morphologically very diverse, they are distinguished by a secreted cuticle that is replaced by molting (ecdysis). The lophotrochozoans include the mollusks, annelids, platyhelminths (flatworms), rotifers, the lophophorates, and a number of other phyla. They are characterized by spiral cleavage (secondarily modified in many lineages) and the presence of either a trochophora larva – a peculiar type of larva with two ciliary bands used in locomotion and feeding – or lophophores – ciliated feeding arms in the adult.

Protostomes and deuterostomes together form the Bilateria defined by their bilateral symmetry (some phylogenetic analyses, as mentioned above, also place the xenacoelomorphs at the base of bilaterians as a sister group to the protostome-deuterostome clade). The sister group to the bilaterians comprises the radially symmetric cnidarians (including sea anemones and jellyfish), which are diploblastic (i.e. they have only two germ layers, ectoderm and endoderm), have no through gut and use very peculiar stinging cells or cnidocysts to catch prey (Fig. 1.8).

The phylogenetic position of another phylum of diploblastic animals, the ctenophores is still controversial. Traditionally, they have been thought to be closely related to cnidarians forming a monophyletic group called the coelenterates which together with their sister group, the bilaterians, form the Eumetazoa (Fig. 1.8). In this view, the poriferans (sponges), which lack true tissues are the most basally branching metazoan lineage, sister group to a clade comprising the very simple placozoans and the eumetazoans (Fig. 1.8). However, recent phylogenomic studies suggest that ctenophores rather than sponges are the most basal metazoan clade (Ryan et al. 2013; Moroz et al. 2014). If true, sponges and placozoans may have been secondarily simplified or some of the cell types found in ctenophores and eumetazoans such as neurons may have independently evolved in both lineages. However, the validity of this view has been called into question for various reasons including the fast evolutionary rate of ctenophores, which make them susceptible to long branch attraction artifacts and the failure to account for heterogeneity of amino acid composition across sites and lineages in previous studies (Telford et al. 2015; Simion et al. 2017; Feuda et al. 2017).

Less controversially, the choanoflagellates are now widely accepted as the unicellular sister group to all multicellular animals (metazoans) (Sebé-Pedrós, Degnan, and Ruiz-Trillo 2017). Cell types with a single cilium surrounded by a collar of microvilli, which strongly resemble choanoflagellates, are found in sponges (choanocytes) and cell types representing variations on this theme, including many sensory cells have been described from several tissues in eumetazoans.

1.2 VERTEBRATES AND FELLOW DEUTEROSTOMES

To understand the evolutionary origin of vertebrates, we'll have to have a closer look at their position in the phylogenetic tree and how they are distinguished from their relatives, the other chordates and deuterostomes. Major defining characters of the chordates include the notochord, a dorsal hollow nerve tube, the endostyle (a ventral ciliated groove in the pharynx that produces mucus for feeding) and a post-anal tail, while they share gill slits and similarities in early development (with the blastopore developing from the anus) with other deuterostomes (Kardong 2018; Gee 2018). In distinction to other chordates, vertebrates also have a vertebral column and a complex head (see Chapter 1 in Schlosser 2021). The latter is equipped with new skeletal and sensory structures derived from neural crest and cranial placodes, with a pharynx ventilated by muscles rather than cilia and with an elaborated forebrain (including the new telencephalon). Vertebrates also have a more complex genome than other chordates due to two rounds (2R) of whole genome duplication, one in the stem lineage of vertebrates and a second one in the stem lineage of gnathostomes (Kasahara 2007; Van de Peer, Maere, and Meyer 2009). A third round of whole genome duplication happened in teleosts (Meyer and Van de Peer 2005). In the following sections, I will first introduce vertebrates and their fossil relatives followed by their fellow chordates (tunicates, amphioxus) and other deuterostomes (hemichordates, echinoderms).

1.2.1 VERTEBRATES AND THEIR FOSSIL RELATIVES

Most living vertebrates have jaws and two paired appendages and belong to the jawed vertebrates or gnathostomes. Only hagfishes and lampreys do not belong to the gnathostomes and lack jaws and appendages (Fig. 1.9). Hagfishes are deep-sea creatures, which live on the carcasses of whales and other large animals, whereas lampreys are blood sucking parasites of fishes in marine and freshwater environments. Both groups lack bones and other mineralized tissues like teeth. Because lampreys but not hagfishes have cartilaginous vertebrae, lampreys were long considered to be the sister group of gnathostomes, both groups together comprising the vertebrates. Hagfishes were thought to be the sister group of vertebrates with both groups together known as craniates (e.g. Janvier 1996). However, vertebra-like skeletal elements have now been described in hagfishes (Ota et al. 2011) and recent molecular phylogenetic studies support the grouping of lampreys and hagfishes as cyclostomes (Kuraku et al. 1999; Furlong and Holland 2002; Takezaki et al. 2003; Heimberg et al. 2010). Many more jawless fishes are known from the fossil record. These are collectively known as "ostracoderms" and form the paraphyletic stem group of gnathostomes (Fig. 1.9). They were small forms heavily armored with a bony skeleton. Bone and other mineralized tissues like dentine and enamel originated probably first in this group of jawless fishes and are primitively absent in cyclostomes (Janvier 1996, 2015; Donoghue and Purnell 2005).

Mineralized tissues have also been found in another group of fossil vertebrates, the enigmatic conodont animals (Donoghue, Forey, and Aldridge 2000). Conodonts were long known as tooth-like microfossils and were widely used to date the age of

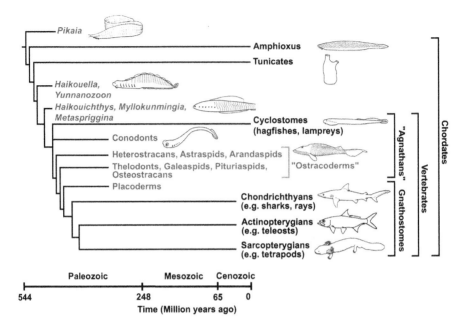

FIGURE 1.9 Chordate phylogeny with some fossil taxa. Fossil taxa are shown in grey. Paraphyletic groups are indicated in quotation marks. See text for details. (Redrawn and modified from Schubert et al. 2006; Cartoons reprinted with permission from Holland 2005; Northcutt 2008; Mallatt and Holland 2013; Janvier 2015.)

rocks although their phylogenetic relationships were unclear. Only in the 1980s were conodonts recognized as part of the feeding apparatus in soft body fossils (Briggs, Clarkson, and Aldridge 1983; Donoghue et al. 2000). These also featured structures interpreted as large paired eyes, v-shaped muscle blocks in the trunk and a notochord suggesting an affinity of conodont animals to vertebrates (Fig. 1.9). However, it is still controversial whether conodont animals (which are now often referred to as "conodonts" and their hard tissues as "conodont elements") are stem gnathostomes or stem cyclostomes and other hypotheses have been advocated as well (Aldridge and Purnell 1996). It is also still debated whether the mineralized tissues of conodonts are homologous to the dentine or enamel of gnathostome teeth although a recent study suggests that the mineralized tissues of conodonts evolved convergently (Murdock et al. 2013). Be that as it may, neither "ostracoderms" nor conodonts are likely to shed much light on how the vertebrate head evolved in a stepwise fashion since they are nested within crown group vertebrates which already have a fully-fledged "New Head".

To get insights into how the vertebrate new head was assembled, we have to focus instead on fossils of stem group vertebrates which branch off the vertebrate lineage after it separated from the last common ancestor with tunicates (Fig. 1.9). Because these stem group vertebrates did not have mineralized tissues, they are only rarely preserved in the fossil record under circumstances that allow the formation of soft-bodied fossils. The richest collections of such soft bodied fossils have been found in two Cambrian deposits, the Canadian Burgess shale (510 million years old), and the Chinese Chengjiang formation (535 million years old) (reviewed in Janvier 2015; Gee, 2018).

However, the structures preserved in these soft body fossils are often open to several different interpretations and their phylogenetic placement is still debated (Donoghue and Purnell 2009; Janvier 2015). Studies on the decay of amphioxus and larval lampreys have further suggested that their more derived characters (shared only among amphioxus or vertebrates, respectively) decay first, whereas more primitive characters (shared among all chordates, e.g. notochord and segmented muscle blocks) decay last (Sansom, Gabbott, and Purnell 2010). This will lead to a bias to place fossils of related animals that were subject to similar decay processes not at their true phylogenetic position but closer to the stem.

Probably most closely related to crown group vertebrates are the Myllokunmingids including *Haikouichthys* and *Myllokunmingia* from the Burgess shale and *Metaspriggina* from the Chengjiang formation (Shu et al. 1999; Shu, Morris, Han et al. 2003; Morris and Caron 2014). These have paired eyes, v-shaped muscle blocks, probably paired olfactory organs and otic capsules, and upper and lower pharyngeal cartilages, whereas a cartilaginous braincase was not well developed or absent. This suggests a placement of these fishlike fossils in the stem lineage of vertebrates (Fig. 1.9). More controversial is the placement of the Yunnanozoans, including *Yunnanozoon* and *Haikouella* from Chengjiang (Fig. 1.9) (Chen et al. 1995; Shu, Zhang, and Chen 1996; Mallatt and Chen 2003; Shu, Morris, Zhang et al. 2003; Cong et al. 2014). Depending on how the preserved structures are interpreted, these have either been considered to be stem vertebrates with paired eyes and pharyngeal cartilages (Mallatt and Chen 2003) or stem deuterostomes related to hemichordates or vetulicolians, another enigmatic fossil group (Shu et al. 1996; Shu, Morris, Zhang et al. 2003). Similar controversy surrounds the famous *Pikaia* from the Burgess Shale (Fig. 1.9). Although widely regarded as a stem group chordate only distantly related to vertebrates, a recent detailed description has revealed many peculiar characters that make a precise placement within the chordates difficult (Morris and Caron 2012; Mallatt and Holland 2013). No brain, sensory organs or skull-like structures have been identified in *Pikaia*.

In summary, the available soft body fossils (of the "rotten squished slug" type), exciting as they are, are still shrouded in mystery and offer only limited information on the evolutionary origin of vertebrates. We, therefore, have to turn to comparisons with other living chordate and deuterostome taxa to better understand vertebrate origins.

1.2.2 TUNICATES

Tunicates are marine animals which get their name from the tunic an extracellular body cover secreted by the epidermis and acting as an exoskeleton (reviewed in Satoh 2009; Lemaire and Piette 2015; Holland 2016a). The tunic is composed of tunicin, a polysaccharide similar to cellulose. The gene encoding the cellulose synthase-like enzyme required for tunicin synthesis is not found in other animals and was most likely acquired by horizontal gene transfer from bacteria (Matthysse et al. 2004). Tunicates are also known as urochordates (tail chordates) due to their tailed, tadpole-like larva exhibiting many of the typical chordate characters such as notochord, dorsal neural tube, and tail.

There are three classes of tunicates, ascidians, thaliaceans, and appendicularians (or larvaceans). The majority of tunicates belong to the ascidians with over 2000 known species, while thaliaceans and appendicularians form smaller groups with a few dozen species each (Satoh 2009). Ascidians are solitary (e.g. *Ciona*, *Halocynthia*) or colonial (e.g. *Botryllus*) sessile filter feeders as adults which use ciliary motion to filter large volumes of water through their pharyngeal basket (Figs. 1.2C, 1.10A). Water is drawn into the pharyngeal basket through the oral siphon, passes through the pharyngeal slits into another space, the atrium, and is expelled through the atrial siphon. Mucus produced in the endostyle is transported by cilia across the pharyngeal basket toward the gut, trapping food particles along the way. Muscles encircling the pharyngeal basket allow quick contraction of the animals when disturbed resulting in expulsion of water and earning them the name "sea squirts". In the thaliaceans, oral and atrial siphons have moved to opposite sides of the bodies allowing them to move through open waters using ciliary or muscle powered jet propulsion while filtering water for food (Fig. 1.10B).

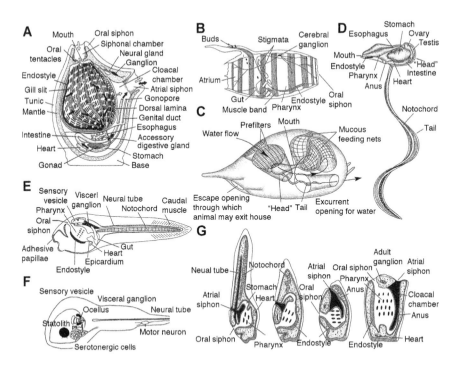

FIGURE 1.10 Tunicates. (**A–D**) Body plans of various tunicates. **A**: Adult ascidian. **B**: Adult thaliacean (*Doliolum*). (**C**) House of an appendicularian. (**D**) Adult appendicularian (*Oikopleura*). (**E**) Ascidian tadpole larva. (**F**) Neural tube of ascidian larva. (**G**) Metamorphosis in ascidians. ([A–E,G] Reprinted with permission from Brusca and Brusca 2003; [F] Reprinted with permission from Stach 2005.)

Thaliaceans comprise the solitary salps and dolioids and the colonial pyrosomes (Piette and Lemaire 2015). The solitary appendicularians or larvaceans, finally, receive their names from maintaining a larva-like body plan with a tail as an adult (Nishino and Satoh 2001; Nishida 2008). In contrast to ascidians and thaliaceans, they never develop a pharyngeal basket for filter feeding (Fig. 1.10C, D). Instead, they produce a secreted covering known as "house" with mesh-like perforations and use their tail to create water current through these meshes. Only very small plankton is permitted through their perforations and these are caught in a secreted mucous feeding net. When meshes are clogged, the house is discarded and a new house made, which happens several times a day. Composed of cellulose fibers and oikosin proteins (only found in appendicularians), the house probably evolved by modification of the tunic (Sagane et al. 2011; Hosp et al. 2012).

The phylogenetic relationships between these different groups are still debated but molecular data provide increasing support for appendicularians being the most basally branching group, sister group of all other tunicates (Wada 1998; Swalla et al. 2000; Delsuc et al. 2018). This is also supported by the structure of gametes since among tunicates only appendicularians have unmodified sperm cells (with an acrosome aiding in fertilization) and eggs without enveloping test cells resembling other deuterostomes (Holland, Gorsky, and Fenaux 1988). Thaliaceans were found nested within the ascidians implying that the latter are a paraphyletic group (i.e. a group that does not contain all descendants from the last common ancestor of its members) and that the thaliacean body plan and life style evolved from a particular group of ascidians (Piette and Lemaire 2015).

Because ascidians are by far the best studied tunicates, I will here focus mostly on them with some brief digression on the other groups in relation to reproductive strategies (reviewed in Satoh 1994; Lemaire, Smith, and Nishida 2008; Sasakura et al. 2012; Kourakis and Smith 2015; Sasakura and Hozumi 2018). Most ascidians start out as a short-lived and non-feeding tadpole larva equipped with a notochord, a dorsal neural tube and a muscular tail (Fig. 1.10E). However, some ascidians have direct development and skip the larval stage (Jeffery 2001). In contrast to other chordates, the tail muscles of the larva are not organized in segmental blocks. The neural tube is enlarged anteriorly to form a sensory vesicle containing a photoreceptive ocellus and a mechanoreceptive otolith (Fig. 1.10F). This is followed by the constricted "neck" region and the visceral ganglion with motor neurons. Based on gene expression domains, the sensory vesicle (expressing Otx) corresponds to the vertebrate fore- and midbrain and the visceral ganglion (expressing Hox genes) to the vertebrate hindbrain (Holland et al. 2013). The motor neurons of the visceral ganglion send their axons down the tail nerve cord, which does not contain any neuronal cell bodies. The sensory vesicle is connected via the so-called neurohypophyseal duct to the oral siphon primordium (further discussed in Chapter 3). The latter is an invagination of the outer ectoderm (also known as stomodeum or buccal cavity) immediately rostral to the oral siphon neural tube and will form the oral siphon and mouth opening in the adult. Further posterior, paired or unpaired atrial siphon primordia (depending on species) will invaginate in older larvae. These will expand during metamorphosis to form the atrium around the pharyngeal basket and the atrial siphon. Several sensory cells allow the larva to probe the environment and find a suitable substrate for settlement. These include various sensory cells in the anterior adhesive papillae (or palps), the rostral ectoderm, and the tail and will be introduced in more detail later in the book (Chapter 3).

Once a suitable substrate is found, larvae attach with their adhesive papillae and enter metamorphosis, during which the larval tail and notochord degenerate. At the same time, the pharynx greatly expands and develops many small stigmata by subdivision of the pharyngeal slits, thereby forming the pharyngeal basket used for filter feeding in the adult (Fig. 1.10G). Oral and atrial siphons also develop during metamorphosis. Moreover, the larval brain gives rise to migratory cells that form the neural complex (a small cerebral ganglion, a collection of neurons known as dorsal strand, and the so-called neural gland) of the adult central nervous system (CNS) (Manni et al. 1999; Manni, Lane et al. 2004; Horie et al. 2011). The function of the neural gland, which connects to the oral siphon with a ciliated duct and funnel, is still enigmatic. It has been proposed to be either an endocrine gland or to be involved in osmoregulation and regulation of blood volume (Ruppert 1990; Joly et al. 2007). Sensory cells of the adult are largely concentrated in the oral and atrial siphons (see Chapter 3). Apart from the photoreceptors of the oral siphon pigment organs and some primary sensory cells, the oral siphon contains the mechanoreceptive secondary sensory cells of the coronal organ, which lines the oral tentacles (Burighel et al. 2003; Manni, Caicci et al. 2004; Manni et al. 2006; Auger et al. 2010). The atrial siphons give rise to the putatively mechanosensory cupular and capsular organs, small sensory organs composed of primary sensory cells (Bone and Ryan 1978; Mackie and Singla 2003, 2004).

In addition to sexual reproduction, many ascidians can also reproduce asexually. Asexual reproduction is particularly frequent in colonial forms, which can form new individuals (or zooids) by budding (Kurn et al. 2011; Brown and Swalla 2012; Kassmer, Rodriguez, and De Tomaso 2016). In many of these colonial forms, adults can be formed by two completely different developmental pathways: sexual reproduction with an intermittent free-living larval stage or asexual reproduction without a larva. Linked to asexual reproduction is an astonishing regenerative capacity of many colonial ascidians although some solitary ascidians, like *Ciona*, do not reproduce asexually, but also have some regenerative capacity (Manni et al. 2007; Jeffery 2015; Kassmer, Nourizadeh, and De Tomaso 2019). The alternation of sexually and asexually reproducing individuals has been elaborated into very complex life cycles in thaliaceans. Thaliaceans either have abandoned the larval stage completely or pass through a rudimentary tailed larval stage (some doliolids), which is never free living and lacks a brain (Piette and Lemaire 2015). In contrast, appendicularians reproduce exclusively sexually and have an extremely rapid life cycle, in which the adult develops from a fertilized egg in 1–2 days (Nishida 2008).

In light of their regenerative capacity (which allows the formation of body parts or even entire adults via developmental trajectories deviating drastically from embryogenesis), it is quite astonishing that ascidians have evolved a very stereotypical development, also known as "mosaic" development, in which the fate of cells is closely linked to the cell division pattern (cell lineage) (reviewed in Venuti and Jeffery 1989; Lemaire et al. 2008; Lemaire 2009). Cell fate in these animals is largely dependent on cell fate determinants, which are invariantly distributed to specific daughter cells during embryonic cell division. In addition, local interactions between neighboring cells play some role. However, the ability of embryos to compensate for damage or deletion of blastomeres (early embryonic cells) is very limited. Experimental removal of blastomeres, therefore, leads to the loss of structures that normally develop from these blastomeres. This was first recognized in experiments by Laurent Chabry (1887). Later, Edward Conklin (1905) provided a detailed description and nomenclature of the cell lineage of the ascidian *Styela*, which turned out to be applicable to other ascidians as well (Fig. 1.11). While the early development of thaliaceans is poorly understood but probably highly aberrant (Piette and Lemaire 2015), the cell lineages in the appendicularian *Oikopleura* have recently been shown to be equally stereotypical and similar to ascidians (Nishida 2008; Stach et al. 2008;) suggesting that it reflects the primitive tunicate mode of development.

Considering the morphology of an adult ascidian or thaliacean, their relationship to vertebrates is far from obvious and it is, thus, not surprising that until the 19th century their classification was contentious. Only after Alexander Kowalevsky (1866) discovered that the ascidian larva has a notochord, dorsal neural tube, and muscular tail were they recognized as chordates. However, based on their deviant morphology tunicates were long thought to be only distant relatives of the vertebrates within the

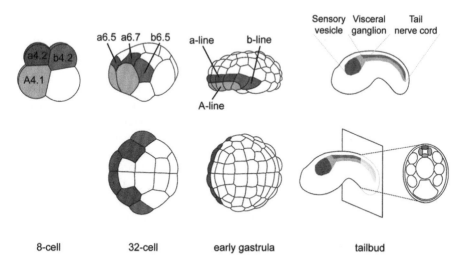

FIGURE 1.11 Tunicate development. The cell lineage of the CNS is highlighted. The cell lineage is stereotypical and is described by a letter followed by numbers. Large letters are used for the vegetal and small letters for the animal hemisphere, the letter A for the anterior half and B for the posterior half. The number that follows indicates the cell generation (increasing by one every division), whereas the last number identifies a particular cell (e.g. cell a4.2 giving rise to a5.3 and a5.4; cell a5.3 to a6.5 and a6.6. and so forth). (Reprinted with permission from Roure, Lemaire, and Darras 2014.)

chordates, whereas amphioxus was thought to be more closely related. As already mentioned above, recent molecular data indicate now that tunicates are more closely related to vertebrates than amphioxus. Apparently, tunicates have evolved a very different adult body plan and different developmental strategies very rapidly, masking these close relationships to vertebrates. The basal position of appendicularians within tunicates further suggests that the tunicate ancestors may have been free-living like amphioxus and vertebrates and that a sessile adult life style was adopted only in ascidians. The rapid evolution of new morphologies and developmental strategies in the tunicate lineage is reflected in their extremely rapid genomic evolution and reduction of genome size with high incidences of gene loss (Tsagkogeorga et al. 2010; Berná and Alvarez-Valin 2014; Holland 2015).

1.2.3 AMPHIOXUS

Like tunicates, amphioxus live in marine environments where most species are found in relatively shallow waters (reviewed in Schubert et al. 2006; Garcia-Fernàndez and Benito-Gutiérrez 2009; Bertrand and Escriva 2011; Holland and Onai 2012). Around 35 species in three genera are known including the well-studied genus *Branchiostoma*. Their name "amphioxus" as well as the English name "lancelet" alludes to their blade-like body form (Figs. 1.2D, 1.12). Alternatively, the animals are also known as cephalochordates (head chordates), because their notochord extends throughout the entire animal until the very tip of the head. Apart from the notochord (composed of muscle cells unlike in other chordates) they are also equipped with other specific chordate traits like a dorsal hollow nerve tube, segmented muscle blocks, which extend into the tail, and an endostyle (Fig. 1.12A). Anteriorly, they have an elaborate perforated pharynx involved in filter feeding. Overall, they are much more vertebrate-like in their body plan than tunicates and are capable of moving by alternating contractions of their muscles blocks on the left and right side of their body similar to fish. Consequently, their chordate affinities are already known since the early 19th century (Garcia-Fernàndez and Benito-Gutiérrez 2009).

The amphioxus life cycle involves a larval and adult form, which are relatively similar in organization and live both as filter feeders (Stokes and Holland 1995; Ruppert 1997). The amphioxus larva exhibits some striking left-right asymmetries in tissues derived from all germ layers (Fig. 1.13) (Soukup 2017). The somites developing on the left side are shifted by a half segment relative to those on the right side of the animal (Schubert et al. 2001). The mouth, Hatschek's pit - a mucus secreting and neurosecretory organ-, and Hatschek's nephridium - an excretory organ - develop on the left side while the endostyle and the club shaped gland - another mucus secreting organ (Holland, Paris, and Koop 2009 - form on the right side. Only the left series of gill slits are present in young larvae but they first appear on the right side of the ventral midline (Stokes and Holland 1995). The peculiar origin of the amphioxus mouth from the left side (Fig. 1.13) suggests that it is not homologous to the midline mouth of other chordates but is probably derived from either a gill slit or a coelomic pore on the left side (Yasui and Kaji 2008; Kaji et al. 2016; Schlosser 2017; Holland 2018).

After spending several months in the plankton, the larva undergoes metamorphosis, during which many of the larval asymmetries disappear. The larval gill slits are

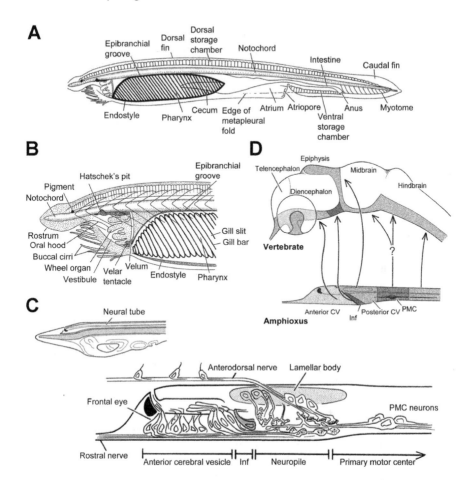

FIGURE 1.12 Amphioxus. (**A, B**) Body plan of an adult amphioxus. (**C**) Organisation of brain in an amphioxus larva. nc: neural cord. (**D**) Correspondence of amphioxus and vertebrate brain divisions. Anterior end of the neural tube shown in detail at the bottom. CV, cerebral vesicle; Inf, infundibular organ; PMC, primary motor center. ([A,B] Reprinted with permission from Brusca and Brusca 2003; [C,D] Reprinted with permission from Wicht and Lacalli 2005 (upper left) and Lacalli 2008 (lower right).)

shifted to their proper position on the left side of the animal and the right series of gill slits develops. Mouth and Hatschek's pit adopt a more rostral and ventral position and additional organs involved in filter feeding such as wheel organ and buccal cirri develop (see below). Two folds of the body wall (metapleural folds) grow ventrally and fuse at the ventral midline to create the atrium, which encloses the pharyngeal region in the adult.

After metamorphosis, amphioxus settles down to an adult life in the benthos (at the bottom of the sea), where it continues filter feeding most of its time, buried in sand with only its anterior-most end sticking out (Fig. 1.2C). Filter feeding in the adult amphioxus is aided by several specialized structures located in front of the mouth in a space enclosed by an oral hood (Fig. 1.12A, B). The ciliated wheel organ creates a current of water and food particles through the mouth into the pharynx. Mucus produced by a groove in the roof of the oral hood named Hatschek's pit helps to trap food particles. The buccal cirri encircling the front end of the animal act as a sieve, which prevents large particles from entering the vestibule in front of the mouth, while the velum around the mouth helps to sort food from other particles at the entrance to the pharynx. From the pharynx water flows through the pharyngeal slits into the atrium and through the atriopore out into the environment. In the pharynx, more food is collected by mucus produced by the endostyle and moved dorsally by cilia over the pharyngeal bars towards the epibranchial groove, from which the food laden mucus is transported by cilia backwards into the gut.

Feeding and swimming in larval and adult amphioxus is controlled by a dorsal nerve cord. Composition and circuitry of the anterior end of this nerve cord in young amphioxus larva have been elucidated in great detail by Thurston Lacalli, who provided 3D-reconstructions of this region based on serial sections analyzed by transmission electron microscopy (Fig. 1.12C) (reviewed in Wicht and Lacalli 2005; Lacalli 2008). The front end of the anterior cerebral vesicle is occupied by the unpaired frontal eye with pigment cells and ciliary photoreceptors. It probably acts as shadow detector but does not have any image forming capacity (Lacalli 2004; Vopalensky et al. 2012; Pergner and Kozmik 2017). Further posterior in the anterior cerebral vesicle there are cells with club-shaped cilia, which have been interpreted as a balance organ, followed by the secretory infundibular cells. Behind the infundibular organ on the dorsal side is the lamellar body, another putative photoreceptive organ, which is prominent in larvae but disintegrates in adults. Its presumptive photoreceptors are characterized by modified cilia, forming expansive lamellae. The increased surface area probably makes these receptors responsive to low light levels and they have been implicated in monitoring diurnal vertical migrations of the larvae in the water column (Lacalli 2004; Pergner and Kozmik 2017). Additional light sensitive cells (the rhabdomeric Joseph cells and Hesse organs) develop within the nerve cord in older larvae and adults (see Chapter 4) (Lacalli 2004; Pergner and Kozmik 2017).

The region below the lamellar body is occupied by a neuropile, a region where various neurons form synaptic connections with each other. Posterior to the neuropile is the primary motor center (PMC) with neurons that send their axons down into the spinal cord. Their dendrites reach forward into the neuropile, where they receive input from the frontal eye, balance organ, and lamellar body as well as from peripheral sensory cells that send their axons into the brain via the rostral and anterodorsal nerves and via the segmental dorsal nerves along the spinal cord (Fig. 1.12C). This circuitry suggests that the PMC neurons probably coordinate locomotor behavior in response to tactile and visual stimuli, mediating for example the escape response after touch or diurnal vertical migrations (Wicht and Lacalli 2005; Lacalli 2008).

Although the amphioxus brain is much simpler than the vertebrate brain, the distribution of cell types and the expression of many marker genes suggest correspondences between regions of the amphioxus brain and subdivisions of the vertebrate brain (Fig. 1.12D) (Wicht and Lacalli 2005; Albuixech-Crespo, López-Blanch et al. 2017;

Albuixech-Crespo, Herrera-Ubeda et al. 2017). The frontal eye is probably homologous to the paired retinae of the vertebrate eyes, and the lamellar body to the pineal gland mediating circadian rhythms in vertebrates. The anterior cerebral vesicle up to the infundibular organ has been proposed to correspond to the anterior-most three prosomeres (hypothalamus and prethalamus) of the vertebrate forebrain; the region of lamellar body and neuropile to the posterior two prosomeres (thalamus and pretectum) and midbrain; and the PMC region to the hindbrain (Albuixech-Crespo, López-Blanch et al. 2017).

In the periphery, sensory stimuli are registered by free nerve endings of sensory neurons located in the dorsal spinal cord. In addition, there are a number of specialized sensory cells that probably respond to mechanical or chemical stimuli although this needs to be verified in physiological studies. These cells will be reviewed in more detail later in the book (Chapter 3) and I will, therefore, mention them here only briefly (reviewed in Holland and Holland 2001; Lacalli 2004; Schlosser 2017). Two major types of cells are found scattered throughout the amphioxus epidermis. Type I cells, which have an axon and a long cilium surrounded by microvilli, are already generated in embryos, while type II cells, which lack an axon and have a short cilium and a peculiar collar studded with microvilli, only appear in late larval stages. Rostrally, there are additional types of sensory cells, for example the oral spine cells associated with the mouth. In the adult, numerous small sensory organs, the corpuscles of Quatrefages, are also found at the rostral tip, which send their axons through the rostral and anterodorsal nerves into the cerebral vesicle.

In contrast to tunicates but similar to vertebrates, amphioxus have an indeterminate or regulative development (Fig. 1.13). The fate of cells is decided relatively late in development in response to long range signaling between different cells and perturbations of early development, for example due to damage or experimental removal of cells, can to some degree be compensated (reviewed in Garcia-Fernàndez and Benito-Gutiérrez 2009; Holland and Onai 2012). Following rapid cleavage divisions, the amphioxus blastula invaginates to form a cup-shaped gastrula (Fig. 1.13). Its dorsal blastopore lip then acts as a signaling center involved in the induction of the neural plate and other dorsal structures similar to the vertebrate organizer (Yu et al. 2007). After gastrulation, the neural plate rolls up to form a neural tube and the first somites develop by outpocketing of the embryonic gut (enterocoely) (Fig. 1.13). Another pair of gut diverticula, which arise anterior to the first somite, develops asymmetrically. The left one will fuse with the adjacent ectoderm to form Hatschek's pit (more about this in Chapter 5), while the right one will become the head coelom. Subsequently, mouth, gill slits, and other anterior structures appear in their peculiar left or right positions described above for amphioxus larvae.

1.2.4 HEMICHORDATES AND ECHINODERMS

Hemichordates are a small group of marine animals comprising around 100 species of enteropneusts (including the model species *Saccoglossus kowalevskii* and *Ptychodera flava*) and 30 species of pterobranchs (genera *Cephalodiscus* and *Rhabdopleura*) (Röttinger and Lowe 2012; Lowe et al. 2015; Tagawa 2016).

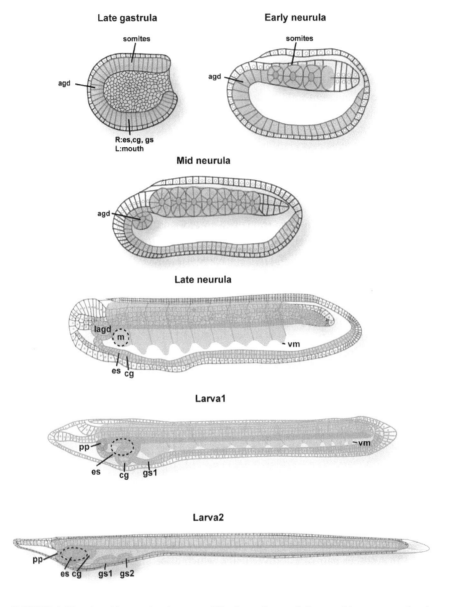

FIGURE 1.13 Amphioxus development. The inner layer of the amphioxus gastrula gives rise to both endoderm and mesoderm. Its dorsal part (red) forms somites by enterocoely. Its ventral part (blue) forms the gut and derivatives such as the anterior gut diverticula (agd), endostyle (es) and club shaped gland (cg), which develop asymmetrically. Es and cg develop on the right hand side and the agd form different structure on the left and right side. The left agd will develop into the preoral pit (pp), the larval precursor of Hatschek's pit. The mouth (m) forms at the late neurula stage on the left side followed by the formation of gill slits (gs) on the right hand side. vm: ventral mesoderm. (Reprinted with permission from Onai, Adachi, and Kuratani 2017. After Hatschek 1881.)

The enteropneusts or acorn worms are benthic animals that either burrow in sediments or live on the surface. They live either as deposit feeders, which ingest sediment, or as filter feeders by using their ciliated epidermis to create water currents trapping small food particles in mucus which is then transported to the mouth by ciliary motion (Figs. 1.2B, 1.14A). Pterobranchs are small sessile and colonial forms that live in secreted tubes and use ciliated tentacles for filter feeding (Fig. 1.14B). Although some phylogenetic analyses suggest that pterobranchs are derived from one particular group of enteropneusts (the Harrimaniids) making enteropneusts paraphyletic (Cameron, Garey, and Swalla 2000; Cannon et al. 2009), a recent study based on increased taxon sampling places pterobranchs as a sister group to the enteropneusts (Cannon et al. 2014).

In distinction to the chordates, hemichordates lack a notochord, a dorsal hollow nerve cord or a postanal tail (Stach and Kaul 2011; Röttinger and Lowe 2012). However, the presence of pharyngeal slits resembling the chordate pharyngeal slits was already noted by Kowalevsky 1866 and later led Bateson (1886) to coin the

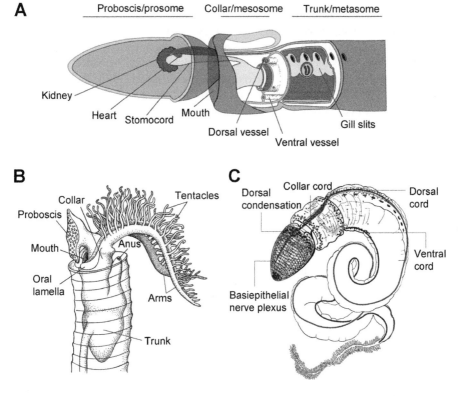

FIGURE 1.14 Hemichordates. (**A**) Body plan of an enteropneust. (**B**) Pterobranch *Rhabdopleura*. (**C**) Nervous system in the adult enteropneust *Saccoglossus kowalevskii*. ([A] Reprinted with permission from Lowe et al. 2006; redrawn after Benito and Pardos 1997; [B] reprinted with permission from Kardong 2009; [C] Reprinted with permission from Lowe et al. 2015.)

name "hemichordates" (half chordates). On the other hand, it was already noted by Metschnikoff (1881) that the larval stages of hemichordates closely resemble the larvae of echinoderms (e.g. sea urchins, starfish). Modern molecular data strongly support Metschnikoff's grouping of echinoderms and hemichordates into a common clade, the Ambulacraria and suggest that these are the sister group of the chordates (e.g. Wada and Satoh 1994; Swalla and Smith 2008; Philippe et al. 2011).

If one considers the disparate anatomy of adult hemichordates and echinoderms, their close relationship is at first very surprising. The adult body plan of echinoderms was highly modified by the evolution of pentaradial symmetry in the ancestors of modern echinoderms and shows little resemblance to hemichordates or chordates (Fig. 1.2A). However, early offshoots of the echinoderm lineage before the adoption of pentaradial symmetry such as the extinct carpoids known from fossils also displayed a series of openings, which may have been pharyngeal slits, and looked much more similar to hemichordates or chordates than the echinoderms living today. The chordate-like appearance of these fossils has even led to a very elaborate proposal that chordates evolved from these extinct groups, which were termed "calcichordates" due to their calcified endoskeleton (Gislén 1930; Jefferies 1986; see Gee 1996 for review). However, uniquely derived chordate traits such as notochord or dorsal hollow nerve cord could not be unequivocally identified in these fossils, while the strong molecular support for an ambulacrarian clade argues against the calcichordate theory.

While echinoderms have evolved a very specialized novel body plan, hemichordates are thought to have retained much more of the characters of the ambulacrarian ancestor and, thus, are of much more interest for understanding the origin of chordates. It has long been debated whether the ancestor of hemichordates was more like the sessile pterobranchs or the motile enteropneusts. Based on the assumption that pterobranchs were nested within a paraphyletic enteropneust group, it was proposed that the hemichordate ancestor was a worm-like creature similar to enteropneusts (Cameron et al. 2000; Brown, Prendergast, and Swalla 2008). Although, the recent study of Cannon (Cannon et al. 2014) has called the premise of this argument into question, both the ancestors of echinoderms and the ancestors of chordates were suggested to be motile forms (see below), suggesting that the ancestral hemichordate most likely also was free-living and that sessility evolved secondarily in the pterobranch lineage.

Here I will describe enteropneusts in a little more detail (Fig. 1.14A, C). Pterobranchs have an overall similar body plan but with some important differences related to their small size and sessile habits (Fig. 1.14B) (e.g. u-shaped gut, reduced number of pharyngeal slits, ciliated tentacles, or lophophores). The enteropneust body can be divided into three different sections, each with an independent coelom or pair of coeloms (fluid filled body cavity). The anterior contractile proboscis (or prosome) used for burrowing is followed by a collar (mesosome) and a long trunk section (metasome) (Fig. 1.14A). At the junction between proboscis and collar there is a ventrally positioned mouth opening into the gut. On the dorsal side of the anterior-most part of the gut, a blind sac branches off into the proboscis. Based on its position and it's make up of vacuolated cells, this so-called stomochord has been homologized with the notochord but this is not supported by recent gene expression studies (Annona, Holland, and D'Aniello 2015). The following section of the gut, the pharynx, is perforated by a paired series of pharyngeal slits in the anterior part of the trunk. The gut then runs through the trunk to end in a terminal anus.

Nerve cells in hemichordates are scattered throughout the epidermis forming a diffuse nerve net (Bullock 1945; Knight-Jones 1952). In addition, nerve cords run along the dorsal and ventral side of the animal (Fig. 1.14C) and part of the dorsal cord in the collar region is internalized in a process resembling vertebrate neurulation to form the collar cord (Kaul and Stach 2010; Miyamoto and Wada 2013). The nerve cords of enteropneusts were originally described as comprising mostly fiber tracts unlike the vertebrate CNS (Bullock 1945). Recent studies now show that at least some regions of the nerve cords contain cellular layers in addition to neuropiles (Brown et al. 2008; Nomaksteinsky et al. 2009; Kaul and Stach 2010; Cunningham and Casey 2014). However, these nerve cords share little more than an agglomeration of nerve cells with the chordate CNS and are patterned differently along the dorsoventral axis (Kaul-Strehlow et al. 2017). Their homology with the vertebrate CNS, thus, remains contentious.

Very little is also known about sensory cells or organs in hemichordates. Ciliated cells with a collar of microvilli are found near the ciliary bands of larvae (see below) and scattered throughout the ectoderm in adults (Jørgensen 1989). Based on morphology, these may be either mechano- or chemoreceptors but this remains to be confirmed. The rostral epidermis is also occupied by so called Mulberry cells, which contain neuropeptides and, thus, may represent a neurosecretory cell (Cameron et al. 1999). However, their function is currently unknown.

The development of hemichordates (Fig. 1.15A) is governed by long-range signaling systems patterning, for example, the anteroposterior and dorsoventral axes similar to amphioxus and vertebrates (Lowe et al. 2003, 2006). However, signals secreted by the dorsal organizer in chordates such as chordin, which block the activity of other signaling molecules of the bone morphogenetic protein (BMP) family on the dorsal side, are localized on the ventral midline in hemichordates as well as in other animal phyla (Lowe et al. 2006). It has therefore been suggested that there was a dorsoventral inversion of the body in the chordate lineage after its divergence from ambulacrarians (Nübler-Jung and Arendt 1996; Gerhart 2006; Lowe et al. 2006; Holland et al. 2013). Hemichordates either undergo indirect development with a larval stage (e.g. in *Ptychodera flava*) or direct development, in which the adult develops without passing through a larva (e.g. in *Saccoglossus kowalevskii*). Indirect development is probably the primitive mode of development for hemichordates and direct development derived. The planktonic tornaria larva of hemichordates is bilaterally symmetrical and has multiple ciliated bands encircling mouth and anus (Fig. 1.15B). The larvae of echinoderms have a similar bauplan, which has been termed a dipleurula type larva. In both indirectly and directly developing species the mesoderm develops by outpocketing from the gut (enterocoely) and forms three sets of coeloms along the body axis: an anterior unpaired protocoel as well as paired meso- and metacoels, which prefigure the coeloms of proboscis, collar and trunk, respectively (Fig. 1.15A, B).

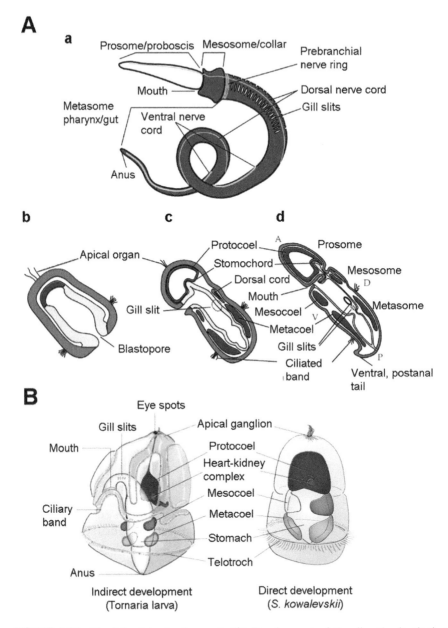

FIGURE 1.15 Hemichordate development. (**A**) Development of the directly developing enteropneust *Saccoglossus*. a: adult with tripartite body plan. b: Late gastrula. c: Late neurula. d: Two gill slit embryo. Dorsal is to the top right; anterior is to the top left. (**B**) Comparison of tornaria larva of an enteropneust with indirect development with an embryo of the direct developer *Saccoglossus kowalevskii*. ([A] Reprinted with permission from Lowe et al. 2003; [B] Reprinted with permission from Lowe et al. 2015.)

1.3   SCENARIOS FOR VERTEBRATE ORIGINS

1.3.1   FROM HAECKEL TO GARSTANG

Being vertebrates themselves, zoologists have always been particularly intrigued by the question of vertebrate origins and a large number of conflicting – and often quite imaginative – theories have been proposed. Here is not the place to give a detailed historical account of these and I will only sketch a few influential views as a backdrop to current scenarios. Readers who want to delve deeper into this fascinating history of ideas are referred to several excellent books and reviews (Nyhart 1995; Bowler 1996; Gee 1996; Lacalli 2005; Holland, Holland, and Holland 2015; Holland 2016b).

After the publication of Charles Darwin's *Origin of species* in 1859, the German zoologist Ernst Haeckel quickly realized that common descent of all living beings as advocated by Darwin required a conceptual overhaul of systematics, the science of classification of organisms. From an evolutionary viewpoint, the only natural way to classify organisms was by their genealogical relationships. These could be depicted as a tree of life, or "phylogeny", a term introduced by Haeckel (Haeckel 1866). According to Haeckel's "biogenetic law", most organisms recapitulate their evolutionary history in a condensed form during their embryonic development or "ontogeny" (another of Haeckel's neologisms) and ontogeny, therefore, helps to elucidate phylogeny.

In the second part of his "Generelle Morphologie", Haeckel presented a classification of all organisms based on these new principles. There he identified amphioxus as the first subphylum of vertebrates (Haeckel 1866). While amphioxus was initially described as a mollusk, other zoologists before Haeckel (e.g. Gegenbaur 1859) had recognized its close affinity to the vertebrates and classified it as a kind of primitive fish (Holland and Holland 2017). However, Haeckel was the first to stress the significant morphological differences of amphioxus to vertebrates such as the lack of a skull and a compact heart, and thus moved it to a separate subphylum within the vertebrates. He also was the first to discuss the evolutionary implications of this grouping and to suggest that amphioxus and vertebrates were derived from the same ancestral form with amphioxus retaining many of the primitive characters of this common ancestor (Haeckel 1866).

Tunicates at this time were still considered mollusks (Gegenbaur 1859; Haeckel 1866). Only when the Russian zoologist Alexander Kowalevsky (also spelled Kowalewski) published a detailed description of the tunicate larva, were its chordate affinities recognized (Kowalevsky 1866). The subsequent description of amphioxus embryology by Kowalevsky supported Haeckel's view of amphioxus as an early offshoot of the lineage leading to vertebrates (Kowalevsky 1867). Taken together, this supported a close evolutionary relationship between tunicates, amphioxus, and vertebrates. During the late 19th century these insights gave rise to the mainstream view, championed by Haeckel and Gegenbaur, that a vermiform ancestor evolved a notochord and a muscular tail to become the first chordate, which may have resembled the tunicate larva (Maienschein 1994). Tunicates then branched off and evolved a very specialized, sessile adult stage, while the main chordate lineage gave rise to amphioxus and vertebrates.

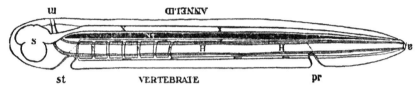

FIG. 140.   Reversible diagram illustrating the Annelid theory.

Reversible designations, applying to both forms: *S*, brain; *X*, nerve cord; *H*, alimentary canal.   Designations applying to Annelid only: *m*, mouth; *a*, anus. Designations applying to Vertebrate only; *st*, stomatodæum; *pr*, proctodæum; *nt*, notochord.

FIGURE 1.16   Origin of vertebrates from inverted annelids. Diagram illustrating the annelid theory of vertebrate origins with dorsoventral inversion. (Reprinted from Wilder 1909.)

However, not everyone agreed. The similarities between the tunicate larva, amphioxus and vertebrates were striking and undeniable and suggested a close relationship between these groups. However, tunicates and amphioxus might just as well represent degenerate and secondarily simplified offshoots of the vertebrates rather than their closest invertebrate relatives. One of the most outspoken defenders of this view was Anton Dohrn, who proposed instead that annelids were the closest relatives of vertebrates, mainly because of the segmental body plan of both groups (Dohrn 1875). However, vertebrates according to Dohrn were upside-down annelids (Fig. 1.16). While extant annelids retained the CNS and dorsal heart of the last common ancestor with vertebrates, vertebrates became dorsoventrally inverted so that the brain ended up on the new dorsal and the heart on the new ventral side (Fig. 1.17A). After a new mouth formed on the new ventral side, the old mouth obliterated. Similar ideas had been proposed already in 1830 by the French zoologist Geoffroy St. Hilaire, who pointed out the similar arrangement of organs in vertebrates and a lobster turned upside down (Appel 1987).

Dohrn's proposal drew a barrage of criticism from Haeckel, Gegenbaur, and others (Maienschein 1994) including the young William Bateson, who later became famous as one of the founding fathers of genetics, but who spend his early career in studying the embryology of enteropneusts (Bateson 1886). Bateson points out that segmentation of some sort is widespread in the animal kingdom affecting different tissues in different groups. Being both a widespread and highly variable character it is a poor guide for uniting vertebrates with annelids. Bateson, thus, rejects Dohrn's arguments for a segmented ancestor of vertebrates and instead proposes enteropneusts as the closest relatives of amphioxus and the vertebrates. This is based on the presence of gill slits, a notochord-like extension of the embryonic gut (stomochord) and the formation of the collar cord resembling the vertebrate neural tube.

All of the theories mentioned so far agreed in one point: the ancestor of vertebrates was a freely moving, worm-like form (Fig. 1.18A). However, not only were the adult ascidians sessile, so were some hemichordates (viz. the pterobranchs) and echinoderms (sea lilies). The latter were now recognized to be closely allied to hemichordates and chordates based on the similarities in embryonic

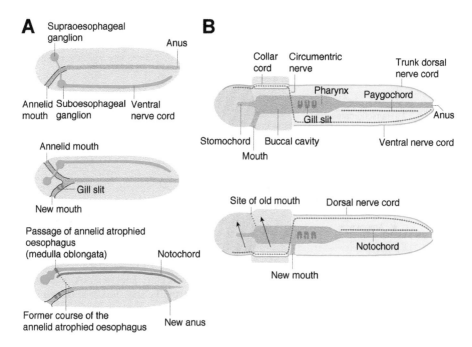

FIGURE 1.17 Dorsoventral inversion during chordate evolution. (**A**) Correspondence of nervous system, gut, and mouth opening between annelids and chordates according to the annelid theory. (**B**) Proposed correspondence of structures between hemichordates and chordates assuming dorsoventral inversion at the base of chordates. The blue dotted line was proposed as the hemichordate CNS by Nübler-Jung and Arendt (Nübler-Jung and Arendt 1996). ([A,B] Reprinted with modification with permission from Holland et al. 2015.)

development such as a tripartite embryonic coelom (Metschnikoff 1881) and deuterostomy (blastopore becoming the anus with mouth breaking through secondarily) (Grobben 1908).

Based on these new insights, Walter Garstang proposed a radically new theory of vertebrate origins (Fig. 1.18B) (Garstang 1928). A few years earlier he had already emphasized that major evolutionary changes may occur at all stages of development and in particular by modifications of larval stages. This invalidated Haeckel's proposal that ontogeny should recapitulate phylogeny, which was based on the assumption that such changes occur mainly at the end of development (Garstang 1922). Applying these ideas to the origin of vertebrates, Garstang now proposed that vertebrates evolved from the larva of ancestors with sessile filter feeding adults. The larva of extant ascidians supposedly maintained many of the characters of this ancestral chordate larva. According to Garstang, vertebrates evolved from this larva by a process termed paedomorphosis, which involves the maintenance of larval characters into the adult, sexually reproductive stage for example due to precocious gonad maturation. A well-known example for paedomorphosis is the Mexican axolotl, which evolved from other salamanders by attaining sexual maturity in the larval stage, foregoing metamorphosis. Already in 1894, Garstang had proposed that

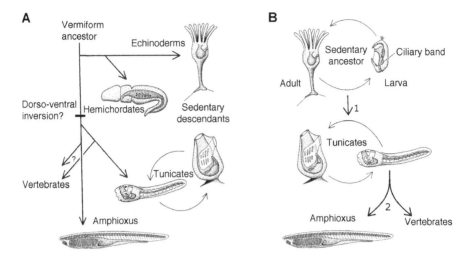

FIGURE 1.18 Origin of vertebrates from wormlike or sedentary ancestors. (**A**) Scenarios that propose the origin of vertebrates from free-living, worm like ancestors assume that sessile forms in hemichordates, echinoderms and tunicates are independently derived. Various such scenarios have been put forward, proposing either tunicates or amphioxus as the closest relative of vertebrates. Some scenarios also propose a dorsoventral inversion at the base of the chordates. (**B**) Scenarios that propose the origin of vertebrates from sedentary ancestors, assume vertebrates evolved in two steps from an ancestor with a motile dipleurula larva and a sessile adult. In step 1, a larva with chordate characteristics (as in extant ascidians) evolved. In step 2, amphioxus and vertebrates evolved by paedomorphosis from this larval stage. ([A,B] Modified with permission from Lacalli 2005.)

this chordate larva arose from a dipleurula-type larva as found in echinoderms and hemichordates by the dorsal movement of the ciliary bands and associated nerve cells, which then coalesced to form the neural tube (Garstang 1894). Vertebrates, thus, may have evolved in a two-step process from animals resembling pterobranchs and sessile echinoderms. In the first step, the dipleurula type larva of these animals evolved into a chordate-type larva similar to the ascidian tadpole larva. In a second step, the sessile adult stage was abandoned and the chordate-type larva became transformed into the vertebrate adult (Fig. 1.18B).

1.3.2 VIEWS OF VERTEBRATE ORIGINS AFTER GARSTANG

Garstang's theory, in particular it's second step, was eagerly discussed and developed further by other zoologists such as Berrill (1955) and Romer (1959, 1972) and dominated thinking about vertebrate origins for much of the 20th century. In the last decades, however, Garstang's theory has increasingly been recognized as untenable in the light of new phylogenetic, molecular, and developmental evidence leading to a revival of older theories of a motile and worm-like vertebrate ancestor. First, tunicates have been recognized as a fast evolving and highly specialized group making them poor models for vertebrate ancestors. Second, new phylogenetic evidence reveals that tunicates and not amphioxus are the closest living relative of vertebrates suggesting that many features of the chordate body plan shared between amphioxus and vertebrates evolved prior to the evolution of a tadpole larva as found in ascidians. Third, evidence for appendicularians being the most basally branching group in tunicates, suggests that a life cycle with a sessile adult as found in ascidians may only have evolved within the tunicate lineage. Similarly, sessility in echinoderms has probably evolved secondarily from a free living ancestor similar to the fossil carpoids, while within the hemichordates, pterobranchs may have evolved a sessile life style secondarily. Taken together, this suggest that a sessile life style evolved several times independently in deuterostomes and that the ancestors of chordates were free-living but presumably quite sluggish and filter feeding animals equipped with a notochord, dorsal neural tube, a postanal tail, and pharyngeal slits similar to amphioxus (Lacalli 2005; Holland 2016a, 2016b).

Whereas notochord, dorsal neural tube, and postanal tail are not found outside the chordates, pharyngeal slits are found in fossil echinoderms (carpoids) as well as in modern hemichordates. Furthermore, recent studies showed that the pharyngeal bars of enteropneusts are composed of acellular cartilage very similar to amphioxus (Rychel et al. 2006) and the developing pharyngeal slits and bars express a similar network of transcription factors than the pharyngeal slits and bars of chordates (Ogasawara et al. 1999; Gillis, Fritzenwanker, and Lowe 2012). This suggests that the pharyngeal slits and bars of ambulacrarians and chordates are homologous and were already present in their last common ancestor, which may thus have been rather enteropneust-like as Bateson originally suggested.

However, in contrast to Bateson and in line with the annelid theory proposed by Dohrn, several observations suggest that the dorsoventral body axis was probably inverted in the stem lineage of chordates after ambulacrarians branched off (Arendt and Nübler-Jung 1994, 1996) (Fig 1.17B). First, several morphological structures found on the ventral side in vertebrates such as the pharyngeal slits, hepatic anlagen and the major body vein carrying blood anteriorly are located on the dorsal side in hemichordates (Benito and Pardos 1997). Second, the dorsoventral BMP patterning system is inverted in hemichordates relative to chordates (with BMP enriched on the dorsal side in the former but on the ventral side in the latter) (Lowe et al. 2006). And third, the left-right patterning module involving the signaling molecule Nodal and the transcription factor Pitx is activated on the right side in echinoderms but on the left side in chordates (Duboc et al. 2005; Blum et al. 2014).

The most comprehensive scenario of vertebrate origins, which ties the proposed changes in development and morphology to functional changes in life style and ecology, is the "New Head hypothesis" of Northcutt and Gans, which was already introduced above (Northcutt and Gans 1983; Gans and Northcutt 1983; Northcutt 2005). It proposed that many of the novel characters of vertebrates are linked to a change in life style from filter feeding to active predation. In a first step, replacement of ciliary by muscular ventilation of the pharynx allowed more efficient gas exchange and suction feeding. This enabled increasingly more active locomotion and predation accompanied by an anterior concentration and elaboration of sensory organs and the formation of a protective skull. The evolution of new embryonic tissues, the neural crest and placodes, enabled the formation of many of these novel cranial structures. While the "New Head hypothesis" was developed in the context of an outdated phylogeny that placed amphioxus and not tunicates as the closest relative of chordates and did not consider chordates as being dorsoventrally inverted relative to hemichordates, its main tenets appear to be still valid even under the revised chordate phylogeny and under the assumption of a dorsoventrally inverted chordate ancestor. This will be elaborated in the remainder of this book.

Several points of contention remain in current scenarios of chordate origins. The first, which will not be discussed here any further, is whether segmentation in chordates and segmentation in annelids and arthropods can be traced back to a segmented last common bilaterian ancestor (reviewed in Tautz 2004; Chipman 2010; Graham et al. 2014). Some similarities in the mechanisms of segmentation between arthropods and vertebrates, such as the involvement of the Notch signaling module, seem to support this scenario. However, even within chordates there appear to be different types of segmentation (pharyngeal arches versus somites; see Chapter 1 in Schlosser 2021) and these affect partly different tissues and organs than in arthropods and annelids. Moreover, segmentation is lacking in most animal phyla. Taken together, this does not provide strong support for a common origin of segmentation in chordates, annelids and arthropods.

A second point of contention is whether the dorsal CNS of chordates is homologous to the ventral CNS of annelids, arthropods and other protostomes both being derived from the CNS in the last common ancestor of protostomes and deuterostome. Alternatively, centralization of the nervous system may have happened several times independently.

Similarities in both dorsoventral and anteroposterior patterning of the nervous system in annelids and arthropods on the one side and chordates on the other side have been cited in favor of the first scenario (D'Alessio and Frasch 1996; Isshiki, Takeichi, and Nose 1997; Hirth et al. 2003; Lichtneckert and Reichert 2005; Holland et al. 2013). It is also supported by the origin of particular neural cell types from regions defined by a similar molecular code of transcription factors subdividing the anteroposterior and dorsoventral axes (Denes et al. 2007; Arendt et al. 2008; Arendt, Tosches, and Marlow 2016) and by shared cytoarchitecture and connectivity (Strausfeld and Hirth 2013; Tomer et al. 2010). However, most of the transcription factors patterning the CNS along the anteroposterior axis are also used to pattern other parts of the ectoderm. This is compatible with the possibility that only a general ectodermal patterning role is evolutionarily conserved, which was then recruited to pattern the CNS several times independently.

In favor of the second scenario (independent centralization of nervous systems), a recent analysis of dorsoventral patterning mechanisms during CNS development in many xenacoelomorphs and lophotrochozoans has revealed substantial differences between taxa (Martin-Durán et al. 2018) although the interpretation of these data has been questioned (Arendt 2018). Furthermore, it has been pointed out that animal phyla with a centralized nervous system (mollusks, annelids, arthropods, chordates) are quite sparsely scattered among a majority of animal phyla equipped only with a diffuse nervous system or with simple ganglia (Roth and Wullimann 1996; Holland

2003; Northcutt 2012). Among the deuterostomes, hemichordates and echinoderms do have very simple nervous systems. As discussed above, hemichordates have a diffuse nerve net and a dorsal and a ventral nerve cord, both of which have been considered as potential homologs of the chordate CNS but currently without strong evidence for either proposal. Taken together, this suggests that centralization of the CNS has either evolved a few times independently (convergently) or has evolved only once and then has been lost many times in different phyla including other deutero-stomes. The first possibility requires a smaller number of evolutionary changes and hence is more economical (parsimonious). The second possibility requires a larger number of changes, but all of these are losses, which are easier to achieve than gains in evolution (Guijarro-Clarke, Holland, and Paps 2020). It is, therefore, not easy to judge, which of these possibilities is inherently more likely. The more complex the similarities between annelid or arthropod and chordate CNS will turn out to be, the less likely it is that they evolved convergently.

2 Teaching Old Cells New Tricks

Although vertebrates have a "New Head" with many novel sense organs and the pituitary as a novel endocrine organ, it will become clear throughout this book that these novel structures are built from sensory and neurosecretory cell types that are much older than vertebrates, have homologs in other lineages and have a long evolutionary history. But to make sense of these sometimes convoluted evolutionary trajectories, some conceptual issues need to be addressed first. This chapter is dedicated to clarify these conceptual issues and illustrate them with examples from sensory evolution. To allow us to trace characters through evolution and to explore the origin of novel characters, I need to explain the general concepts of character identity, homology, and evolutionary innovation in the first part of this chapter. In the second part, I will then apply these general concepts to the evolution of cell types, after introducing cell types as particular, independently evolving units.

2.1 CHARACTER IDENTITY, HOMOLOGY, AND EVOLUTIONARY INNOVATION

2.1.1 CHARACTERS AS INDEPENDENTLY EVOLVING UNITS

The concept of homology was initially coined by the British comparative anatomist Richard Owen who defined a homologous organ as "the same organ in different animals under every variety of form and function" (Owen 1843). Owen formulated this concept in a pre-evolutionary framework. He thought that corresponding organs could be identified in different animals because they were constructed as variations on a common theme, a bauplan or "archetype" (Owen 1848). After Darwin's publication of the *Origin of Species* (1859), homology was then re-interpreted from an evolutionary perspective as reflecting descent from a common ancestor (Lankester 1870). We now consider characters in different species as homologous if they are derived from the same character in their last common ancestor (Wagner 2007a) (Figs. 2.1, 2.2A). Similarities in homologous characters are due to inheritance from a common ancestor. In contrast, characters may have evolved similarities independently, for example, as responses to similar environmental challenges. Owen termed this "analogy" but it is now usually referred to as "homoplasy" (Lankester 1870; Wake, Wake, and Specht 2011; Hall 2013) (Fig. 2.2B). Sometimes two different types of homoplasy are distinguished, convergence and parallelism. Convergence refers to the independent evolution of similar characters in different lineages, while parallelism refers to the independent evolution of similar characters in different descendants of a common ancestor (Hall 2013).

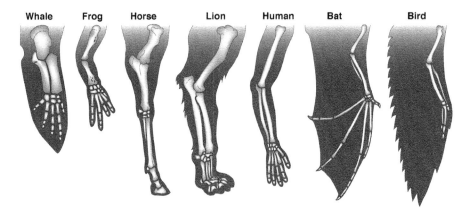

FIGURE 2.1 Homology. Characters such as the tetrapod forelimbs shown here are homologous if derived from the same character in their last common ancestor. Despite adaptations to different types of locomotion, there are similarities in the underlying bauplan inherited from a common ancestor: from proximal to distal (top to bottom), a single long bone (humerus) is followed by a pair of long bones (radius and ulna), which are linked to the hand with five (or less) digits. (Reprinted with permission from Wagner 2007b.)

Since characters are subject to heritable variation, a major conceptual problem for this historical (or genealogical) notion of homology is how the identity of a character can be determined in different species (Wagner 1989, 2014). Genealogical continuity is easy to establish for systems that directly reproduce such as organisms. There is genealogical continuity between A and B if both are linked by an uninterrupted chain of generations (e.g. A being parent of X, X being parent of Y, Y being parent of B). A similar chain can be established for parts of organisms that are directly copied from one generation to the next such as genes. However, most phenotypic characters of organisms (e.g. cell types, organs) are not simply copied from a parental template but instead are constructed anew during development of each generation. Given that

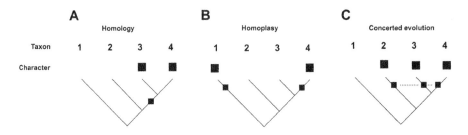

FIGURE 2.2 Homology, homoplasy and concerted evolution. Phenotypic similarities between characters in different lineages may be due to (**A**) inheritance from a common ancestor, in which the character originated (homology) or (**B**) independent evolution in two different lineages (homoplasy). (**C**) Characters with a partly shared genetic basis may undergo correlated evolutionary changes (concerted evolution).

those characters are not directly transmitted from generation to generation, how can corresponding characters be identified across generation (and ultimately species) boundaries (Wagner 2014)? This problem may not be apparent when comparing slowly evolving characters in closely related species where there is little ambiguity of character identification (e.g. the forelimbs of different vertebrates). However, it becomes evident in comparisons of quickly evolving characters and/of distantly related species (e.g. the ocellus in the sensory vesicle of tunicates and the eyes of vertebrates). In these cases, homology may only be possible to establish if transitional forms are preserved in other species that have branched off intermittently from the same lineages.

Genealogical continuity can meaningfully only be established for characters that clearly retain their identity in subsequent generations. They, thus, need to remain stable in evolution and distinct from other characters in spite of the fact that they are subject to heritable variation. Homology statements can only be attached to such characters or their character states. It has therefore been proposed that only individualized parts of the organism, i.e. those parts that evolve as an integrated unit but relatively independently from other parts should be recognized as proper characters reflecting real units in nature (also referred to as units or modules of evolution) (Wagner 2001, 2014; Schlosser 2002c). Such individualized characters have at least a partially different genetic basis from other characters allowing them to be subject to heritable variations that affect only those and no other characters (genetic individuation or individualization). In order to take an independent evolutionary trajectory, genetically individuated characters also have to have a separable function from other characters and make an independent (or neutral) fitness contribution (Wagner 1995; Schlosser 2002c, 2004). Should this not be the case, changes in one character will induce selection on the other resulting in the co-evolution of both characters over time (Schlosser 2002c; Weinreich, Watson, and Chao 2005; Hansen 2013).

For example, inner ear and lateral line system in vertebrates arise from a common embryonic precursor, but different genes are ultimately required to specify inner ear versus lateral line placodes (Schlosser 2002a, 2002b; Schlosser and Ahrens 2004). Whereas the genes *Pax8*, *Sox9*, *Sox10*, and *Tbx1* are required for ear development, these genes are not expressed in the developing lateral line placodes, which instead express *Tbx3* (Schlosser 2006). Moreover, inner ear and lateral line fulfill different functions in detecting body movements or sound and water movements, respectively (Chagnaud et al. 2017). Consequently, inner ear and lateral line are different characters with an independent evolutionary history. Both systems have acquired independent adaptations and the lateral line system has been lost, for example, several times independently in several vertebrate lineages such as amniotes and some frogs without affecting the inner ear (Northcutt 1997; Schlosser 2002b).

In contrast, although some regional specializations of individual lateral line placodes have evolved, shared gene expression patterns suggest that the genetic basis for lateral line development is largely shared between different lateral line placodes (Schlosser and Ahrens 2004; Piotrowski and Baker 2014). Moreover, the mechanosensory organs (neuromasts) derived from the different lateral line placodes must functionally cooperate to allow the extraction of information on the size and direction of moving objects from the displacement of hair bundles in neuromasts (Dijkgraaf 1962; Bleckmann 2008). Consequently, the entire lateral line system tends to evolve as a coherent unit and many evolutionary changes (e.g. regarding the structure and function of lateral line neuromasts) affect the entire lateral line system rather than individual lateral line placodes (Dijkgraaf 1962; Coombs, Janssen, and Webb 1988).

This last example illustrates a more general principle. In many organisms we find repeated parts, for example, multiple cells belonging to the same cell type (e.g. muscle cells, neurons), multiple segments containing the same tissues and organs, or multiple appendages. Repeated structures in one organism are also known by Owen's term of "serial homologs", which he introduced in distinction to "special homologs" referring to structures corresponding between different individuals or species (Owen 1848; Wagner 2014). Such structures often are built by the same genes activated repeatedly at different locations and time points during development. There are large numbers of muscle cells in our body, and by and large the same set of genes is required to build each of these cells. The same is true for many other cell types (e.g. liver or red blood cells) or for many structures repeated along the body axis in segmented animals (e.g. the ribs and vertebrae in vertebrates). The different instances of such repeated structures are, therefore, not individualized and do not have independent evolutionary histories. Serial homologs of this type have been called "homomorphs" (Riedl 1975; Wagner 2014). The similarity between homomorphs reflects shared developmental mechanisms rather than inheritance from a common ancestor. However, if there are at least some genetic differences between, for example, anterior or posterior segments, forelimbs and hindlimbs, or between a dorsal and a ventral group of neurons or other cell types, they will acquire independent character identity and will be able to follow separate evolutionary trajectories. Serial homologs that are thus genetically individualized have been called "paramorphs" in analogy to paralogous gene, as will be discussed further below (Minelli 2000; Wagner 2014).

However, the individualization of characters is a matter of degree. Characters that are partly genetically individualized, but share some of their genetic basis, may sometimes undergo correlated evolutionary changes due to mutations of genes shared by both characters. This has been termed "concerted evolution" (Musser and Wagner 2015) (Fig. 2.2C). The forelimbs and hindlimbs of vertebrates or segmented structures at different levels of the body axis as discussed above are particularly obvious examples. However, characters may also be linked genetically to other characters, even if they are not serial homologs. Two structures may, for example, develop from a common genetically specified precursor before their developmental fate separates and is directed by different genes. The lateral line and the inner ear, which are both derived from a common region, the pre-placodal ectoderm (PPE) defined by expression of *Six1* and *Eya1* are a case in point (e.g. Schlosser 2006; Moody and LaMantia 2015; Streit 2018; see Chapter 3 in Schlosser 2021). Or the same gene may be used for the construction of different and developmentally unrelated structures. The *Six1* and *Eya1* genes just mentioned, for example, play essential roles not only for PPE specification but also for kidney development (e.g. Xu 2013). Characters may also be incompletely individualized and linked to other characters if their function is only partially separable. Occasional functional interactions may promote some co-evolution between different characters in such cases (e.g. between the ear and lateral line system, if otic modulation of lateral line information contributes to prey localization in some environments).

Because characters are complex, composite structures and because parts of these structures may be reshuffled during evolution, character identity or homology has to be established always for a particular level of comparison (Roth 1991; Striedter and Northcutt 1991). For example, a character may be homologous between two species because it forms a conserved part of a larger structure inherited from a common ancestor, even though its components have been substituted by other (non-homologous) components or the way it is built developmentally has changed in evolution. For example, arthropod segments and the regulatory network of segment-polarity genes establishing these segments have been evolutionarily conserved in spite of wide divergence of upstream generative mechanisms and genes involved (Peel, Chipman, and Akam 2005), a phenomenon termed "genetic piracy" or "developmental system drift" (Roth 1988, True and Haag 2001). Conversely, a character may be homologous between two species because it inherited its compositional structure or development from their common ancestor but has become co-opted into new (non-homologous) developmental contexts. For example, many signaling pathways have adopted roles in new developmental contexts during evolution; a well-studied case is the new role of Hedgehog signaling in butterfly eye spot development (Pires-daSilva and Sommer 2003).

2.1.2 CHARACTER IDENTITY NETWORKS

It has recently been proposed that the evolutionary stability of characters despite pervasive changes in their properties (or character states) is enabled by a hierarchical organization of metazoan development. Eric Davidson and colleagues have first suggested that evolutionarily conserved core networks of regulatory genes or "kernels" are linked to flexible "differentiation gene batteries" (Davidson and Erwin 2006). Günter Wagner has subsequently proposed, more generally, that core networks of regulatory genes form so-called character identity networks (ChINs) that determine character identity, while a battery of "realizer genes" downstream of these core networks is responsible for establishing the detailed properties (Wagner 2007b, 2014). For example, the retinal determination network between the *Eyeless*, *Sine oculis*, *Eyes absent*, and *Dachshund* genes, determines eye identity in insects and other metazoans, while the downstream genes activated by this network, determine size, shape, color, sensitivity, and other properties of the eye (Wawersik and Maas 2000; Silver and Rebay 2005; Kumar 2009; Xu 2013). Evolutionary conservation of the retinal determination network preserves the identity of eyes and allow us to recognize their homology through evolutionary lineages, while changes in the realizer genes account for their diversification and adaptation.

Regulatory genes encoding transcription factors (e.g. *Eyeless*, *Sine oculis*), play a particular central role as members of these kernels or ChINs because they are able to turn transcription (expression) of other genes on or off (Fig. 2.3). A gene is only transcribed into messenger RNA when the basal transcriptional apparatus (BTA),

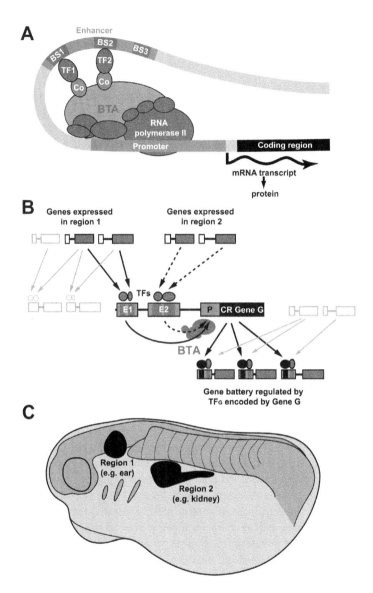

FIGURE 2.3 Combinatorial regulation of gene transcription. (**A**) A gene is transcribed into mRNA when the basal transcriptional apparatus (BTA) is recruited to the promoter by stabilizing interactions with transcription factors (TF) and cofactors (Co) that bind to their binding sites (BS) in enhancers. The mRNA is then translated into protein. (**B, C**) A gene G may be regulated by multiple enhancers (E1, E2), each with a particular combination of binding sites (shaded boxes) for TFs. Therefore, each enhancer responds to a different combination of TFs (encoded by genes expressed in different regions) (**B**) and regulates transcription of gene G in a different region (E1 for region1; E2 for region 2) (**C**). If gene G itself encodes a TF, it can bind to enhancers of multiple other genes and thus activate (in cooperation with other TFs) a whole gene battery in each region (**B**). ([A] Redrawn from Wolpert et al. 2019; [B,C] Modified from Schlosser 2004.)

a complex of the enzyme RNA polymerase II with multiple other proteins, stably binds to the promoter region located in front of the coding region of the gene (Fig. 2.3A) (Spitz and Furlong 2012; Peter and Davidson 2015; Barresi and Gilbert 2019; Wolpert, Tickle, and Martinez-Arias 2019). This requires interactions with specific transcription factors, proteins equipped with a DNA-binding domain that recognizes particular short nucleotide sequences. With a second protein domain, transcription factors are able to bind directly or indirectly (via cofactors) to the BTA. For some transcription factors (activators) this results in the stabilization of BTA binding to the promoter, while for others (repressors) it leads to de-stabilization. Binding sites for different transcription factors are clustered together in so-called enhancers. Like promoters, enhancers are located on the same chromosome as the coding region and therefore are referred to as cis-regulatory regions; however, in contrast to promoters, enhancers are often found at some distance from the coding region. Activation of gene transcription requires that enhancers are occupied by a combination of different activating transcription factors and that repressive transcription factors are absent. The combination of binding sites within each enhancer essentially defines the combination of transcription factors that can activate gene expression. Because this particular cocktail of transcription factors may be present only in a particular region or stage of the developing embryo, enhancers regulate the region- and stage-specific expression of a gene. Often genes are regulated by multiple enhancers, each driving transcription of the gene in a different region or developmental stage (Fig. 2.3B, C).

When genes encoding transcription factors (or co-factors and other regulatory proteins) cross-regulate each other (either directly or indirectly, e.g. mediated by signaling between cells) in a ChIN, they define a regulatory state characterized by the persistent presence and regulatory activity of a combination of transcription factors. These will then be able to activate batteries of realizer or differentiation genes (all those genes with enhancers responding to this particular combination). I will use the term "regulatory state" here to refer to a relatively stable, non-transitory state or "attractor" of a gene regulatory network (Graf and Enver 2009 – or valleys in the "epigenetic landscapes" of Waddington 1957, for those readers familiar with this model). ChINs establishing a particular regulatory state may be maintained, even if downstream target genes change over time (by losing or gaining responsive enhancers). An attractive feature of this view is that it allows us to understand how we can recognize identity (homology) of structures across taxa even if their properties have changed drastically: structures are homologous as long as they share the same ChIN, even if they have a very different phenotype due to highly diverged or different realizer/differentiation genes.

The view that character identity in evolution is guaranteed by invariant ChINs provides a useful criterion to trace the evolutionary fate for some types of characters such as cell types (see below). However, we should bear in mind that it is probably an oversimplification. The maintenance of character identity and individuality throughout evolution may be possible even in the absence of ChINs or when ChINs themselves change during evolution (e.g. due to developmental system drift) as long as a series of plausible transformations between character-defining regulatory networks in different taxa can be established. I only want to mention this here as a caveat; this is not the place to develop these ideas further.

2.1.3 EVOLUTIONARY INNOVATION

Many characters show a surprising resilience in the face of evolutionary changes and are preserved over vast periods of time and speciation events. We are, therefore able to recognize the same "bauplan" composed of multiple homologous structures, for large clades of organisms often comprising thousands of different species. However, occasionally novel characters appear in evolution that have no clear homolog in ancestors or sister taxa (Moczek 2008; Hallgrimsson et al. 2012; Peterson and Müller 2013). There is still a lively debate among evolutionary biologists on what, precisely, constitutes an evolutionary novelty (e.g. Brigandt and Love 2012; Hall and Kerney 2012; Peterson and Müller 2013). Here I adopt the view that characters that merely differ in their properties or character states while retaining their identity (e.g. by retaining a ChIN) do not count as evolutionary innovations or novelties (Wagner 2014). Only characters that form a new independently evolving, individualized unit qualify.

However, evolution is notoriously opportunistic. It can only "tinker" with what is already there rather than creating new structures out of thin air (Jacob 1977). Moreover, it has to rebuild the organism while keeping it alive and without interfering with any essential functions. Evolution is thus like rebuilding ships while at sea: novel characters are fashioned from cobbling together old parts while keeping the vessel afloat. Therefore, any new structure is typically built out of old parts (such as pre-existing cell types, signaling pathways, or local patterning mechanisms). What is new in a novelty are not the parts but the way they interact with each other. Thus, even though novelties do not have homologs in other lineages (Müller and Wagner 1991), their parts often do. Moreover, novel structures may be embedded into pre-existing higher order structures (such as global patterning systems). Even for novel structures, we therefore expect to find homologous components and possibly a homologous address in a conserved global coordinate system in other species that are derived from common ancestors preceding the origin of the novelty. The phenomenon that novel structures are built from components that have clear homologs in other taxa has been termed "deep homology" (Shubin, Tabin, and Carroll 2009), but I will avoid this terminology here because it is misleading and obscures the insight that homology is level specific and that non-homologous structures can be built from homologous parts.

Functionality of the organism has always to be maintained when old parts are being redeployed to form novel structures. One way to achieve this is by duplication (or multiplication) of structures followed by at least partial individualization. One of the structures can then continue to fulfill its original function, while the other one is free to diverge. Duplication and divergence is, therefore, one of the major routes to innovation (Wagner and Lynch 2010). A second major route to innovation is the recombination (redeployment) of old parts in new ways, thereby forming new networks of interaction. If the parts themselves are present multiple times – for example, repeated organs in segmented animals or multiple cells of the same cell type – this may be possible without compromising previous functions. Once a novel structure has originated it may accumulate various adaptive changes without losing its identity. The following examples of novel sense organs that evolved within vertebrates shall briefly illustrate these two pathways to innovation.

2.1.3.1 Novel Sense Organs Originating by Duplication and Divergence

As introduced in Chapter 1 of Volume 1 (Schlosser 2021), the vertebrate inner ear contains multiple sensory areas with specialized functions in detection of gravity, angular acceleration, or hearing. However, the number of these sensory areas is quite variable between different vertebrate groups (Fig. 2.4A). While some areas have been lost in certain taxa, other areas have probably split into two (Fritzsch et al. 2013; Fritzsch and Elliott 2017). For example, shared innervation patterns suggest that the amphibian papilla, which is unique to amphibians and contributes to hearing, has most likely evolved by the splitting of the evolutionarily ancient neglected papilla. The latter is present in many fishes and is maintained together with the amphibian papilla in one group of amphibians, the caecilians. Similarly, the basilar papilla of tetrapods, which is involved in hearing, has probably evolved by the splitting of the lagenar macula, a sensory area involved in gravity reception in many fishes and amphibians (Fig. 2.4A). Evolutionary persistence of these new sensory areas and the adoption of a new function can only have been possible after they were at least partly individualized and acquired a distinct genetic identity. Although, the regulatory networks defining the identity of different sensory areas are still not well understood, different sensory areas in the mammalian inner ear do indeed express different combinations of transcription factors, some of which have been shown to be essential for their identity (e.g. Fekete and Wu 2002; Groves and Fekete 2012). Once sensory areas are individualized they may evolve adaptations to peculiar environments or life styles. The basilar papilla, for example has evolved into the organ of Corti in mammals, a highly specialized and structured hearing device employing modified sensory hair cells – the so-called outer hair cells – equipped with the motor protein prestin for stimulus amplification (Fritzsch and Elliott 2017; see Chapter 6 in Schlosser 2021).

The evolutionary origin of the vomeronasal organ, involved in detection of pheromones and other large molecules in tetrapods, provides a second example for the role of duplication and divergence in sensory innovation (Grus and Zhang 2009; Silva and Antunes 2017) (Fig. 2.4B). Lampreys, cartilaginous and most bony fishes lack a vomeronasal organ. Instead the so-called ORA and OlfC receptor proteins, related to the tetrapod's vomeronasal V1R and V2R receptor families (Fig. 2.4B), respectively, are expressed in receptor cells within the main olfactory epithelium (for details see Chapter 6 of Schlosser 2021). The differential expression of transcription factors Fezf1 and Fezf2 in the olfactory and vomeronasal epithelia, respectively, may contribute to the individualization of both sensory epithelia probably in cooperation with other factors that remain to be identified (Shimizu and Hibi 2009; Chang and Parrilla 2016; Parrilla et al. 2016).

The four-eyed teleost fishes of the genus *Anableps,* which swim along the water line, may serve as a final example here (Perez et al. 2017). During development of these fishes the eye becomes subdivided by the iris growing inwards along the equator splitting the pupil and cornea. The dorsal and ventral parts of the eye, including cornea, lens, and retina then acquire different specializations for vision in aerial or aquatic environments. The ventral part of the retina, which receives visual input from the air, for example, has higher visual acuity and a thicker inner nuclear layer

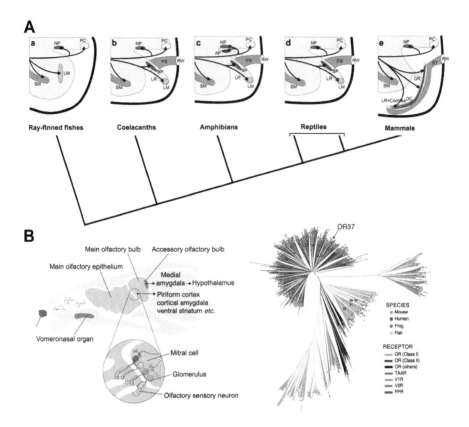

FIGURE 2.4 Evolutionary innovation in sense organs arising by duplication and divergence. (**A**) Diagrams of the sensory areas in the inner ear of various vertebrates. Innervation is shown in black. The primitive condition (**a**) with an undivided lagenar macula (LM) and neglected papilla (NP) is found in basal actinopterygians, some cartilaginous fishes (chimeras) and lungfishes. In the derived condition (**b**), found in coelacanths and most amniotes (except therian mammals), the lagenar macula has moved into the lagenar recess (LR) and a basilar papilla (BP) has split off. In therian mammals (**e**), the lagenar macula was lost and probably incorporated into the elongated basilar papilla, which forms the organ of Corti (OC). In amphibians (**c**), the amphibian papilla (AP) has separated from the neglected papilla. The situation illustrated here is typical for many caecilians; frogs and salamanders (not shown) have lost the neglected papilla. DR, ductus reuniens; PC, posterior canal crista; PS, perilymphatic sac; RW, round window; SM, saccular macula; ST, scala tympani. (**B**) Origin of the vomeronasal organ. Left side: Main olfactory epithelium (projecting to the main olfactory bulb) and vomeronasal organ (projecting to the accessory olfactory bulb) in a rodent. Sensory neurons expressing the same odorant receptor (OR) (same color) project to the same glomerulus in the olfactory bulb. Right side: Gene tree based on a sample of 2700 olfactory GPCR genes from zebrafish (*Danio rerio*), frog (*Xenopus laevis*), mouse (*Mus musculus*), and human (*Homo sapiens*). Note the presence of receptors related to V1R and V2R already in fishes; the large expansions of Class II ORs (red) in tetrapods, V2Rs (purple) in amphibians, TAARs (blue) in fishes, and V1Rs (teal) in mammals; and significant losses of functional ORs, V1Rs, and V2Rs in humans (blue dots) compared to mice (green dots). ([A] Reprinted with modification with permission from Fritzsch et al. 2013; [B] reprinted with permission from Bear et al. 2016.)

probably reflecting an increased number of bipolar cells. It also expresses a different opsin (Rh2-1, responding to medium wavelengths) than the dorsal retina (LWS, responding to long wavelengths). Which regulatory networks control the identity of aerial and aquatic eye in this case is currently not known.

2.1.3.2 Novel Sense Organs Originating by Recombination/Redeployment

Novel sensory organs mediating different sensory modalities evolved multiple times independently in several vertebrate lineages, when somatosensory neurons of the trigeminal nerve, which innervate the mouth cavity and snout, became closely associated with other cell types. I will present a few of those examples here to illustrate how new combinations of pre-existing structures can result in the formation of novel sense organs. In mammals, the whisker hairs have evolved as a new tactile organ, which may have facilitated the adoption of a primarily nocturnal life style (Gerkema et al. 2013; Maor et al. 2017). The development of whiskers is controlled by partly different transcription factors than other body hair (e.g. Fantauzzo and Christiano 2012) allowing whiskers to evolve to some extent independently from other body hair and to acquire independent adaptations. Whisker hairs can fulfill their peculiar sensory function due to their innervation by several types of mechanosensory trigeminal neurons that form specialized nerve endings (Takatoh et al. 2018). These trigeminal neurons are themselves characterized by the expression of a peculiar combination of transcription factors including Tbx3 and Onecut2, allowing them to evolve independently from other trigeminal neurons (Erzurumlu et al. 2010).

Further specialized mechanosensory organs evolved in some groups of mammals, but we know very little about the genetic control of their development. Rorqual whales are a group of baleen whales with expandable throats that feed by taking in large gulps of water and filtering out prey organisms, a strategy known as lunge feeding. To monitor this, a new sense organ has evolved in this group of whales by repurposing some of the innervation of teeth, which do not develop in baleen whales. Blood vessels and trigeminal nerve fibers from either the right or the left anterior-most tooth socket grow into the fibrous connection between their unfused lower jaws, where they presumably respond to the mechanical stresses in this tissue during lunge feeding (Pyenson et al. 2012) (Fig. 2.5A,B). Another sense organ has evolved in moles. In this group of insectivore mammals, which live underground in self-made tunnels, tactile stimuli have become particularly important for orientation and predation. For this purpose, moles use specialized organs, termed Eimer's organs, which cover the tip of their noses. Each Eimer's organ is a small swelling produced by epidermal cells, which are stacked up in a columnar fashion and are associated with specialized touch receptor cells (Merkel cells and lamellated corpuscles) and free nerve endings of trigeminal neurons (Marasco et al. 2006) (Fig. 2.5C–E).

Other groups of mammals have evolved electroreceptive sense organs by redeploying somatosensory neurons of the trigeminal nerve. This has happened at least twice independently, in the egg-laying mammals (monotremes) and in some dolphins. In monotremes such as the platypus, both mucous and serous glands of the snout are innervated by trigeminal neurons with specialized terminals and form

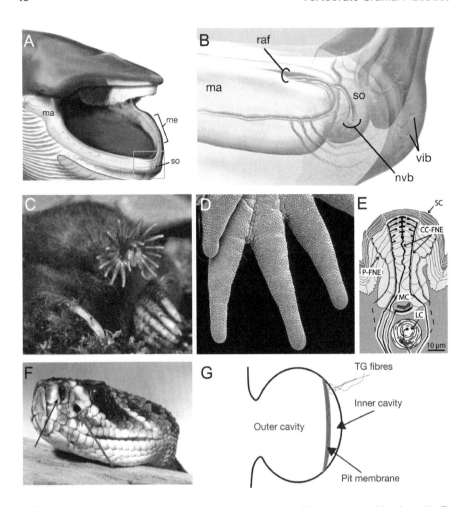

FIGURE 2.5 Evolutionary innovation in sense organs arising by recombination. (**A, B**) Novel mechanosensory organ in the mandibular symphysis of rorqual whales. ma, mandibles; me, mental foramina; nvb, neurovascular bundles; raf, relictual alveolar foramen; so, sensory organ; vib, vibrissae. (**C–E**) Eimer's organs as novel mechanoreceptive organs in the star-nosed mole (*Condylura cristata*). Appendages on the snout tip (rhinarium) (**C**) are covered with Eimer's organs, apparent as small bulges in a transmission electron micrograph (**D**). (**E**) Schematic view of Eimer's organ. A central epidermal cell column is associated with intraepidermal free nerve endings (CC-FNE) and surrounded by smaller peripheral free nerve endings (P-FNE). Merkel cell–neurite complexes (MC) and lamellated corpuscles (LC) are located at the base of each organ. SC: stratum corneum. (**F, G**) Novel pit organ for infrared sensing in pit vipers. (**F**) Rattlesnake head with location of nostril (black arrow) and pit organ (red arrow) indicated. (**G**) Schematic of pit organ structure showing innervation of pit membrane by trigeminal (TG) fibers. ([A,B] Reprinted with permission from Pyenson et al. 2012; [C-E] reprinted with permission from Marasco et al. 2006; [F,G] reprinted with permission from Gracheva et al. 2010.)

electroreceptive organs responding to positive (cathodal) stimuli (Proske, Gregory, and Iggo 1998; Pettigrew 1999). In the Guiana dolphin, the crypts of the whisker hairs, which have been lost, are also innervated by trigeminal neurons and serve a similar purpose (Czech-Damal et al. 2012, 2013). In both cases, the mucus or gel present in the gland or crypt serves as an electrically conductive substrate. Unfortunately, nothing is currently known about how the development and function of these mammalian electroreceptors is genetically controlled.

A final example of a novel vertebrate sense organ involving trigeminal neurons is the infrared organ in snakes (Fig. 2.5F, G). This is particularly sensitive in venomous pit vipers such as rattlesnakes, where it forms a characteristic pit organ below the eye. The pit organ acts as an infrared sensor, which produces a thermal image of warm-blooded prey organisms and together with the eye helps to localize the prey (Gracheva et al. 2010). The narrow opening of the pit organ produces a rough image on the pit membrane, which is innervated by free nerve endings of trigeminal neurons (Fig. 2.5G). It has been shown that these fibers respond to thermal stimuli and not photons and do so by using TRPA1 ion channels, which respond to chemical irritants or cold temperatures in other vertebrates (Gracheva et al. 2010).

2.1.4   THE PHYLOGENY OF CHARACTERS: TREES VERSUS NETWORKS

Because individualized characters form independent evolutionary units, their evolutionary history or phylogeny can be traced like the phylogeny of species and genes (Geeta 2003; Serb and Oakley 2005; Wagner 2014). However, while the evolution of species and genes usually can be depicted as a tree (except for cases like hybridization between species), the evolutionary relationships between characters are not necessarily tree-like (Oakley 2017; Schlosser 2019). This is only the case for those characters that give rise to new characters by duplication and divergence. If novel characters originate by the duplication and divergence of ancestral characters their evolutionary lineage can be represented as a branching tree (Fig. 2.6A). The character tree – like a gene tree – is embedded within the species tree but can have additional branching points due to duplication and divergence. Since characters or genes that are derived from a common ancestor are considered homologous regardless of whether they are derived from an ancestral character or gene through duplication or through speciation, two types of homologs can be distinguished. This was first pointed out by Fitch who introduced a new terminology for gene trees (Fitch 1970): Genes are called orthologs, if they are derived from a common ancestor through speciation and paralogs, if they are derived from a common ancestor through gene duplication (Fig. 2.6A). In analogy, characters may be called orthomorphs, if they are derived from a common ancestor through speciation and paramorphs (Minelli 2000; Wagner 2014), if they arise through duplication and divergence. Like paralogous genes, paramorphous characters can evolve independently from each other. Characters, thus, only qualify as paramorphs, if they have acquired a certain genetic individuality (e.g. forelimbs versus hindlimbs of vertebrates). As explained above, paramorphs have to be distinguished from homomorphs, which are simply repeated structures with the same genetic basis (e.g. different muscle cells in the same organism). A genetic equivalent of homomorphs would be genes only duplicated in the somatic cells but not in the germ line of an organism.

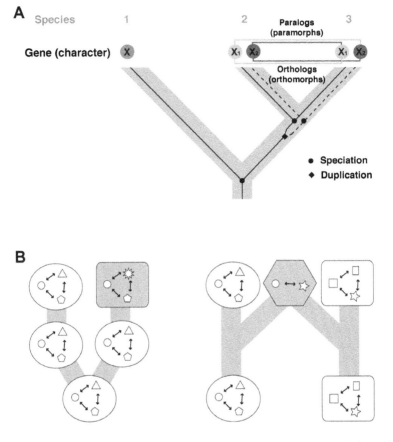

FIGURE 2.6   Character phylogeny. (**A**) Genes or characters can duplicate and diverge independent of speciation events. The resulting gene or character trees (black) are embedded within the species tree (grey). Genes (or characters) are homologs if they are derived from a common ancestor through speciation (orthologs; for characters: orthomorphs) or through duplication (paralogs; for characters: paramorphs). Thus X_1 in species 2 and 3 are orthologs as are X_2 in species 2 and 3, while X_1 and X_2 are paralogs. To describe the relationships of genes before and after duplication, gene X is called the pro-ortholog of X_1 and X_2, while X_1 and X_2 are called the semi-ortholog of X. (**B**) Novel characters may arise either by duplication and divergence (left) or by recombination (right). Characters are depicted as large symbols, their components that interact to form their core regulatory network (CoRN) as small symbols. (Reprinted from Schlosser 2019. Modified from Oakley 2017.)

However, novel characters do not necessarily arise by duplication and divergence. If novel characters originate instead by the recombination of components of two or more pre-existing characters, this may be depicted as fusion rather than a divergence of evolutionary lineages resulting in a network-like pattern of evolutionary lineages (Oakley 2017) (Fig. 2.6B). However, the depiction of character origination as a "fusion" event has to be taken with some grains of salt. First, novel characters often originate not by the recombination of individualized pre-existing characters, but by the recombination of parts that may not be individualized, independently evolving units themselves. Second, the origination of a novel character involves the establishment of specific, new interactions between pre-existing components rather than their mere juxtaposition. The whisker hairs of mammals only become a new character, which is independently evolving from other body hair, due to their association with a peculiar subset of trigeminal neurons mediating their heightened tactile responsiveness (Takatoh, Prevosto, and Wang 2018).

Character phylogenies are also distinguished from phylogenies of species or genes by another important difference. Because characters are often only partially individualized but still share some of their genetic basis with other characters, they will act as independently evolving unit only for some genetic changes, whereas for other they will change in coordination with other characters (concerted evolution; Fig. 2.2C). Changes in the shared genetic basis of characters may therefore lead to correlated changes in different branches of a tree or network depicting character phylogeny. Evolutionary changes in the *Eda*, *Sox9*, and *NF-κB* genes involved in hair follicle development (Kapsimali 2017), for example, are likely to affect development of body hair and whiskers. In addition, characters may also co-evolve with other characters due to functional interactions between them (see above), a topic that I will not further explore here.

2.2 HOMOLOGY AND INNOVATION IN CELL TYPE EVOLUTION

2.2.1 WHAT IS A CELL TYPE?

As for any other character, homology and innovation in cell types can only be recognized if cell types are conceived as independently evolving units. Therefore, an evolutionary concept of cell types has been recently proposed according to which a cell type is "a set of cells in an organism that change in evolution together, partially independent of other cells, and are evolutionarily more closely related to each other than to other cells" (Arendt et al. 2016). In contrast to traditional definitions of cell types, which are based on similarities in morphology, physiology, or gene expression, this concept allows us to distinguish homology from homoplasy in cell type evolution and to trace the evolutionary lineages of cells even in the face of evolutionary changes of phenotype or function.

Like other characters, cell types can evolve by duplication and divergence of pre-existing cell types or by recombination and re-deployment of some of their components. In the first case, an ancestral cell type gives rise to two sister cells (Arendt 2008) and cell type evolution can be depicted as a tree (Fig. 2.7). By repeated duplication events, one cell type can give rise to a cell type family, for example, the family of neurons (Fig. 2.7) (Arendt et al. 2019). Because cell types (like genes) may duplicate in a lineage independent of speciation events, the cell lineage tree (like a gene tree) can be different from but is nested within the species tree (Fig. 2.8A, B). After duplication, the sister cell types originating from a duplication event may both persist in individuals of the same species (forming paramorphs) just as paralogous genes may persist in the same genome. The evolution of various types of photoreceptors (Fig. 2.8C) discussed below illustrate this mode of cell type evolution.

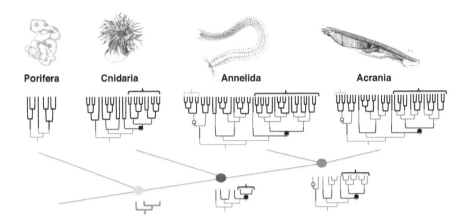

FIGURE 2.7 Cell types and cell type families. Hypothetical scheme showing cell type diversification during animal evolution. Colored branches of cell type tree show cell diversification events that had taken place already in the last common ancestor of a group of extant animals with their sister group. Black branches show additional, lineage-specific events of cell type diversification. Black box highlights origin of one particular cell type (e.g. neurons) and black brackets indicate members of the cell type family arising from its diversification. Grey box and brackets show another cell type family. Because a cell type (e.g. neuron) may diversify differently in each of two animal groups A and B after these separate in a phylogenetic branching event, there is no one to one correspondence between members of the cell type family arising in lineage A and members of the cell type family arising in lineage B (e.g. the different types of neurons in cnidarians and annelids). (Reprinted with modification with permission from Arendt et al. 2019.)

However, in the second mode, when cell types arise by the recombination of components of older cell types, depictions of their evolutionary history will show the fusion rather than divergence of evolutionary lineages. As discussed above, their phylogeny will then resemble a network more than a tree. It is generally accepted that novel cell types can originate in both ways and I will discuss some examples below. However, it is currently controversial whether both modes of cell type evolution are equally prevalent or whether one of these modes – either duplication and divergence or recombination – predominates (Arendt et al. 2016; Oakley 2017). In the last section of this chapter, I will show that the two modes of cell type evolution are not mutually exclusive and may typically intertwine.

The identity of cell types like that of other individualized, independently evolving characters is defined by a core regulatory network of genes, often genes encoding transcription factors (Fig. 2.9A). Of particular importance are so-called "terminal selector genes" (Hobert 2008, 2011), which control the expression of large batteries of genes encoding the proteins responsible for the cell-type specific morphology and physiology of a differentiated cell. In addition, they usually activate their own expression, often by direct transcriptional autoregulation and repress alternative cell fates. Consequently, overexpression of one or a small number of terminal selectors in other cell types is often sufficient to stably convert cell type

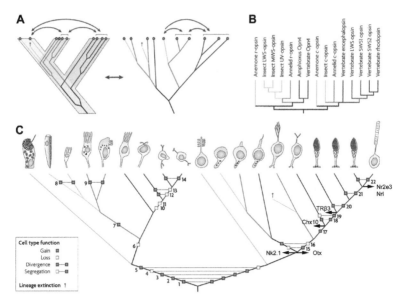

FIGURE 2.8 Cell type evolution. (**A**) A cell type tree shown as it is nested within the species tree (left) or as a separate tree (right). Evolution of sister cell types within a species (brackets) and of homologous cell types between species (arrows) is indicated. (**B**) Gene tree for the *opsin* family of G-protein coupled receptors. *C-opsin*: ciliary opsin; *MWS*, medium wavelength-sensitive; *Opn4*, opsin4; *R-opsin*, rhabdomeric opsin; UV, ultra violet. (**C**) Cell type tree for photoreceptors. Yellow, blue, red, and grey represent the evolution of cell types in Ecdysozoa, Lophotrochozoa, Deuterostomia, and Cnidaria, respectively. Green represents evolution in the protostome stem line. Black represents evolution in the metazoan and bilaterian stem lines. Depicted from left to right are photoreceptors from the cnidarian rhabdomeric eye, insect rhabdomeric compound eye, insect rhabdomeric larval eye, polychaete rhabdomeric adult eye, polychaete rhabdomeric larval eye, amphioxus rhabdomeric Hesse eyecup, vertebrate retinal ganglion cell, vertebrate amacrine cell, vertebrate horizontal cell, cnidarian ciliary photoreceptor, insect light-sensitive neurosecretory cell, annelid light-sensitive neurosecretory cell, vertebrate light-sensitive neurosecretory cell, annelid brain ciliary photoreceptor, vertebrate bipolar cell, vertebrate long wavelength-sensitive (LWS) cone, vertebrate short wavelength-sensitive 1 (SWS1) cone, vertebrate SWS2 cone and vertebrate rod. Changes in core transcription factors are indicated (arrows) as are the following changes in character states: 1, opsin storage in the surface-extended apical membrane (rhabdomeric photoreceptor) or in the membrane of the cilium (ciliary photoreceptor); 2, deployment of the r-opsin of the c-opsin duplicate; 3, use of the $G\alpha_q$ and phospholipase C or of the $G\alpha_i$ and cGMP phototransduction cascade; 4, control of reproduction via neurosecretion; 5,6, locomotor cilium; 7, assembly of rhabdomeric photoreceptor cells into an eye: visual function; 8, specialization into larval or adult eye photoreceptors (for example, in arthropods); 9, specialization into larval or adult eye photoreceptors (for example, in annelids); 10, rhabdomeric photoreceptor structure; 11, direct association with pigment cell; 12, light sensitivity; 13, axonal projection to brain; 14, specialization into horizontal or vertical interneuron; 15, surface-extended cilium; 16, control of reproduction via neurosecretion; 17, visual function; 18, axonal projection or interneuron function; 19, light sensitivity via c-opsin; 20, deployment of the LWS or of the SWS1–2–rhodopsin duplicate. 21, deployment of the SWS1 or of the SWS27–rhodopsin duplicate; 22, deployment of the SWS2 or of the rhodopsin duplicate. ([A-C] Reprinted with modification with permission from Arendt 2008.)

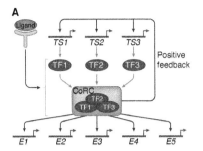
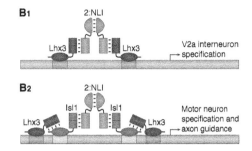

FIGURE 2.9   Core regulatory networks (CoRNs) and complexes (CoRCs) for cell type identity. (**A**) A model of cell type identity determination. A small set of terminal selector genes (*TS1* to *TS3*) encode transcription factors (TF1 to TF3), which are modified through the activation of signaling pathways upon binding of a ligand and form a core regulatory complex (CoRC). The CoRC regulates downstream effector genes (*E1* to *E5*) and maintains its own expression. Grey arrows indicate translation and complex formation. Black arrows indicate transcriptional regulation. (**B**) Combinatorial and cooperative interactions of transcription factors Lhx3 and Islet1 and their cofactor (NLI) determine the identity of V2a interneurons and motor neurons in the spinal cord. Based on Lee et al. 2008. ([A,B] Reprinted with permission from Arendt et al. 2016.)

identities (Graf and Enver 2009). The conversion of fibroblasts into muscle cells by overexpression of the muscle selector gene *MyoD* is a well-known example (Davis, Weintraub, and Lassar 1987). Similarly, the transcription factor PET-1, directly controls expression of multiple proteins of serotonergic neurons in the vertebrate nervous system. These include the enzymes involved in serotonin synthesis, proteins involved in serotonin transport, and reuptake and many others (Deneris and Wyler 2012; Spencer and Deneris 2017). PET-1 also directly activates its own expression (Spencer and Deneris 2017) and in combination with other transcription factors is able to transform fibroblasts into serotonergic cells (Vadodaria et al. 2016; Xu et al. 2016). Two other transcription factors, Nurr1 and Pitx3 play comparable roles in regulating the identity of dopaminergic neurons in the midbrain (reviewed in Smidt and Burbach 2009).

In many cases, the terminal selectors controlling cell identity have been shown to engage in protein-protein interactions with other transcription factors and cofactors when binding to the enhancers of their target genes (Fig. 2.9A). Such regulatory protein complexes that regulate the transcription of target genes in a cooperative fashion have been termed "core regulatory complexes (CoRCs)" (Arendt et al. 2016). The interaction of the LIM homeodomain proteins Lhx3 and Islet1 with each other and with the cofactor nuclear LIM receptor (NLI) during specification of different cell types in the amniote spinal cord illustrates this concept nicely (Fig. 2.9B). While a tetrameric complex of two copies each of Lhx3 and NLI specifies V2a interneuron identity, a hexameric complex of Lhx3, Islet1, and NLI specifies motor neurons (Lee et al. 2008). More generally, and to also include cases where cross-regulating transcription factors control cell type identity without direct protein-protein interactions, I will refer to a network of transcriptional regulators of cell type identity as a "core regulatory network" (CoRN).

Although terminal selectors are crucial components of a CoRN, they are often not sufficient to specify a particular cell type and may specify different cell types in different contexts. For example, the serotonergic selector PET-1 discussed above is also expressed in non-serotonergic neurons of the brain as well as in pancreas and kidney, where it does not promote serotonergic differentiation (Pelosi et al. 2014). Only tissues that also express another transcription factor, Nkx2.2, such as the ventral hindbrain respond to PET-1 activation with serotonergic differentiation because Nkx2.2 and its immediate downstream targets GATA2/3 confer the competence allowing Pet-1 to drive serotonergic cell fate (Cheng et al. 2003; Deneris and Wyler 2012; Spencer and Deneris 2017). Competence here refers to the ability of cells to adopt a particular cell fate when challenged, for example, by exposing it to certain signals or by up-or down-regulation of certain transcription factors. Competence to adopt a particular fate in response to some signals requires that a cell is able to recognize the signals (e.g. by expressing the proper receptors), that it expresses all required transcription factors and cofactors for activation of terminal differentiation genes for this cell fate, and that these terminal differentiation genes are ready for activation (e.g. Servetnick and Grainger 1991; Groves and LaBonne 2014; Singh and Groves 2016). The last condition requires that genes are accessible to activating transcription factors and are not located in regions of highly condensed chromatin or associated with histones carrying repressive posttranslational modifications (e.g. trimethylation of lysine 27 in histone 3, abbreviated as H3K27me3).

Thus, cell type identity is not specified by terminal selectors alone (e.g. PET-1), but also depends on the expression of additional transcription factors promoting competence (e.g. Nkx2.2 and GATA2/3). These act in combination with terminal selectors and allow the same terminal selectors to specify different cell fates in different contexts. Therefore, such competence factors need also be considered to be part of the CoRN defining cell type identity. Competence factors are not sufficient to activate transcription of their target genes but are required (sometimes only transiently, paving the way for subsequent binding of other transcription factors) for activation of their target genes in combination with other transcription factors. Typically, competence factors bind to a large number of target genes, only a fraction of which will be finally activated in a particular differentiated cell after binding of additional transcription factors with narrower binding specificities, viz. the terminal selectors. Competence factors, thereby define a developmental potential for cells that is larger than the fate they ultimately adopt.

A particularly important category of competence factors are pioneer factors. These are able to attach to their DNA binding sites even when methylated (repressed) and/or located in condensed chromatin and make these regions accessible for other transcription factors (Cirillo et al. 2002; Zaret and Carroll 2011; Iwafuchi-Doi and Zaret 2016; Iwafuchi-Doi 2019). Once bound, pioneer transcription factors recruit DNA demethylases and histone modifying enzymes and establish a "poised" state of chromatin characterized by the presence of activating histone marks (e.g. H3K4me3, H3K4me1, H3K27ac) facilitating subsequent activation of their target genes by other transcription factors.

Cell types can be recognized as homologous between different evolutionary lineages as long as their CoRNs (or CoRCs) remain intact even if they acquire different adaptations in different lineages. These adaptations may involve the divergence of existing cellular parts or pathways or the recombination and integration of cellular components into new functional parts or pathways downstream of CoRNs. Duplication and divergence of differentiation genes or the acquisition of new regulatory elements establishing new expression domains of differentiation genes are often involved in these processes. In contrast to these adaptive changes of persistent cell types, novel cell types only originate when new CoRNs themselves evolve (followed by the accumulation of changes in their downstream targets). Like for the evolution of novelties in general, there are two major routes to cell type innovation. First, novel cell types may arise by duplication and divergence, where two sister cell types originate from one ancestral cell type. This involves the evolution of two new CoRNs by rewiring or duplication and divergence of an ancestral CoRN as illustrated schematically in Fig. 2.10 (Arendt et al. 2016). Or second, novel cell types may originate by recombination. This involves the establishment of a novel CoRN due to the formation of new interactions between regulatory genes previously active in different cell types.

Whereas cell types specified by the same CoRN are usually homologous (derived from a common ancestral cell type), the possibility of homoplasy has to be acknowledged. Under particular circumstances, expression of similar sets of transcription factors may even be selected for in different cell types, which do not share a common evolutionary history. For example, the establishment of cell to cell contacts between different cell types may be beneficial, there by allowing the formation of synaptic connections between particular neurons during evolution of neuronal circuits or facilitating the fusion of two epithelia to form an opening. Because transcription factors control the activation of genes promoting cell contacts (e.g. cell adhesion molecules, chemoattractants, digestive enzymes to remove basal lamina etc.), the expression of the same transcription factor in two different cell types may be favored in certain circumstances by natural selection and may lead to the independent establishment of similar CoRNs in different cell types. A well-studied case is the expression of Phox2 transcription factors in all types of neurons (sensory, motor, and interneurons) connected with each other in reflex circuits of the autonomic nervous system (Brunet and Pattyn 2002).

In the next sections, I will illustrate the concept of cell type identity introduced here with some examples from vertebrate sensory evolution. After considering the CoRNs defining the identity of sensory cells, I will discuss both the acquisition of adaptive changes in cell types and the origin of novel cell types by duplication and divergence or by recombination. I will then end the chapter by presenting a general model of cell type evolution.

2.2.2 Core Regulatory Networks Defining Sensory Cell Type Identity

For many cell types, the regulatory basis of cell type identity is still unknown. However, two sensory cell types of vertebrates, the photoreceptors of the retina and the hair cells of the inner ear, have been particularly well studied allowing us to draft preliminary CoRNs defining their cell type identity. In the retina, the homeodomain transcription factor Crx forms the hub of a network of auto- and cross-regulating

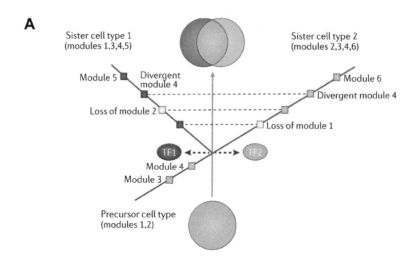

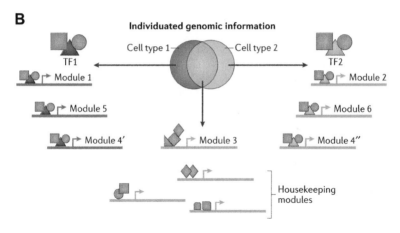

FIGURE 2.10   A model for the origin of novel cell types by duplication and divergence. (**A**) The phenotype of the precursor cell type shown here is characterized by two functional building blocks or modules (1, 2). Before diversification, two new modules arise (3, 4). A key step in the formation of sister cell types is the evolution of two distinct core regulatory networks (CoRNs) or complexes (CoRCs) employing transcription factors TF1 and TF2, respectively. The two sister cell types then diverge phenotypically by a number of adaptive changes including division of labour events (modules 1, 2), module divergence (module 4) and the acquisition of new modules (5, 6). Corresponding modules in the sister cells are connected by dashed lines. Venn diagram shows shared and cell type-specific genomic information of sister cell types. (**B**) In this example, cell types 1 and 2 gain evolutionary independence through the duplication of an ancestral transcription factor, resulting in related trimeric CoRCs. The cell type-specific genetic information (red and green) allows cell types to evolve with partial independence. Shared genetic information (orange), including housekeeping modules and the *cis*-regulatory elements driving their expression, leads to concerted evolution of the two cell types when altered. 4′ and 4″ represent different genes belonging to module 4 that have arisen through module divergence. (Reprinted with modification with permission from Arendt et al. 2016.)

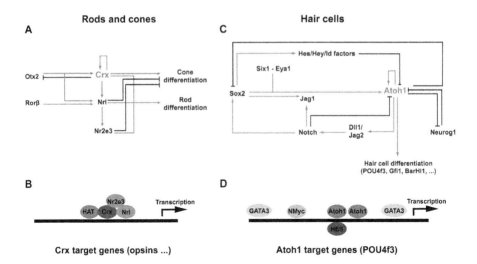

FIGURE 2.11 Core regulatory networks (CoRNs) and complexes (CoRCs) of sensory cells. (**A**) CoRN regulating cell type identity in vertebrate photoreceptors. (**B**) Regulation of enhancers of photoreceptor differentiation genes by a CoRC comprising Crx and other transcription factors. (**C**) CoRN regulating cell type identity in vertebrate hair cells. (**D**) Regulation of the hair cell enhancers of *POU4f3* by Atoh1 and other transcription factors. ([A] Based on Hennig, Peng, and Chen 2008; [B] Based on Peng and Chen 2007; [C] Modified from Raft and Groves 2015; [D] Based on Masuda et al. 2017.)

transcription factors essential for the specification of both rods and cones (Hennig et al. 2008; see Chapter 7 of Schlosser 2021) (Fig. 2.11A). In conjunction with the transcription factors Nrl and Nr2e3, which are themselves upregulated by Crx, Crx activates rod specific differentiation genes, while Nrl and Nr2e3 repress cone genes. In the absence of Nrl and Nr2e3, Crx instead synergizes with Rorβ to promote cone differentiation. The regulation of rod or cone specific target genes involves the formation of cooperative protein complexes by direct protein-protein interactions between Crx and several members of this network such as Nrl and Nr2e3 (Fig. 2.11B). Other transcription factors including Rx also play important roles for photoreceptor specification but how they integrate with Crx is not well understood (Pan et al. 2010).

In the inner ear, the transcription factor Atoh1 plays a similarly central role for conferring hair cell identity (Bermingham et al. 1999; Cai and Groves 2015; Raft and Groves 2015; see Chapter 6 of Schlosser 2021). Atoh1 is activated in hair cell precursors by two other transcription factors, Sox2 and Six1, in cooperation with its cofactor Eya1 (Ahmed et al. 2012). Once activated, Atoh1 maintains expression by positive autoregulation and cross-regulates Sox2 (Raft et al. 2007; Dabdoub et al. 2008). Atoh1 then directly activates a suite of hair cell-specific target genes and prevents adoption of the alternative neuronal fate by repression of Neurog1 transcription (Raft et al. 2007; Cai and Groves 2015) (Fig. 2.11C, D). Recent evidence indicates that Atoh1 synergizes with some of its target genes, most notably *POU4f3* (also known as *Brn3c* or *Brn-3.1*) and *Gfi1* in hair cell specification (Costa et al. 2015, 2017; 2019). It has been suggested that direct protein-protein interactions between Atoh1, POU4f3, and Gfi1 may underlie this synergy but this remains to be confirmed experimentally (Costa et al. 2017). In addition, Atoh1 probably cooperates with other, ear-specific factors (possibly competence factors) for activating hair cell-specific genes, since it is also involved in the specification of other neuronal subtypes with different gene expression profiles elsewhere in the nervous system (Cai and Groves 2015).

The regulatory networks centered on Crx and Atoh1 have been evolutionarily well conserved in vertebrates and are prime examples for CoRNs conferring cell type identity of photoreceptors and hair cells, respectively. The specification of rods versus cones by different cooperative interactions between Crx and other transcription factors furthermore suggests, how two distinct sister cell types can evolve from a common ancestral cell type by the rewiring of such CoRNs.

2.2.3 ADAPTIVE EVOLUTION OF SENSORY CELL TYPES

Once different sister cell types have their own identity and are defined by distinct CoRNs they can evolve independently and acquire different adaptations. This can again be illustrated by the evolution of rod versus cone photoreceptors. Rods have become specialized for mediating vision at low light levels. They are highly sensitive and respond broadly to light of intermediate wavelengths. This maximizes their ability to respond to weak light. In contrast, cones are less sensitive but occur in several subtypes, each with a photopigment tuned to light of different wavelengths, making them suitable for differentiating between different colors. These differences evolved after the two rounds of genome duplication at the base of vertebrates, which led to the duplication of most components of the ancestral phototransduction pathway (Hisatomi and Tokunaga 2002; Lagman et al. 2012, 2013). Rods and cones use different duplicates allowing the accumulation of different properties in the two lineages (Lamb 2013) (see Chapter 4 for more details).

While the adaptive changes of rod and cone photoreceptors happened in association with the origin of two distinct sister cell types, many cell types undergo adaptive changes in different lineages without giving rise to new cell types. The capacity to smell, for example has been highly evolutionarily flexible in vertebrates. As briefly introduced in Chapter 1 of the first volume (Schlosser 2021), vertebrate olfaction is mediated by volatile or water-soluble molecules binding to membrane bound odorant receptors (OR) in the olfactory receptor cells of the nose (for more details see Chapter 6 of Schlosser 2021). Vertebrate genomes encode a large variety of *OR* genes and each olfactory receptor cell expresses only a single OR. Different odorants can be distinguished due to the activation of different combinations of olfactory receptor cells. Importantly, however, the differences between different olfactory receptor cells are not genetically hardwired. Instead expression of a single *OR* gene is randomly assigned during development of each olfactory receptor cell (reviewed in Degl'Innocenti and D'Errico 2017; Monahan and Lomvardas 2015; see Chapter 6 of Schlosser 2021). Initially all *OR* genes are epigenetically silenced by histone methylation but low-level demethylation activity later stochastically removes repressive histone marks from one of the *OR* genes (Magklara et al. 2011). Activation of a single *OR* gene is then stabilized by an OR activated feedback signal, which keeps all other OR genes repressed (Serizawa et al. 2003; Dalton, Lyons, and Lomvardas 2013).

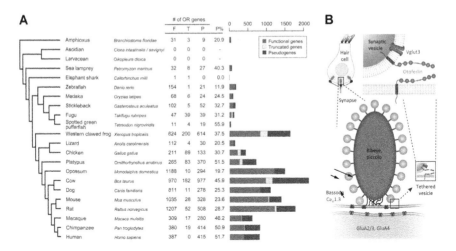

FIGURE 2.12 Adaptive changes of sensory cell types. (**A**) Numbers of different odorant receptors expressed in olfactory receptor cells are highly variable in vertebrates reflecting differences in life style and environment. (**B**) The ribbon synapse of vertebrate inner hair cells allows faithful transmission of a large range of stimulus intensities. The magnified view of a ribbon shows synaptic vesicles docked to the ribbon made up mostly of Ribeye protein. ([A] Reprinted with permission from Niimura 2009; [B] Reprinted with permission from Pangršič et al 2012.)

Therefore, olfactory receptor cells expressing different *OR* genes are not governed by different CoRNs and are merely different variants of a single cell type.

Now the repertoire of *OR* genes is highly variable in vertebrates (Niimura 2009). In mammals, *OR* genes comprise the largest gene family in the genome, with around 1000 genes in rodents. However, significant losses or increases of *OR* gene numbers by gene duplication have happened many times independently in different mammalian lineages (Figs. 2.4B, 2.12A). Primates, which are mostly diurnal and rely more heavily on vision than other vertebrates, have a much smaller complement of functional *OR* genes (300–400) and a higher proportion of pseudogenes indicating a substantial loss of *OR* genes (Figs. 2.4B, 2.12A) (Niimura 2009). Relatively low numbers of functional *OR*s and high proportions of pseudogenes have also been reported for several aquatic mammals including whales and the platypus reflecting the insignificance of airborne odorants in their aquatic habitat (Fig. 2.12A). However, other lineages of mammals, such as elephants, have greatly expanded their *OR* repertoire (Niimura, Matsui, and Touhara 2014). These various changes in the *OR* complement reflect adaptive changes of the vertebrate olfactory receptor cell by duplication and divergence (or loss) of differentiation genes without involving the origin of novel cell types.

The evolution of photoreceptors and olfactory receptor cells reviewed so far illustrate the importance of duplication and divergence for the adaptive evolution of cell types. I now want to discuss the evolution of the so-called ribbon synapse of the vertebrate hair cell as an example of a different process, where new adaptive structures originate by the recombination and integration of cellular components into new functional parts. Ribbon synapses are a specialized form of synapse found in vertebrate photoreceptors, bipolar cells of the retina and hair cells (Fig. 2.12B) (Safieddine, El-Amraoui, and Petit 2012; Wichmann and Moser 2015). They are characterized by their potentially very high release rates of synaptic vesicles and the ability to maintain a constant rate of release in response to sustained stimulation. This allows receptor cells equipped with ribbon synapses to faithfully encode a large range of stimulus intensities – up to six orders of magnitude in the case of hair cells (Matthews and Fuchs 2010).

Ribbon synapses are distinguished from regular synapses by several unique components most notably the name-giving "ribbon", an aggregate of multiple Ribeye proteins that serves as a docking station for synaptic vesicles (Schmitz, Königstorfer, and Südhof 2000). Interactions of the Ribeye protein with other proteins also present in conventional synapses such as Bassoon anchor it to the presynaptic membrane. Interestingly, Ribeye is encoded by a vertebrate-specific gene that has evolved by the fusion of a new exon to the ancient *CTBP2* gene (Schmitz et al. 2000).

The ribbon synapses of hair cells also employ a special membrane bound protein, Otoferlin, for the calcium-dependent release and recycling of synaptic vesicles (Pangršič, Reisinger, and Moser 2012). Neurotransmitter release at the ribbon synapse of photoreceptors instead relies on the SNARE complex (composed of the proteins Synaptobrevin 1/2, SNAP25, and Syntaxin) similar to other synapses (Matthews and Fuchs 2010; Safieddine et al. 2012; Wichmann and Moser 2015). Otoferlin is a member of the Ferlin family, involved in the regulation of various calcium-activated events throughout the animal kingdom (Lek et al. 2010, 2012). During the two rounds of genome duplication at the base of vertebrates, each of the two Ferlins found in invertebrates triplicated. Otoferlin, as one of the three paralogs of type 2 Ferlin, subsequently acquired its hair cell-specific function at the ribbon synapse. Finally, the ribbon synapses of hair cells use L-type voltage-gated $Ca_v 1.4$ channels and Vglut3 vesicular glutamate transporters, paralogs of the $Ca_v 1.3$ channels, and Vglut1/2 transporters used at other glutamatergic synapses or the ribbon synapses of photoreceptors (Pangršič et al. 2012).

During the evolution of the ribbon synapse in vertebrate hair cells, therefore, various cellular components were integrated into a novel functional unit. Protein-protein interactions of the Ribeye protein with itself allowed the formation of a new structure, the ribbon, whereas interactions of Ribeye with pre-existing synaptic proteins like Bassoon ensured its positioning at the presynaptic membrane. Both interactions were probably essential for the evolution of sustained high vesicle release rates. Interestingly, such ribbons are shared between photoreceptors and hair cells despite the fact that vertebrate photoreceptors and hair cells most likely did not evolve from a common cell type in the last common vertebrate ancestor but from distinct pre-existing photoreceptive and mechanoreceptive cell types (see Chapters 3 and 4). A convergent evolution of synaptic ribbons in photoreceptors and hair cells is extremely unlikely given their extensive molecular similarities. The synaptic ribbon, therefore, may have either arisen first in one of these sensory cell types followed by redeployment into the other sensory cell type or by concerted evolution in photoreceptors and hair cells. The functional replacement of SNARE proteins at the ribbon synapse of hair cells involved the integration of another protein, Otoferlin, into the ribbon synapse. It is currently unclear to what extent this involved regulatory rewiring or the evolution of novel protein-protein interactions. However, molecular interactions of otoferlin with calcium, phospolipids, and other proteins have subsequently contributed to further functional adaptations of vesicle release and recycling mechanisms in hair cells (Pangršič et al. 2012; Lek et al. 2012).

2.2.4 ORIGIN OF NOVEL CELL TYPES

The examples discussed in the last section show that sensory cell types often have a very long evolutionary history, during which they acquire different adaptations and change their phenotype, sometimes quite drastically. However, as long as cells remain the same independently evolving units across generation boundaries and speciation events and are defined by identical CoRNs, all of these differently adapted cells are homologous and belong to the same cell type. Occasionally, however, novel cell types originate, when new CoRNs emerge. A famous example is the cnidarian stinging cells or cnidocyte, which is found exclusively in cnidarians (Tardent 1995; Babonis and Martindale 2014; see Chapter 3). It uses a modified cilium – the cnidocil – as trigger to release the contents of a special venom filled capsule – the cnidocyst – equipped with a protrusible tubule for predation. Various cnidaria-specific proteins such as minicollagens, NOWA, spinalin, and poly-γ-glutamate synthase acquired by horizontal gene transfer from bacteria combine with other evolutionarily more conserved proteins to build the cnidocyst (Denker et al. 2008; Babonis and Martindale 2014). Although cnidocytes share developmental and probably also evolutionary precursors with cnidarian neurons, they are specified by different transcription factors including PaxA and Mef2, suggesting that they have acquired a new CoRN (Babonis and Martindale 2017). Other examples of novel cell types are bone-forming cells (osteocytes) in gnathostomes and neurons in eumetazoans (e.g. Donoghue and Sansom 2002; Moroz 2009; Burkhardt and Sprecher 2017).

Novel cell types like other novel characters can originate either by duplication and divergence from an ancestral cell type or by recombination. I will briefly illustrate these two modes of cell type origin with some examples from sensory evolution. Because many of these sensory cell types will be described in more detail later in the book, I will only introduce the bare essentials here to make my point.

2.2.4.1 Novel Sensory Cell Types Originating by Duplication and Divergence

The origin of novel cell types by duplication and divergence was first proposed by Arendt and others (Arendt and Wittbrodt 2001; Arendt 2003) as a model to explain the evolution of two types of photoreceptors – ciliary and rhabdomeric photoreceptors – which differ with respect to the location of membrane expansions containing the opsin photopigments (see Chapter 4 for details) (Fig. 2.8). While it was originally thought that ciliary photoreceptors are confined to deuterostomes and cnidarians and rhabdomeric photoreceptors to protostomes, both types of photoreceptors have now been found in some protostomes and deuterostomes (Arendt et al. 2004; Pergner and Kozmik 2017; Schlosser 2018). Apart from their morphological differences, ciliary and rhabdomeric photoreceptors also use different but evolutionarily related phototransduction mechanisms (Arendt 2003; Plachetzki, Fong, and Oakley 2010). Whereas phototransduction in ciliary photoreceptors (e.g. in the vertebrate retina) depend on c-opsins, Gα-proteins of the Gα_i or Gα_o type and cyclic nucleotide gated (CNG) ion channels, rhabdomeric photoreceptors use r-opsin, Gα_q and transient receptor potential (TRP) channels instead. In addition, the identity of these two types of photoreceptors appears to be specified by different transcription factors indicating that they are regulated by different CoRNs. The identity of rhabdomeric photoreceptors is controlled by Atonal, POU4, Gfi, and Six1/2, whereas ciliary photoreceptors depend on a different set of transcription factors including Rx (Arendt 2003; Lamb, Collin, and Pugh 2007; Schlosser 2018). However, Pax6 plays a role for both types of photoreceptors, and the related PaxB is important for photoreceptor identity in cnidarians (Piatigorsky and Kozmik 2004). Taken together, the observations that rhabdomeric and ciliary photoreceptors of bilaterians share expression of some transcription factors (Pax6) and use related components of the phototransduction cascade but depend on distinct transcription factors (Atonal vs. Rx) for specification suggest that they evolved as sister cell types from a distinct type of photoreceptor in the cnidarian–bilaterian ancestor.

Duplication and divergence of an ancestral cell type may also have led to the evolution of the axonless hair cell and the sensory neuron of the vestibulocochlear ganglion that transmits the signal from the hair cell to the brain (see Chapter 3 for details) (Fig. 2.13). Bernd Fritzsch and colleagues have suggested that both of these cell types have arisen as sister cells by division of labor from primary mechanoreceptor cells possessing an axon (Fritzsch, Beisel, and Bermingham 2000). Such primary mechanoreceptors are found in many invertebrate taxa including other chordates and some of them also express Atonal, POU4, and Gfi, orthologous to the core transcription factors specifying hair cell identity (Fritzsch et al. 2005; Schlosser 2005; Fritzsch et al. 2007; Schlosser 2018; Arendt et al. 2015). The expression of different but related genes encoding transcription factors in hair cells and sensory neurons supports this sister cell scenario. Vestibulocochlear neurons are specified by *Neurogenin1* (*Neurog1*). This belongs to a gene family that arose together with the *Atonal*-family by duplication of a gene in early bilaterians (Simionato et al. 2008). In addition, differentiation of vestibulocochlear neurons also depends on *POU4f1* (also known as *Brn3a* or *Brn-3.0*) and *POU4f2* (also known as *Brn3b* or *Brn-3.2*) (McEvilly et al. 1996; Smith, Dawson, and Latchman 1997). These arose together with *POU4f3* involved in hair cell differentiation from a single ancestral *POU4* gene during the two rounds of genome duplications in early vertebrates (Gold, Gates, and Jacobs 2014). The common clonal origin of hair cells and vestibulocochlear sensory neurons in at least some sensory areas of the inner ear (Satoh and Fekete 2005) may also reflect their common evolutionary origin. As a caveat, the picture sketched here briefly is almost certainly oversimplified and glosses over several difficulties (e.g. the expression of similar transcription factors in other sensory cell types; the fact that hair cells originated much later than the *Atonal* and *Neurogenin* gene families). A more detailed account will be presented in the next chapter.

Once a hair cell-like secondary sensory cell had evolved, it probably gave rise to further cell types by duplication and divergence. In many cartilaginous fishes, ray-finned fishes and lobe-finned fishes, the lateral line placodes give rise to

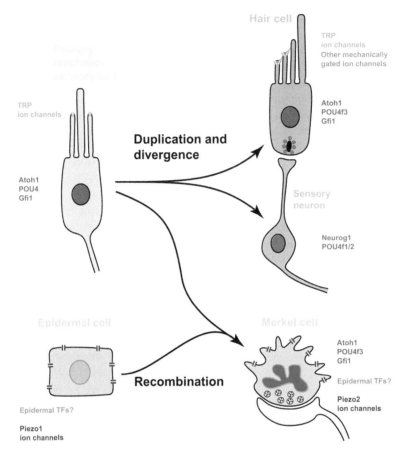

FIGURE 2.13 Novelty in cell type evolution. Novel cell types may arise by duplication and divergence or by recombination as illustrated here for mechanoreceptor evolution. The vertebrate hair cell (specified by transcription factors Atoh1, POU4f3, and Gfi1) and the sensory neuron innervating it (specified by the paralogous transcription factors Neurog1 and POU4f1/2) arose probably as sister cell types from primary mechanosensory cells. TRP channels are used for mechanotransduction in most primary mechanosensory cells. These are still expressed in hair cells, which have, however, recruited other channels for mechanotransduction. The Merkel cell may instead have arisen by recruitment of a gene regulatory network (involving transcription factors Atoh1, POU4f3, and Gfi1) from mechanoreceptors into epidermal cells, which expressed stretch sensitive Piezo channels under control of epidermal transcription factors. (Reprinted with modification with permission from Arendt et al. 2016.)

electroreceptors in addition to the mechanoreceptive hair cells although electroreceptors have been lost in teleosts and amniotes (Baker and Modrell 2018). Apart from their embryonic origin, these electroreceptors resemble hair cells in sporting a ribbon synapse and in most taxa carry an apical cilium associated with microvilli (Jørgensen 2005). A recent study in the basal actinopterygian paddlefish indicates that the ribbon synapses of its electroreceptors also employ otoferlin and L-type calcium channels and the cells express Atoh1 and POU4f3 similar to hair cells (Modrell et al. 2017). However, expression of the transcription factor NeuroD4 is confined to electroreceptors suggesting that these are genetically individualized from hair cells (Modrell et al. 2017). Taken together, these findings support the model that hair cells and electroreceptors evolved as sister cell types.

Although teleosts have lost electroreception, electroreceptors re-evolved at least two times independently in some teleost fishes (Baker, Modrell, and Gillis 2013; see Chapter 6 in Schlosser 2021). These electroreceptors differ in several respects from electroreceptors in other vertebrates: teleost ampullary electroreceptors respond to anodal stimuli (positive relative to the interior of the cell) on their basal membrane, while the electroreceptors of other vertebrates respond to cathodal stimuli on the apical membrane (Bodznick and Montgomery 2005; Baker et al. 2013). These differences suggest that teleost electroreceptors are not homologous to electroreceptors of other vertebrates. Instead, they may have evolved independently by duplication and divergence of teleost hair cells, which also respond weakly to anodal stimuli (Baker et al. 2013).

2.2.4.2 Novel Sensory Cell Types Originating by Recombination/Redeployment

In all examples discussed so far, novel sensory cell types originated by duplication and divergence. The second mode of origin for novel cell types – recombination and redeployment of components of several cell-types to generate a new CoRN – is much less well understood although it may play a far more important role for cell type evolution than is currently recognized (Oakley 2017; Schlosser 2018). It probably was essential during the evolutionary expansion of the first cell types in early eumetazoans and bilaterians, a topic I will return to later (Chapter 3). And it continued to play important roles for subsequent cell type diversification as I will briefly discuss below.

Before I proceed, however, I wish to point out that some novel cell types, which have been suggested to have originated by duplication and divergence, were most likely established not by the mere duplication and divergence of an ancestral CoRN but in a process involving the recruitment of transcription factors or small regulatory networks from other cell types (see also Oakley 2017). This is most likely the case for the ciliary and rhabdomeric photoreceptors discussed above, because some transcription factors considered to be main CoRN members for rhabdomeric versus ciliary photoreceptors are not encoded by paralogous genes (potentially arisen by gene duplication of ancestral CoRN genes) but by unrelated genes – the bHLH gene *Atoh1* for rhabdomeric and the homeobox gene *Rx* for ciliary photoreceptors.

The vertebrate Merkel cell provides a putative example for the origin of a novel sensory cell type by recombination, even though we currently don't have sufficient information about the CoRN guiding Merkel cell development to fully support this (Fig. 2.13). As briefly mentioned in Chapter 1 of the first volume (Schlosser 2021), Merkel cells are one of several specialized mechanoreceptor cells mediating touch perception in the vertebrate skin. Although derived from the epidermis, Merkel cells depend for their differentiation on many of the same transcription factors involved in the specification of the placode-derived hair cells, notably Atoh1, POU4f3, and Gfi1 (Whitear 1989; Wallis et al. 2003; Morrison et al. 2009; Van Keymeulen et al.

2009; Masuda et al. 2011; Woo, Lumpkin, and Patapoutian 2015). However, Merkel cells use the Piezo2 channel for mechanotransduction, different from hair cells, which use TMC channels (possibly in cooperation with TRP channels) instead (Eijkelkamp, Quick, and Wood 2013; Pan et al. 2013; Woo et al. 2015). Although Piezo2 is also present in the apical membrane of hair cells, it does not seem to play a major role in hair cell mechanotransduction (Wu et al. 2017). In contrast, Piezo2 and its paralog Piezo1 are activated by membrane stretching in the epidermis and other epithelial cells, where they control cell extrusion in response to cell crowding (Eisenhoffer and Rosenblatt 2013; Honoré et al. 2015). Because Merkel cells are epidermally derived, their Piezo channels may be under similar transcriptional control than those of other epidermal cells. If confirmed in future studies this would suggest that Merkel cells may have evolved by the recombination of CoRN components from epidermal and hair cells.

2.2.5   A New Model of Cell Type Evolution

Current models of cell type evolution mostly discuss evolution of novel cell types by duplication and divergence, which can be represented by cell family trees (Arendt 2008; Arendt et al. 2016, 2019). This is based on the assumption that this is a frequent or even the predominant mode of cell type evolution, while evolution of novel cell types by recombination is rare. In this last section of Chapter 2, I would like to argue that this is a misleadingly dichotomous view of cell type evolution and that both modes of cell type evolution typically go hand in hand.

Duplication of a cell type requires that it is able to follow its independent evolutionary trajectory and this is only possible, when each of the two sister cell types acquires at least a partially different genetic basis. In special cases, this may be achieved without any rewiring of existing CoRNs by duplication of at least some of the genes involved in the CoRN (Fig. 2.14A). However, alternatively, duplication and divergence of cell types may also be achieved by the differential modulation of the regulatory state defined by the CoRN due to recruitment of different sets of additional transcription factors A and B (or the replacement of some CoRN members by different transcription factors) in the two sister cell types, which I will also refer to as cell subtypes (Fig. 2.14B).

These additional transcription factors A and B, may differ from each other in regional or temporal expression as well as in the set of target genes that they are able to activate (since each transcription factor is able to bind to a different subset of enhancers). Regarding expression patterns, these modulating transcription factors may act very locally and allow the subdivision of a group of cells expressing a particular CoRN into local subtypes (Fig. $2.14B_1$). Alternatively, they may be expressed in broad body regions containing multiple cell types and may then confer regional identity on many different cell types, by modulating their respective CoRNs resulting in regional subtypes (Fig. $2.14B_2$). Regarding sets of target genes, modulating transcription factors may act very narrowly on a small set of target genes and will then only affect expression of a few genes in a subtype specific manner. Alternatively, they may affect a large fraction of genes in the genome and act as competence factors (e.g. pioneer factors) for their expression, potentially leading to very different gene expression profiles in different subtypes.

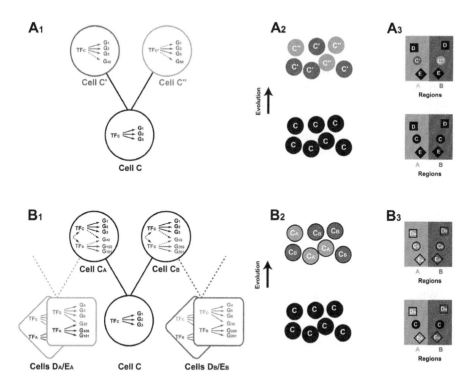

FIGURE 2.14 A new model of cell type evolution. Evolutionary changes in cell types are depicted with a time axis proceeding from bottom to top in each panel. (**A**) Diversification of a cell type C may result from duplication of genes encoding members of its CoRN such as transcription factor TF_C (**A₁**), which original regulates genes G_1, G_2, and G_3. Two genetically individuated sister cell types C' and C", may result by silencing complementary copies ($TF_{C''}$ silenced in C'; $TF_{C'}$ silenced in C") of the duplicated gene, thereby acquiring a different genetic basis. $TF_{C'}$ and $TF_{C''}$ may subsequently diverge in function, e.g. by gaining control over different new target genes (e.g. G_{40} for $TF_{C'}$ and G_{50} for $TF_{C''}$) or by losing some of their old target genes (e.g. G_1 by $TF_{C''}$). If expression of $TF_{C'}$ and $TF_{C''}$ depends on different local (**A₂**) or regional cues (**A₃**), C' and C" will come to develop in different local subsets of cells or body regions, respectively. D and E depict other cell types. (**B**) Alternatively, diversification of a cell type C may result from recruitment (e.g. due to co-regulation by some upstream regulator as indicated by broken arrows) of additional, modulating transcription factors (TF_A, TF_B) to the CoRN of cell type C (**B₁**). TF_A and TF_B may already function as local or regional modulators of other cell types D and E. They now impart partial genetic individuality to subtypes C_A and C_B, which differ in expression of genes controlled by TF_A (G_{100}, G_{101}) and TF_B (G_{200}, G_{201}), respectively, and may gain differential expression of additional target genes (G_{40} in C_A and G_{50} in C_B) due to co-regulation of these genes by TF_C/TF_A and TF_C/TF_B, respectively. If expression of TF_A and TF_B depends on different local (**B₂**) or regional cues (**B₃**), C_A and C_B will come to develop in different local subsets of cells or body regions, respectively.

In any case, the recruitment of these modulating transcription factors will allow the combinatorial regulation of different target genes in the two sister cell types and will thus enable the generation of region- or stage-specific sister cell types (or subtypes) of an ancestral cell type. For example, a neuronal cell type such as the motor neurons of the vertebrate hindbrain innervating cranial muscles, may have diversified into different subtypes in the anterior and posterior part of the hindbrain due to cooperation between the CoRN defining motor neuron identity with different Hox-type transcription factors expressed in more anterior and posterior part of the hindbrain, respectively (Guthrie 2007). Consequently, motor neurons of the trigeminal nerve (which express HoxA2) now innervate different cranial muscles than motor neurons of the facial nerve (which express HoxB1) because motor neurons expressing different Hox transcription factors also activate different downstream genes encoding proteins that guide outgrowing axons of neurons to their target muscles. Regional differences along other dimensions, such as the dorso-ventral position of motor neurons in the hindbrain and spinal cord may control other aspects of motor neuron identity (as suggested by the "quasi-cartesian model of character identity" proposed in Wagner 2014).

Whenever duplication and divergence of cell types involves the recruitment of modulating transcription factors (rather than gene duplication), it involves some rewiring of CoRNs (for example, the establishment of new regulatory links between upstream regulators so that these now co-activate the new transcription factors together with ancestral CoRN members). It also involves the recombination of pre-existing regulatory networks, since the new modulating transcription factors likely were already functioning – and may continue to function – in the regulation of other networks. Transcription factors expressed in broad body regions (e.g. Hox proteins) that were recruited to confer regional identity to subtypes of a particular cell type e.g. neurons) often also confer regional identity to subtypes of other cell types (e.g. muscles and skeletal cells).

As a consequence, there is no sharp divide between the two modes of generating novel cell types introduced above – duplication and divergence versus recombination/redeployment. Since duplication of cell types may typically involve some degree of recombination of regulatory networks, cell type evolution will often not be completely tree-like (divergence of solid lines in Fig. 2.14A$_1$, B$_1$) but will involve some degree of coalescence (convergence of hatched and solid lines in Fig. 2.14B$_1$). Occasionally, this may even lead to complete fusion of cell types when the CoRN of one cell type gains control over the CoRN of another cell type or both CoRNs get activated by a common upstream regulator leading to the co-activation and co-regulation of both CoRNs in a single cell.

3 Evolution of Mechano- and Chemosensory Cell Types

In the first volume (Schlosser 2021), I introduced the specialized sense and endocrine organs of the vertebrate head and reviewed how these cell types arise from the cranial placodes during vertebrate embryonic development. In the first two chapters of the present volume, I briefly discussed the phylogenetic relationships of vertebrates and laid out what defines the identity of sensory and neurosecretory cell types and how we can track them through evolution. In the remainder of this book, I will now compare the sensory and neurosecretory cell types of the vertebrate head with similar cell types in other animals to get insights into their evolutionary origins.

In this chapter, I will review the evolution of mechano- and chemosensory cells throughout the animal kingdom. I will consider these together because most of the evidence presented in this chapter suggests that chemosensory cells have evolved repeatedly from mechanosensory cells. In the following two chapters, I will consider the evolutionary history of photosensory cells (Chapter 4) and neurosecretory cells (Chapter 5). In the final chapter, I will then address how cranial placodes and the novel sensory organs and neurosecretory glands derived from them, may have originated in early vertebrates by the evolution of new regulatory connections between "old" cell types and ectodermal patterning mechanisms and will draw some general conclusions (Chapter 6).

Chapters 3–5 will trace the evolutionary history of vertebrate sensory and neurosecretory cell types in a manner borrowed from Richard Dawkins' book *The Ancestor's Tale* (Dawkins 2004) using the comparison between extant sister taxa and more distantly related taxa (outgroups) to gain insights into the cell types present in increasingly distant vertebrate ancestors. The logic of such an "outgroup comparison" was explained in Chapter 1 and is summarized in Fig. 3.1. For each class of cell types, I will start by discussing cell type differentiation in the tunicates, the sister group of vertebrates (Figs. 1.8 and 1.9). For cell types that are shared between tunicates and vertebrates, comparison with an outgroup (e.g. amphioxus) will indicate whether these are most likely shared derived characters which originated in the last common ancestor of the tunicate-vertebrate clade, or whether they are shared primitive characters which were inherited from earlier ancestors (e.g. the last common ancestor of chordates). For cell types that differ between tunicates and vertebrates, outgroup comparison will help to clarify which of these cell types (if any) is primitive and was inherited from a common ancestor, and which cell type is derived and originated as novelty in the tunicate or vertebrate lineage, respectively.

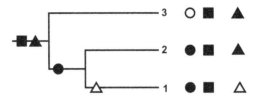

FIGURE 3.1 Outgroup comparison. When characters are shared between two sister groups 1 and 2 (circles and squares), comparison with the outgroup 3 allows to determine whether these are shared derived characters, which evolved in the last common ancestor of the [1,2] clade (circle), or whether they are shared primitive characters, which evolved already in earlier ancestors (square). For characters that differ between two sister groups 1 and 2 (triangles), outgroup comparison may help to clarify which character is primitive (closed triangle) and was inherited from a common ancestor, and which character is derived (open triangle). (Reprinted from Schlosser 2015.)

I will then use similar comparisons with increasingly distantly related groups (e.g. amphioxus, hemichordates/echinoderms, protostomes, cnidarians, sponges) to shed light on the cell types present in ancestors of more and more inclusive groups (e.g. chordates, deuterostomes, bilaterians, eumetazoans, and metazoans; Fig. 1.8). In principle, this comparative approach should allow us to gain insights into which cell types were present even in very distant ancestors. However, since all of these inferences to the past are necessarily incomplete and often ambiguous, our picture of the cell types present will become more and more uncertain, sketchy, and speculative for more distant ancestors.

3.1 MECHANO- AND CHEMOSENSORY CELLS IN THE LAST COMMON TUNICATE-VERTEBRATE ANCESTOR

Our review of sensory cells in the first volume revealed a diversity of placodally derived mechano- and chemosensory cells in vertebrates (Chapter 6 in Schlosser 2021), which I will very briefly recapitulate here. The chemosensory cells comprise the primary sensory cells of the olfactory and vomeronasal epithelia, derived from the olfactory placode. These are primary sensory cells (with an axon) which carry cilia and/or microvilli on their apical surface. Olfactory sensory cells employ proteins of the odorant receptor (OR) family of G-protein-coupled receptors (GPCR), which activate a cyclic nucleotide gated (CNG) channel to detect odors. In contrast, vomeronasal sensory cells use different types of GPCRs, which activate transient receptor potential (TRP) channels.

The specialized mechanosensory cells of vertebrates are known as hair cells (see Chapter 6 in Schlosser 2021). These are secondary sensory cells (without an axon) and are associated with special somatosensory neurons (SSN) that transmit information from the hair cells to the brain. Hair cells and SSNs are derived from the otic and lateral line placodes. On the apical side, hair cells are equipped with one cilium (kinocilium) and a bundle of specialized microvilli (stereovilli). Stereovilli are connected via tip links (a complex of the proteins Protocadherin-15 and Cadherin-23), which open ion channels (including proteins of the transmembrane channel or TMC family) when stereovilli are bent by mechanical stimuli. In addition, vertebrates also have sensory neurons that have free nerve endings or are associated with epithelium-derived receptor cells (e.g. Merkel cells) (see Chapter 6 in Schlosser 2021). These mediate pain, temperature, touch, and information from muscle spindles. They include the general somatosensory neurons (GSNs) derived from profundal and trigeminal placodes and neural crest, which innervate the skin, as well as the viscerosensory neurons (VSN) derived from epibranchial placodes, which innervate the viscera and taste buds. To infer which of these cell types were present in the last common ancestor of vertebrates and tunicates, we now have to have a closer look at the tunicates.

3.1.1 MECHANO- AND CHEMOSENSORY CELL TYPES IN TUNICATES

Several different types of mechano- and/or chemosensory cells have been described in the vertebrates' closest living relatives, the tunicates (reviewed in Holland and Holland 2001; Mackie and Burighel 2005; Burighel, Caicci, and Manni 2011; Ryan and Meinertzhagen 2019). The most detailed description of sensory cells and their developmental origins is available for tunicate larvae, in particular the larva of the ascidian *Ciona* (Figs. 3.2 and 3.3), in which the entire complement of neurons in the peripheral and central nervous system (CNS), as well as their synaptic connections, was recently described (Imai and Meinertzhagen 2007; Ryan, Lu, and Meinertzhagen 2016, 2018). To facilitate the comparison between ascidian, vertebrate, and amphioxus body plans, corresponding ectodermal regions are illustrated in Fig. 3.2.

3.1.1.1 Sensory Neurons of the Larva

A very peculiar mechanoreceptor of tunicates is the statocyte (also known as otolith). It is found on the left side of the sensory vesicle (i.e., the anterior brain) in the larva of solitary ascidians like *Ciona* and has been experimentally shown to sense gravity (Ohtsuki 1990; Tsuda, Sakurai, and Goda 2003; Sakurai et al. 2004; Esposito et al. 2015). The statocyte is formed by a single specialized pigment cell in which a large melanin granule serves as a weight. It moves in relation to gravity and presses down on two dendrites of neighboring neurons. In contrast to the mechanoreceptor cells of the peripheral nervous system (PNS), which we will discuss in the remainder of this chapter, the statocyte is a derivative of the CNS. It develops from the same blastomere (a9.49; see Fig. 3.2B) on the left side than the photoreceptive ocellus (see Chapter 4) on the right side of the ascidian sensory vesicle and is associated with a group of unpigmented photoreceptors (Horie et al. 2008; Esposito et al. 2015). This strongly suggests that the statocyte has evolved from a modified pigment cell of an ancient ocellus. In line with this hypothesis, many colonial ascidians (e.g. *Botryllus*) have only a single sensory organ known as photolith that combines a pigment cell serving as statocyte with photoreceptor cells and perceives both gravity and light (Grave and Riley 1935; Sorrentino et al. 2000).

The ascidian statocyte with its associated neurons represents the unsual case of a mechanosensory organ derived from the CNS. Other mechanosensory cells in tunicates (and in most other animals) are found in the PNS. Typically, these cells

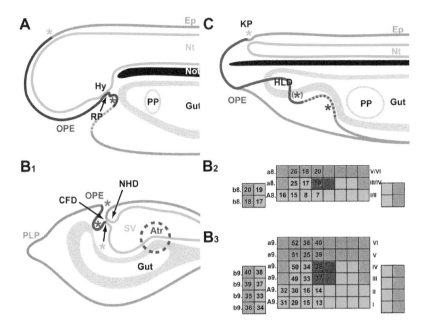

FIGURE 3.2 Comparison of ectodermal regions in vertebrates, tunicates, and amphioxus. Different regions of the ectoderm are distinguished by different colors (green, neural tube; orange, general epidermis; red, oral and preoral part of epidermis; pink, palp-forming region of epidermis in tunicates). The position of anterior neuropore (green asterisk) and mouth (red asterisk) is marked. The oral part of the epidermis is located around the mouth, while its preoral part is located between mouth and anterior neuropore. (**A**) Vertebrates. The preoral ectoderm (including precursors of adenohypophyseal, olfactory, and lens placodes) is expanded due to elaboration of the forebrain. The adenohypophyseal placode buds off the stomodeum as Rathke's pouch (RP) to form the anterior pituitary. The hatched region below the mouth corresponds to the palp forming region in tunicates. (**B**) Tunicates (*Ciona*). The oral and preoral ectoderm (OPE) comprises the oral siphon primordium and participates in neurulation. As a consequence, the external "neuropore" (purple asterisk) is different from the proper neuropore (green asterisk). The connection between anterior neural tube and the OPE may persist giving rise to a neurally derived neurohypophyseal duct (NHD) and a ciliated funnel and duct (CFD). It is still contentious whether the CFD is entirely derived from the oral siphon primordium as shown here or receives a contribution from NHD. The atrial siphon primordia (Atr; hatched brown line) invaginate only at late larval stages. The origin of the various ectodermal territories from the so-called neural plate of ascidians at 110-cell stage (**B₂**) and mid-gastrula stage (**B₃**) are shown to the right (Nishida 1987). (**C**) Amphioxus. Hatschek's left diverticulum (HLD), an endomesodermal pouch will fuse with the preoral ectoderm to give rise to Hatschek's pit. The mouth of amphioxus (red asterisk) is probably not homologous to the vertebrate mouth. The approximate position of the region corresponding to the vertebrate mouth is indicated (red asterisk in parentheses). The hatched part of the ectoderm between this point and the amphioxus mouth corresponds to the palp forming region in tunicates. Atr, atrial siphon primordium; CFD, ciliary funnel and duct; Ep, epidermis; HLD, Hatschek's left diverticulum; Hy, hypothalamus; KP, Kölliker's pit; NHD, neurohypophyseal duct, Not, notochord; Nt, neural tube; OPE, oral–preoral ectoderm; PLP, palps; PP, pharyngeal pouches; RP, Rathke's pouch; SV, sensory vesicle. (Modified from Schlosser 2005 and Schlosser et al. 2014.)

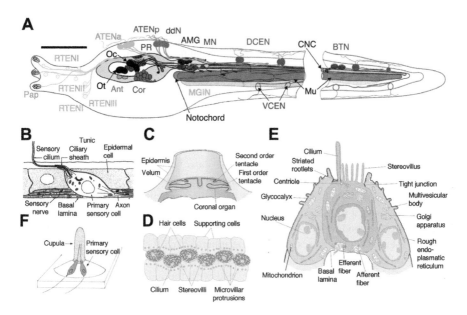

FIGURE 3.3 Sensory cells in ascidians. (**A**) Diagram of whole larva of Ciona intestinalis with sensory and motor cell types indicated. Pap, papilla neuron; RTEN, rostral trunk epidermal neurons; ATENa, anterior apical trunk epidermal neurons; ATENp, posterior apical trunk epidermal neurons; Oc, ocellus; Ot, otolith; Ant, antenna neuron; Cor, coronet cell; PR, photoreceptor; ddN, descending decussating neuron; AMG, ascending motor ganglion interneuron; MGIN, motor ganglion interneuron; MN, motor neuron; DCEN, dorsal caudal epidermal neuron; VCEN, ventral caudal epidermal neuron; BTN, bipolar tail neuron; Mu, muscle; BV, brain vesicle; MG, motor ganglion; CNC, caudal nerve cord; Not, notochord; IN, interneuron; BVINs, brain vesicle interneurons. Scale bar: 10 μm. (**B**) Primary sensory cell in larva of *Diplosoma macdonaldi* with cilium extending into the tunic. (**C–E**) Coronal organ in oral siphon of *Botryllus schlosseri* showing location in oral siphon (**C**), arrangement of sensory cells (**D**) and schematic diagram of sensory cell, nerve endings and supporting cells (**E**). (**F**) Cupular organ in atrial siphon of *Ciona intestinalis*. ([A] Reprinted with permission from Ryan and Meinertzhagen 2019, after Ryan et al. 2016; [B] Reprinted with permission from Torrence and Cloney 1982; [C-E] Reprinted with permission from Burighel et al. 2003; [F] Reprinted with permission from Mackie and Singla 2004.)

have one or several sensory cilia, which may be surrounded by microvilli. Number and arrangement of cilia and microvilli differ between different cell types and species. We currently can only tentatively guess which of these sensory cells respond to mechanical or chemical stimuli or both based on their structure and location, since direct experimental evidence to firmly assign sensory modality is lacking.

The ascidian tadpole larva has clusters of sensory neurons in its adhesive palps (palp sensory neurons [PSNs]) and scattered sensory cells concentrated in several patches of ectoderm in its main body and in dorsal and ventral rows along the tail (Fig. 3.3) (Torrence and Cloney 1982, 1983; Takamura 1998; Imai and Meinertzhagen 2007; Caicci et al. 2010; Ryan et al. 2016, 2018). These neurons are known as rostral trunk epidermal neurons (RTENs), anterior apical trunk epidermal neurons (aATENs), posterior apical trunk epidermal neurons (pATENs), and in the tail as caudal epidermal sensory neurons (CESNs) (Fig. 3.3A). The sensory neurons of the ascidian larva have a single cilium, which is surrounded by a membrane sheath and projects into the tunic (Fig. 3.3B) (Torrence and Cloney 1982; Konno et al. 2010; Terakubo et al. 2010; Yokoyama, Hotta, and Oka 2014). All larval sensory neurons are primary sensory cells. Most of them send their axon directly into the brain. However, the trunk sensory neurons (CESNs) only have short axons, which do not reach the brain. Instead these form synapses with other peripheral neurons, the bipolar tail neurons (BTNs), which then project into the brain (Stolfi et al. 2015). CESNs, therefore, present an interesting intermediate case between primary sensory cells (with an axon that usually reaches the brain) and secondary sensory cells (without an axon). Interestingly, the Langerhans cells of the appendicularian *Oikopleura*, which are equipped with a single stiff cilium and also function as caudal mechanosensory cells, are true secondary sensory cells without an axon; they are connected to the CNS with sensory neurons residing in a peripheral caudal ganglion (Bone and Ryan 1979; Holmberg 1986; Bassham and Postlethwait 2005).

Except for the sensory neurons associated with the palps (discussed in the next paragraph), the cilia of the primary sensory cells in the ascidian larva extend into long and branched dendrites within the tunic (Fig. 3.3 B) (Terakubo et al. 2010; Yokoyama et al. 2014). How they function and whether they respond to mechano- and/or chemosensory stimuli is currently not known. The co-expression of a CNG channel with the neuropeptide GnRH2 and GPCRs involved in neurosecretion (RXFP3, SOG) in the *Ciona* aATEN sensory cells suggests that this subtype of sensory cells at least may be chemosensory and neurosecretory rather than mechanosensory (Abitua et al. 2015). However, this needs to be confirmed experimentally and the sensory modalities mediated by the other clusters of primary sensory cells need to be resolved.

In contrast to other sensory neurons, the palp sensory neurons form tight clusters with each other and with two other cell types, the collocytes, which produce adhesive substances, and another, microvillous cell type of unknown function (Torrence and Cloney 1983; Dolcemascolo et al. 2009; Caicci et al. 2010; Zeng et al. 2019). The palps of the ascidian larva mediate attachment to the substrate in response to mechanical and chemical cues that induce metamorphosis. Two types of palp sensory neurons have been described: a larger, elongated cell type in which the cilium is accompanied by microvilli ("anchor cells" of Torrence and Cloney 1983), and a smaller, rounder cell type ("basal cell" of Torrence and Cloney 1983) with a cilium surrounded by a sheath closely resembling the caudal epidermal neurons. It is currently unclear which, if any, of these cell types is specialized for mechano- or chemosensory functions (Torrence and Cloney 1983; Caicci et al. 2010; Zeng et al. 2019). The axons of PSN and RTEN cells enter the sensory vesicle via a rostral nerve, reminiscent of the olfactory nerve (Fig. 3.3A) (Imai and Meinertzhagen 2007).

The permanently pelagic appendicularians like *Oikopleura* lack palps and instead display a ventral sense organ composed of primary sensory cells probably also serving mechano- and/or chemosensory functions (Bollner, Holmberg, and Olsson 1986; Bassham and Postlethwait 2005). In addition, *Oikopleura* has secondary sensory cells surrounding the mouth as circumoral cells (Olsson, Holmberg, and Lilliemarck 1990; Bassham and Postlethwait 2005).

3.1.1.2 Sensory Neurons of the Adult

In *Ciona* and other ascidians, primary sensory cells, which send their cilia into the tunic, continue to be present after metamorphosis and mediate responses to gentle touch and vibration (Fig. 3.3B) (Mackie et al. 2006; Manni et al. 2006). In addition, new types of sensory cells develop in tunic-free regions of the oral and atrial siphons which form by invaginations from the dorsal non-neural ectoderm (Manni, Lane et al. 2004; Manni et al. 2005; Mazet et al. 2005; Bassham and Postlethwait 2005; Kourakis, Newman-Smith, and Smith 2010; Veeman et al. 2010). The oral siphon already begins to develop at neural plate stages from non-neural ectoderm located at the anterior rim of the "neural plate" (but not yet committed to a neural fate; see discussion in Schlosser, Patthey, and Shimeld 2014), while the atrial siphon only forms at the end of the larval stage from anterolateral non-neural ectoderm (Fig. 3.2).

In the oral siphon, there are secondary sensory cells (without an axon) which resemble vertebrate hair cells in carrying one or several cilia on their apical side that are associated with a group of microvilli in many species (Fig. 3.3C–E) (Burighel et al. 2003, 2011; Manni, Caicci et al. 2004; Manni, Lane et al. 2004; Bassham and Postlethwait 2005; Manni et al. 2006; Caicci, Burighel, and Manni 2007, Caicci et al. 2013; Rigon et al. 2013). In *Ciona*, these cells also express the TRPA channel like stereovilli in hair cells (Rigon et al. 2018). They have, therefore, been proposed to be homologs of vertebrate hair cells (Burighel et al. 2003; Manni, Caicci et al. 2004). However, in contrast to hair cells, their microvilli, if present, are not arranged as a staircase and are positioned to one side of the cilium only in some species, while they surround the cilium in others (Fig. 3.3D, E) (Burighel et al. 2011; Rigon et al. 2013). These mechanosensory cells together with flanking supporting cells are arranged in a long strip which encircles the tentacles inside the oral siphon thereby forming the coronal (ascidians and thaliaceans) or circumoral organ (appendicularians) (Fig. 3.3C, D). Experimental evidence shows that the coronal organ mediates reflexes that promote contraction of sphincter muscles as well as ciliary arrest in oral and atrial siphons in response to mechanical stimulation by large particles, suggesting that its main function is in particle rejection (Mackie et al. 2006).

In the atrial siphons, primary sensory cells (with an axon) cluster together with supporting cells to form different types of mechanosensory organs, known as cupular organs, cupular strands, or capsular organs (Bone and Ryan 1978; Mackie and Singla 2003, 2004) (Fig. 3.3F). They probably also mediate closure of siphons in response to mechanical stimuli although this remains to be confirmed experimentally.

3.1.2 TRANSCRIPTION FACTORS INVOLVED IN SPECIFYING TUNICATE MECHANO- AND CHEMOSENSORY CELL TYPES

To address questions of cell type homology between the sensory cells or neurons of vertebrates and those of tunicates we have to consider whether core regulatory networks (CoRNs) of transcription factors that determine cell type identity are shared between different classes of vertebrate and tunicate sensory cells/neurons. The transcription factors involved in specifying the various mechano- and chemosensory neurons in vertebrates were discussed in Chapter 6 of Volume 1 (Schlosser 2021) and are depicted schematically in Fig. 3.4. This overview reveals that many transcription factors are shared between the different mechano- and sensory-cell types of vertebrates, pointing to a common evolutionary origin as will be discussed below. The specification of GSNs developing from the profundal and trigeminal placodes depends on a CoRN of Neurog1/2, POU4f1, Islet1, and Tlx1/3. The VSNs derived from epibranchial placodes, instead require Phox2b (which represses POU4f1), but otherwise share many transcription factors with GSNs (e.g. Neurog1/2, Islet1, Tlx1/3). The CoRNs of the SSNs derived from the otic and lateral line placodes likewise overlap with those of GSNs (sharing Neurog1, POU4f1, and Islet1). Specification of hair cells (also from otic and lateral line placodes) instead depends on paralogs of bHLH and POU4 transcription factors that are used in SSNs (Atoh1 instead of Neurog1; POU4f3 instead of POU4f1) in addition to Gfi1 and BarHl1. Furthermore, transcription factors such as Pax2/8, Hmx2/3, and Prox1 that are specifically expressed in hair cells and SSNs help to distinguish these from other sensory neurons. In contrast to other sensory neurons, the olfactory and vomeronasal receptor neurons originating from the olfactory placode appear to be regulated by a different CoRN (e.g. Ascl1, Lhx2, DMRT4/5, FoxG, Sp8).

In tunicates, previous reports of transcription factor expression based on in situ hybridization (Giuliano et al. 1998; Imai et al. 2004, 2006; Bassham and Postlethwait 2005; Candiani et al. 2005; Mazet et al. 2005; Joyce Tang, Chen, and Zeller 2013; Wagner et al. 2014; Waki, Imai, and Satou 2015), together with recent single-cell RNA sequencing studies of *Ciona* embryos and larvae have now also provided us with a detailed profile of transcription expression in different sets of neurons (Fig. 3.5) (Horie et al. 2018; Cao et al. 2019; Sharma, Wang, and Stolfi 2019). Unfortunately, the lack of functional data currently does not allow us to determine conclusively which of these transcription factors play central roles for cell type specification and, thus, belong to the CoRNs for particular cell types in tunicates. However, comparison of these expression profiles with vertebrates invites tentative hypotheses regarding cell type homologies.

Overall, the profiles of transcription factor expression of the different sensory cell types as well as BTNs in *Ciona* are strikingly similar although sensory cells arise from multiple lineages of the non-neural ectoderm (Figs. 3.4 and 3.5). In particular, all tunicate sensory cells and BTNs express POU4 and Islet and a combination of bHLH transcription factors (either Neurog1 or various combinations of Neurog1, Atonal, and Ascl) and most of them co-express Pax3 and Pax2/5/8. This suggests that all of these cells are evolutionarily related and may have originated as sister cells by diversification of a common ancestral cell type. Recent experiments indeed show that the identity of some types of tunicate sensory cells can be easily transformed into other types of sensory cells by modulating expression of just a single transcription factor (Horie et al. 2018). BTNs, for example, adopt PSN identity after overexpression of FoxC (Horie et al. 2018). The recent single-cell RNA sequencing studies of *Ciona* also reveal an extensive overlap between the transcription factors expressed in *Ciona* sensory neurons and the core transcription factors of vertebrate

	Tunicates							Vertebrates					
	Bipolar tail neurons (BTN)	Caudal epidermal sensory neurons (CESN)	Posterior apical trunk epidermal neurons (pATEN)	Anterior apical trunk epidermal neurons (aATEN)	Rostral trunk epidermal neurons (RTEN)	Palp sensory neurons (PSN)	Collocytes	Hair cells	Special somatosensory neurons (SSN)	General somatosensory neurons (GSN)	Viscerosensory neurons (VSN)	Olfactory and vomeronasal sensory neurons	Neurosecretory cells in adenohypophysis
Six1/2					?								
POU4													
Islet1/2													
Prox													
Pax2/5/8													
GATA1/2/3	?	?	?	?	?	?	?						
Hmx2/3													
Atoh1													
Gfi1													
BarH													
Neurog1/2													
Tlx1/3	?	?	?	?	?	?	?						
DRG11	?	?	?	?	?	?	?						
Phox2													
Pax3/7													
Ascl													
DMRT4/5													
Lhx2/9													
EBF2/3													
Pax4/6													
FoxG													
Sp6-9													
Pitx1/2/3													
Lhx3/4													
Foxa1/2/3													
POU1	?	?	?	?	?	?	?						

FIGURE 3.4 Transcription factor expression in sensory cells of tunicates and vertebrates. Expression of transcription factors in *Ciona intestinalis* is indicated based on data from Horie et al. 2018; Sharma et al. 2019; and Cao et al. 2019 and compared with transcription factor expression in vertebrate sensory cells, sensory neurons and neurosecretory cells (see Chapters 3–8 in Schlosser 2021 for references). Faint colors indicate weak expression. Question marks indicate insufficient data. Pax3 is only expressed in general somatosensory neurons derived from the profundal placode in vertebrates. For the ascidian GATA1/2/3 ortholog GATAb, a recent study suggests expression in the entire non-neural ectoderm, but blastomeres of origin for the various sensory cells were not unequivocally identified (Ogura and Sasakura 2016).

hair cells, SSNs and GSNs and/or VSNs (Fig. 3.4). Shared transcription factors include POU4, Islet, Prox, Pax2/5/8 and/or Pax3, Hmx2/3, Gfi, BarH and Atoh, Ascl, and/or Neurogenin (the expression of Tlx1/3 and DRG11 has not been described). Let us first consider this in more detail for the sensory neurons of the head/trunk region of the *Ciona* tadpole before considering the tail neurons of the larva and the adult sensory neurons.

3.1.2.1 Sensory Neurons of the Larval Head and Trunk

The most rostral group of larval sensory neurons are the primary sensory PSNs of the palps, putatively mechanosensory neurons that express Prox and Six1/2, in addition to POU4, Islet1, and other transcription factors defining the identity of hair cells and of SSNs and GSNs in vertebrates. PSNs arise together with aATENs from the region of non-neural ectoderm bordering the anterior neural plate that expresses Six1/2 in addition to DMRT4/5 (DMRTa) (rows V and VI in Fig. 3.2B). Subsequently, only the more lateral aATEN precursors (derived from cell a8.26, Fig. 3.2B) maintain Six1/2 expression, while FoxC suppresses Six1/2 in more medial cells (derived from cells a8.18 and a8.20, Fig. 3.2B) and promotes PSN over aATEN differentiation (Horie et al. 2018). Another group of primary sensory cells that share a similar transcription

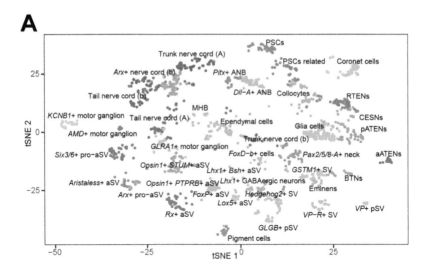

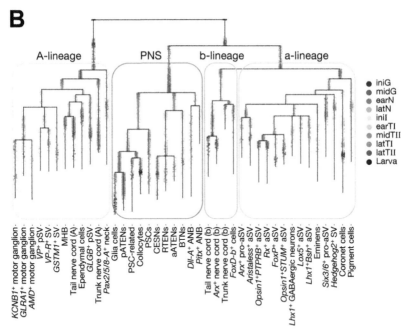

FIGURE 3.5 Single-cell RNA seq of neural cells in *Ciona*. (**A**) So-called *t*-distributed stochastic neighbour-embedding (*t*-SNE) projections of transcriptomes allow the identification of coherent clusters of individual neural cells. Shown is the *t*-SNE plot of all neural cells recovered from the larval stage (*n* = 1,704 cells). Identified cell types are labeled. (**B**) Transcriptome trajectories for individual neurons were reconstructed from transcriptome data of many developmental stages and are shown in pseudotemporal ordering. Cells are color coded by developmental stages ranging from early gastrula stages (iniG) to larval stages as shown to the right. The a-lineage, b-lineage, and A-lineage branches of the central nervous system and peripheral nervous system (PNS) are identified. (Reprinted with permission from Cao et al. 2019.)

profile (Figs. 3.4 and 3.5) and arise from non-neural ectoderm immediately adjacent to the anterior neural plate (Horie et al. 2018) are the mechano- and/or chemoreceptive RTENs (which have not been mapped but most likely originate from the same cells as PSNs or aATENs) and pATENs (originating from b8.20, Fig. 3.2B). The shared pattern of transcription factor expression between all these cell types suggest that they are all related and may be derived from one (or several related) cell types in the last common ancestor of vertebrates and tunicates.

In addition to the many similarities with vertebrate hair cells, SSNs, GSNs, and VSNs, some of the rostral-most *Ciona* sensory neurons share some transcription factors specifically with olfactory and vomeronasal receptor neurons of vertebrates that are characterized by co-expression of Ascl1 with Lhx2/9 as well as DMRT4/5, Sp8 and FoxG1 (Fig. 3.4) (see Chapter 6 in Schlosser 2021). While Lhx2/9 expression is not found in any sensory neurons of *Ciona* (being confined to the anterior CNS) and Ascl is always co-expressed with Neurog and/or Atonal, DMRT4/5, Sp8, and FoxG are co-expressed in PSNs, RTENs, and aATENs as well as collocytes (see Chapter 5) suggesting that they may be related to vertebrate olfactory and vomeronasal receptor neurons (Tresser et al. 2010; Wagner and Levine 2012; Horie et al. 2018; Cao et al. 2019). The axons of PSNs and RTENs also enter the brain of *Ciona* rostrally like the olfactory nerve in vertebrates.

However, different from these rostral sensory neurons of *Ciona* (PSNs, RTENs, and aATENs), vertebrate olfactory and vomeronasal receptor neurons do not rely on POU4 and Islet and are specified by a CoRN that is very different from somatosensory neurons (Figs. 3.4 and 3.6; for details see Chapter 6 in Schlosser 2021). Moreover, olfactory and vomeronasal receptor neurons in vertebrates use chordate-type *OR* genes, which are also present in amphioxus but have been lost from the tunicate genome (Niimura 2009; Churcher and Taylor 2011). Other classes of receptors involved in vertebrate olfaction (e.g. vomeronasal or TAAR receptors) have only evolved in the vertebrate lineage. This indicates that tunicates and vertebrates use different mechanisms for chemosensation. The candidate chemosensory aATEN cells do, however, express CNG receptors suggesting that the cyclic nucleotide mediated signaling pathways may play a role for the transduction of chemical stimuli in tunicates (Abitua et al. 2015).

Taken together, this suggests that rostral chemosensory (and possibly OR expressing) neurons in the tunicate-vertebrate ancestor may have been specified in a POU4 and Islet depending manner, having evolved as sister cells of POU4 and Islet expressing mechanosensory neurons. In tunicates, these chemosensory neurons may have maintained their POU4- and Islet-dependent CoRN, while losing OR receptors. In vertebrates, these chemosensory neurons may have evolved into olfactory and vomeronasal neurons, which lost POU4 and Islet dependency and significantly rewired their CoRN (Fig. 3.6). In an alternative scenario, vertebrate olfactory and vomeronasal sensory neurons may have an independent evolutionary origin from POU4- and Islet-independent sensory cells. However, given the extensive similarities in transcription factor expression between PSNs and the olfactory/vomeronasal sensory neurons and hair cells/SSN/GSN/VSNs of vertebrates and the transient expression of POU4 and Islet1 in some olfactory/vomeronasal sensory neurons (chapter 6 in Schlosser 2021), this seems rather unlikely.

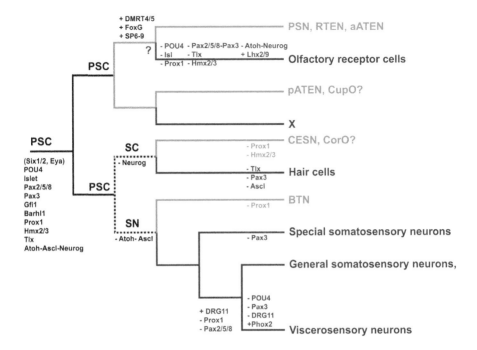

FIGURE 3.6 Scenario for the evolution of mechano- and chemosensory cells in tunicates and vertebrates. Based on shared patterns of transcription factor expression between sensory cells in tunicates and vertebrates it is proposed that a primary sensory cell (PSC) gave rise to two additional sister cells, a dedicated sensory cell (SC; either a secondary sensory cell or a primary sensory cell with a short axon that does not reach the brain) and a sensory neuron (SN). As indicated by the hatched grey-black line, this may have happened in the last common ancestor of chordates or in the last common ancestor of tunicates and vertebrates. PSC, SC, and SN then gave rise to the different sensory cells in the tunicate (blue) and vertebrate lineages (red) as indicated. Proposed changes in transcription factor expression are indicated. Although Tlx expression in tunicates has not been described, Tlx expression in the ancestral primary sensory cell is inferred from shared expression in amphioxus and vertebrate sensory cells. Six1/2 and Eya are shown in parentheses because they may have been expressed only in a subset of sensory cells. The red question mark indicates that derivation of olfactory receptor cells from this ancestral primary sensory cell is uncertain due to the highly divergent transcription factor expression profile of olfactory receptor cells. The placement of coronal and cupular organs of tunicates in this tree is also indicated with question marks since little is known about expression of transcription factors in these cells. aATENs, anterior apical trunk epidermal neurons; BTN, bipolar tail neurons; CESN, caudal epidermal sensory neurons; CorO, coronal organ; CupO, cupular organ; pATENs, posterior apical trunk epidermal neurons; PSNs, palp sensory neurons; RTENs, rostral trunk epidermal neurons.

3.1.2.2 Sensory Neurons of the Larval Tail

In contrast to the sensory neurons, discussed so far, the primary sensory cells of the tail (CESNs) do not send their axon into the brain but synapse onto sensory neurons (BTNs) whose axon reaches the brain. CESNs and BTNs, thus, form a functional unit similar to the hair cells and SSNs of vertebrates. Also like in vertebrates, both CESNs and BTNs arise from non-neural ectoderm that immediately borders the neural plate. Indeed, both cell types arise from the b8.18 lineage (see Fig. 3.2B) (Pasini et al. 2006; Stolfi et al. 2015; Horie et al. 2018) and are, thus, lineage related like the hair cells of the vertebrate ear or lateral line and the neurons of the respective vestibulocochlear or lateral line ganglia. With some exceptions, the profile of transcription factor expression is also very similar between BTNs, CESNs, and other primary sensory cells (sharing POU4, Islet, Pax2/5/8, Pax3, Gfi, BarH), suggesting that these are related cell types (Fig. 3.4). However, CESNs, like hair cells, strongly express Atonal (but not Neurog) expression, whereas BTNs, like somato- and viscerosensory neurons, strongly express Neurog but not Atonal (Tang et al. Joyce Tang, Chen, and Zeller 2013; Cao et al. 2019). Additional transcription factor differences between these cells (e.g. regarding expression of Hmx2/3, Phox2a, and Ascl) suggest that they represent genetically well-individualized subtypes (Fig. 3.4).

Taken together, this raises the possibility that the last common ancestor of tunicates and vertebrates may have already been equipped – like vertebrates – with a functional unit of an Atoh1-dependent sensory cell and a Neurog1-dependent sensory neuron, mediating mechanosensory information from the skin. Consequently, CESNs and hair cells may be evolutionarily related cell types (as may be BTNs and SSNs), possibly derived from a common ancestral cell type in the tunicate-vertebrate ancestor.

The fact that CESNs are morphologically quite unlike hair cells – displaying a cilium without microvilli and a short axon – and are located in the tail, may seem to be at odds with this proposal. However, in contrast to shared CoRNs (as tentatively inferred here from shared profiles of transcription factor expression), both morphology and location are relatively unreliable guides to infer cell type homologies. First, the morphology of sensory cells is notoriously variable. We have seen examples of this in our survey of the olfactory and vomeronasal sensory neurons derived from the olfactory placode in Chapter 6 of the first volume (Schlosser 2021). While most of these sensory cells contain apical cilia and/or microvilli, one or the other of these structures may be lost and their number and configuration is highly variable between different subtypes within one species as well as between species. Nevertheless, the various subtypes of olfactory and vomeronasal sensory neurons express largely the same set of transcription factors, suggesting that they are most likely evolutionarily closely related. Similarly, the fact that CESNs have a short axon while hair cells are axonless does not argue against cell type homology. The most likely scenario for the evolution of a chain of sensory cell and sensory neuron from a primary sensory cell assumes that duplicated ancestral primary sensory cells initially cross-talked to one another before one of them specialized for a role as a sensory cell by reducing its connection to the CNS, while the other specialized as a sensory neuron by losing its responsiveness

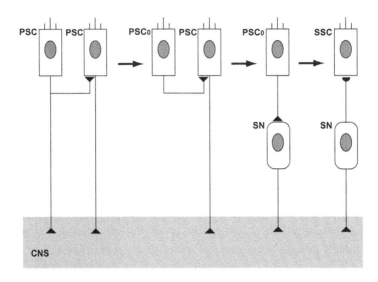

FIGURE 3.7 Model for evolution of secondary sensory cells. Black triangle: presynaptic ending of axon. Black half circle: postsynaptic ending of dendrite. PSC, primary sensory cell with connection to CNS; PSC_0, primary sensory cell without connection to CNS; SN, sensory neuron; SSC, secondary sensory cell. See text for details.

to sensory stimuli (Fig. 3.7). Sensory cells that retain a short axon like CESNs may correspond to an early stage of specialization for a role as dedicated sensory cell, whereas axonless secondary cells such as cells of the coronal organ or hair cells may represent a later stage.

Second, the same cell type can develop at multiple locations (and stages) in the embryo and these locations (and stages) may change during evolution due to alterations in the wiring of gene regulatory networks that lead to the activation of CoRNs for cell type identity in new locations. For example, in *Ciona*, tail sensory neurons (CESNs) are found both along the dorsal and ventral midline. While these two populations develop from different precursors and by different upstream mechanisms, they ultimately both activate the same profile of POU4 and other transcription factors (Waki et al. 2015). As a consequence, the gene expression profile of differentiated dorsal and ventral neurons is indistinguishable and they cluster together in single cell transcriptome analyses (Cao et al. 2019). Therefore, the fact that cell types develop at different locations, does not per se argue against cell type homology either within or between species.

Contrary to my suggestion here, it has been previously proposed that BTNs are neural-crest like cells resembling the neurons of the dorsal root ganglia because they express Pax3/7, Snail and Neurog, are migratory and have similar central and peripheral connections, whereas the Atoh1 expressing CESNs have been proposed to be Merkel cell-like (Stolfi et al. 2015). However, their similarities of transcription factor expression with other primary sensory cells (e.g. PSNs, aATENs, pATENs) (Figs. 3.4 and 3.5) together with the joint origin of BTNs and CESNs from the same precursor cell (b8.18) located just outside of the Zic expressing neural plate

(Wada and Saiga 2002; Imai et al. 2004, 2006), which indicates that these cells arise from non-neural ectoderm rather than from the lateral neural plate, does not support this previous proposal.

3.1.2.3 Sensory Neurons in the Adult

Unfortunately, we know very little about the transcription factors expressed in the sensory neurons of the cupular and coronal organ, which develop from the atrial and oral siphon primordia, respectively, after metamorphosis. Both oral and atrial siphon primordia express pan-placodal transcription factors (Six1/2, Eya, Six4/5 in atrial primordium) and Pax2/5/8 which demarcates the posterior placodal area in vertebrates (Wada et al. 1998; Bassham and Postlethwait 2005; Mazet et al. 2005; Gasparini et al. 2013). The atrial primordium also expresses additional posterior placodal transcription factors (FoxI, Irx) (Imai et al. 2004; Mazet et al. 2005). Together with the expression of other transcription factors implicated in overall regionalization of the ectoderm, this suggests that the atrial primordium corresponds in position to the posterior placodal area, from which hair cells and sensory neurons of the ear and lateral line will develop in vertebrates (see Chapter 6).

However, Atoh1, the central regulator of hair cell specification in vertebrates, was shown to be expressed in the tentacles of the oral siphon (Rigon et al. 2018), which contain the secondary sensory cells of the coronal organ. Conversely, Neurog and Phox2, involved in the specification of sensory neurons in vertebrates, were upregulated in the atrial primordium, which give rise to the primary sensory cells of the cupular organ (Mazet et al. 2005). This provides some support for the proposal that in spite of their rostral location the secondary sensory cells of the coronal organ but not the primary sensory cells of the cupular organs may be homologs of vertebrate hair cells (Burighel et al. 2003; Manni, Caicci et al. 2004). It is, thus, tempting to speculate that the secondary cells of the coronal organ in adult ascidians and the CESNs in their larva despite their different developmental origins may be evolutionarily related cell types that diverged from a common ancestor in the tunicate lineage (Fig. 3.6). However, a more thorough analysis of the CoRNs of these two cell types in future studies will be necessary to confirm this.

3.1.3 The Last Common Tunicate-Vertebrate Ancestor

In summary, comparison of vertebrate with tunicate sensory cell types suggests that the last common ancestor of the tunicate-vertebrate clade possessed mostly POU4- and Islet-dependent sensory cells and neurons. These probably also expressed Pax2/5/8, BarH, Prox, Gfi, Hmx2/3, and other transcription factors possibly in different combinations defining several mechano- and/or chemosensory subtypes (Fig. 3.6). Some of them were primary sensory cells which send their axons into the brain. Such primary sensory cells were retained and further diversified in the tunicate lineage. The vertebrate olfactory and vomeronasal neurons, which express POU4 and Islet1 only transiently or in small subpopulations and are specified by a completely different CoRN, may have evolved as strongly divergent sister cells of POU4 and Islet1 expressing primary sensory cells that lost expression of these and other transcription factors and evolved a new CoRN (Fig. 3.6).

Other cells in the tunicate-vertebrate ancestor may have been primary sensory cells with a short axon that made contact with a sensory neuron relaying its sensory information to the brain. These two cells, which formed a functional unit, may have originated by duplication of a common precursor and subsequently became specifically dependent on Atoh and Neurogenin, respectively, for cell identity. This is in accordance with a previous proposal that vertebrate hair cells and the SSNs that innervate them arose as sister cells from a common precursor, which acquired differential dependence on Atoh1 and Neurog1, respectively (Fritzsch, Beisel, and Bermingham 2000; Fritzsch, Eberl, and Beisel 2010).

The presence of functional units comprising a dedicated sensory cell (CESN or receptor cell of coronal organ) and a sensory neuron in tunicates suggests that these two sister cells may have originated before the separation of the tunicate and vertebrate lineages by duplication of a POU4-Islet-dependent primary sensory cell and subsequent divergence into sensory cell and sensory neuron. Subsequently, the dedicated sensory cells may have evolved into CESNs and possibly coronal cells in tunicates and into hair cells in vertebrates, while the sensory neuron evolved into BTNs and SSNs, respectively (Fig. 3.6). This suggests that the complete loss of the axon in hair cells and coronal cells may have possibly been acquired twice independently in the vertebrate and tunicate lineages, respectively. The duplication of *POU4* genes in the vertebrate lineage (Gold, Gates, and Jacobs 2014) probably contributed to the increased genetic individuation of hair cells versus sensory neurons, which became specifically dependent on *POU4f3* and *POU4f1*, respectively. However, we can currently not rule out the alternative possibility that functional units of a dedicated mechanosensory cell and a sensory neuron evolved convergently from primary mechanosensory cells in tunicates and vertebrates.

There is no evidence in tunicates or amphioxus (see Section 3.2) for the existence of GSNs and VSNs comparable to the neurons that innervate skin and viscera of vertebrates without being associated with a dedicated sensory cell (reviewed in Chapter 6 of Schlosser 2021). This suggests that these two populations of neurons originated only in the vertebrate lineage. However, the last common ancestor of tunicates and vertebrates may have had sensory neurons similar to the BTNs of *Ciona* that had free nerve endings in addition to its innervation by dedicated receptor neurons (like CESNs). It may, thus, have also responded to stimuli not mediated by these dedicated receptors, facilitating the subsequent evolution of stand-alone sensory neurons. GSNs express Islet1 and POU4, while VSNs express Islet1 and Phox2, which represses POU4 (D'Autreaux et al. 2011) suggesting that these neurons are also evolutionarily derived from the lineage of POU4-Islet dependent sensory cells (Fig. 3.6). Furthermore, GSNs and VSNs share Neurog1/2 and Tlx1/3 specifically with SSNs but not hair cells, indicating that they may be offshoots of the SSN lineage and that the associated sensory cells have been lost (Fig. 3.6) (Baker, O'Neill, and McCole 2008; Patthey, Schlosser, and Shimeld 2014).

During the evolution of different subtypes of sensory cells, some transcription factors that were co-expressed in an ancestral vertebrate sensory cell type became differentially expressed in these subtypes, whereas there was probably subtype specific upregulation of others. For example, the co-expression of Pax2/5/8 and Pax3 in many tunicate sensory cells suggests that these transcription factors may also have been co-expressed in mechanosensory neurons in the last common tunicate-vertebrate ancestor. Subsequently, they may have been differentially retained in hair cells/SSNs (Pax2/5/8) and GSNs (Pax3) in the vertebrate lineage (Fig. 3.6). GSNs and VSNs may then have adopted different subtype identity by downregulation of Pax3 and upregulation of Phox2 in VSNs (Fig. 3.6).

Finally, Six1/2 and Eya may have been important only for the specification of a subset of sensory cells in the tunicate-vertebrate ancestor (Fig. 3.6). In tunicates, only PSNs, aATENs and, possibly, RTENs and the sensory cells of the adult coronal organs, but not pATENs, CESNs, and BTNs arise from Six1/2 expressing non-neural ectoderm at the anterior neural border. The adult cupular organs arise from another domain of non-neural ectoderm adjacent to the lateral part of the neural plate, which upregulates Six1/2 during later larval stages. Although functional experiments have not been performed yet, this suggests that while Six1/2 may be required for some subtypes of POU4-dependent sensory cells/neurons, it is dispensable for others (pATENs, CESNs, and BTNs) in the PNS of tunicates. This contrasts with vertebrates, where differentiation of all cranial placode-derived POU4-dependent sensory cells and neurons is inhibited after Six1 loss of function. Taken together, it suggests that the central role of Six1/2 as an upstream regulator of all cranial sensory cells and sensory neurons of the PNS evolved only after vertebrates branched off from their last common ancestor with tunicates. I will return to this point in Chapter 6, when discussing the origin of cranial placodes in vertebrates.

3.2 MECHANO- AND CHEMOSENSORY CELLS IN THE LAST COMMON CHORDATE ANCESTOR

3.2.1 MECHANO- AND CHEMOSENSORY CELL TYPES IN AMPHIOXUS

Lancelets (amphioxus) are the closest living relatives of the tunicate-vertebrate lineage; together they comprise the chordates. Comparisons between these taxa and relevant outgroups (e.g. other deuterostomes) should, therefore, provide us with insights regarding the mechano- and/or chemosensory cells present in the last common ancestor of chordates. In amphioxus, such sensory cells are found scattered throughout the non-neural ectoderm (reviewed in Holland 2005; Holland and Holland 2001; Lacalli 2004). Based on ultrastructural differences, two major types have been described (Schulte and Riehl 1977; Bone and Best 1978; Baatrup 1981; Stokes and Holland 1995; Lacalli and Hou 1999). Type I cells have a single cilium surrounded by several rings of microvilli (Fig. 3.8A). There are two subtypes, a subtype Ia with a long cilium surrounded by two different types of microvilli and a subtype Ib with a shorter cilium surrounded by microvilli of a single type (Lacalli and Hou 1999). Type II cells have a single cilium that is recessed below the surface and is surrounded by a peculiar collar covered with microvilli (Fig. 3.8A).

Type I and type II cells also differ with respect to their central projections. Most but possibly not all type I cells appear to be primary sensory cells with an axon, which probably projects into the CNS (Lacalli and Hou 1999). In contrast, most but not all type II cells are secondary sensory cells (Lacalli and Hou 1999; Zieger, Garbarino et al. 2018). They are innervated by dendrites of sensory neurons that

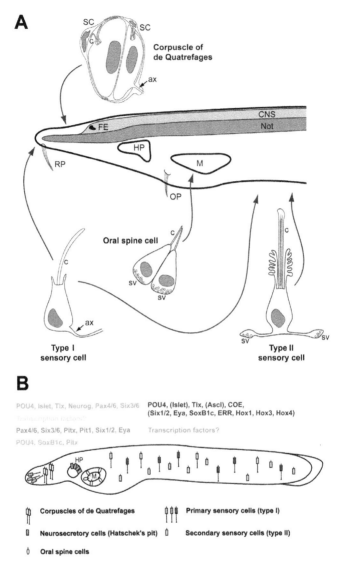

FIGURE 3.8 Sensory cells in amphioxus. (**A**) A selection of sensory cells in amphioxus. Red arrows indicate locations of these cells in an amphioxus larva (left side of head region shown). See text for details. ax, axon; c, cilium; CNS, central nervous system; FE, frontal eye; HP, Hatschek's pit; M, mouth; Not, notochord; OP, oral papilla; RP, rostral papilla; SC, sheath cells; sv, synaptic vesicles. (**B**) Distribution of different types of sensory cells in amphioxus and their pattern of transcription factor expression. Type I sensory cells are heterogeneous as indicated by different colors. Transcription factors listed in parentheses are only expressed in subsets of sensory cells. The transcription factors expressed in type II sensory cells and corpuscles of de Quatrefages, which develop later, are currently not known. ([A] Reprinted from Schlosser 2017 after Meulemans and Bronner-Fraser 2007; [B] reprinted from Schlosser 2017, adapted from Baatrup 1982; Lacalli and Hou 1999; Lacalli, 2004.)

probably have their cell body residing in the CNS, although this remains to be confirmed. Many type I cells originate already during early embryonic stages in the ventral ectoderm, where they delaminate and migrate dorsally before reinserting into the epidermis and differentiating into sensory neurons (Stokes and Holland 1995; Benito-Gutiérrez, Nake et al. 2005; Benito-Gutiérrez, Illas et al. 2005; Kaltenbach, Yu, and Holland 2009). The majority of type II cells develop later, just prior to metamorphosis (Stokes and Holland 1995).

In addition to type I and II cells, which are found widely distributed in the ectoderm, there are other cell types with a more localized distribution. Ventral pit cells with a recessed cilium and no microvilli are found on folds of the body wall (metapleural folds, which enclose the atrium), while cells with stiff ciliary spines or with recessed cilia in irregular pits occur on the rostral tip of the animal (Stokes and Holland 1995; Lacalli, Gilmour, and Kelly 1999; Lacalli and Hou 1999; Lacalli 2002). In addition, there are rostral epithelial cells without obvious sensory specializations that contain synaptic vesicles and are innervated (Lacalli 2002). The mouth is surrounded by several spines, each of which is formed by a cluster of oral spine cells (Fig. 3.8A) (Lacalli et al. 1999). These cells carry a single cilium, but no microvilli and the cilia of several adjacent cells are fused together to form the spine. Oral spine cells are secondary sensory cells, which are innervated by sensory neurons, with cell bodies located nearby in a circumoral plexus (Lacalli et al. 1999; Zieger, Candiani et al. 2018; Zieger, Garbarino et al. 2018). In adult amphioxus, the rostral ectoderm also harbors small sensory organs, the so-called corpuscles of de Quatrefages (de Quatrefages 1845; Baatrup 1982). They contain a small number (1–4) of primary sensory cells each equipped with two cilia and surrounded by up to seven sheath cells (Fig. 3.8A). Finally, there are ciliated cells that form the tufts of the rostral and oral papillae (Fig. 3.8A). However, based on their lack of synaptic specializations and the movements of the oral papillae during feeding they were suggested to be more likely ciliary effectors than sensory cells (Lacalli et al. 1999).

Like in tunicates, the sensory modalities mediated by these different types of sensory cells have been tentatively assigned based on their morphology but only few supporting behavioral or physiological data are available. Oral spine cells are most likely mechanoreceptors, which, when touched, initiate a "cough" response expelling excess debris (Lacalli et al. 1999). The structure of the corpuscles of de Quatrefages suggests that they may be mechanoreceptors responding to deformation, but this needs to be confirmed by physiological studies (Baatrup 1982). Type Ia cells have been proposed to be mechanosensory cells, based on the length of their cilia, whereas type II and possibly type Ib cells have been suggested to also or alternatively serve chemosensory functions judged from the surface increase of their elaborate microvillar collars (Lacalli and Hou 1999). Again, this needs still to be confirmed by physiological studies.

However, the expression of a vertebrate-type odorant receptor (OR) in some type I sensory cells located in rostral ectoderm suggests that these cells are in fact chemosensory (Satoh 2005). Amphioxus has approximately 50 genes encoding vertebrate-type ORs in its genome, which are, however, not related one to one to vertebrate ORs but rather represent an independent expansion of the OR family of GPCRs in the amphioxus lineage (Churcher and Taylor 2009; Niimura 2009). In vertebrates, different OR family members are expressed in different olfactory sensory neurons, with each neuron expressing only a single OR (see Chapter 6 in Schlosser 2021). Whether this is also true for the chemosensory neurons in amphioxus still needs to be resolved. Resembling the olfactory nerve in vertebrates, the rostral primary sensory cells of amphioxus send their axons via a rostral nerve into the anterior-most part of the brain (Fig. 1.12C). Here they converge with the axons of other primary sensory cells onto neurons, which control the amphioxus escape response (Lacalli 2002).

3.2.2 TRANSCRIPTION FACTORS INVOLVED IN SPECIFYING AMPHIOXUS MECHANO- AND CHEMOSENSORY CELL TYPES

To understand how the various sensory cell types of amphioxus are related to those of vertebrates and tunicates, we need to consider whether they express transcription factors implicated in the specification of sensory neurons in vertebrates and tunicates. However, while the expression of several of these transcription factors including POU4, Islet, Tlx, Pax2/5/8, Pax3/7, Neurog, and Ascl has been investigated in amphioxus by in situ hybridization and immunostaining (Holland et al. 1999, 2000; Kozmik, Holland et al. 1999; Jackman, Langeland, and Kimmel 2000; Schubert et al. 2004; Candiani et al. 2006; Kozmik et al. 2007; Kaltenbach et al. 2009; Lu, Luo, and Yu 2012; Barton-Owen, Ferrier, and Somorjai 2018; Zieger, Garbarino et al. 2018), no double-labeling studies have been done and no single cell RNA-sequencing data are currently available. We, therefore, currently know very little about which transcription factors are co-expressed in individual cells. Furthermore, the expression of many transcription factors with important roles in conferring sensory identity, including Atonal, Prox, Hmx2/3, Gfi, BarH, DRG11, Phox2, and DMRT4/5 has not yet been described in amphioxus, or has only been reported for early embryonic stages (e.g. Lhx2/9; Albuixech-Crespo et al. 2017).

At least some of the type I neurons, which originate from the ventral ectoderm in amphioxus embryos, express POU4, Islet, and Tlx, in addition to the progenitor factors SoxB1c, Six1/2, Six4/5, Eya, the pan-neuronal differentiation markers Hu/Elav and Trk (encoding a neurotrophin receptor), and the Notch ligand Delta, which represses neuronal fates in adjacent cells via lateral inhibition (Fig. 3.8B) (Satoh et al. 2001; Schubert et al. 2004; Mazet et al. 2004; Benito-Gutiérrez, Nake et al. 2005; Benito-Gutiérrez, Illas et al. 2005; Candiani et al. 2006; Rasmussen et al. 2007; Meulemans and Bronner-Fraser 2007; Kozmik et al. 2007; Kaltenbach et al. 2009; Lu et al. 2012). However, in contrast to vertebrates and tunicates, neither Neurogenin nor Pax3 or Pax2/5/8 expression were detected in these neurons by in situ hybridization, although low-level expression may have been missed (Holland et al. 1999, 2000; Kozmik, Pfeffer et al. 1999; Barton-Owen et al. 2018). Amphioxus also lacks the horseshoe-shaped domain of Six1/2 and Eya expression found in the dorsal-most non-neural ectoderm surrounding the anterior neural plate in vertebrates and tunicates.

Neurogenin is, however, co-expressed with POU4, Islet, and Tlx as well as the anterior transcription factors Pax4/ and Six3/6 in the rostral ectoderm of amphioxus containing type I sensory cells whose axons enter the brain via the rostral nerve (Fig. 3.8B) (Glardon et al. 1998; Holland et al. 2000; Jackman et al. 2000; Candiani et al. 2006; Kozmik et al. 2007; Kaltenbach 2009; Lu et al. 2012). Lhx2/9 is expressed in the adjacent anterior neural plate but has not been reported in the non-neural ectoderm (Albuixech-Crespo et al. 2017). POU4 is also expressed around the mouth, where oral spine cells develop (Candiani et al. 2006). Which transcription factors are expressed in type II receptors and the neurons that innervate them is currently unknown due to their late development just prior to metamorphosis.

Although we lack conclusive experimental evidence regarding the co-expression of transcription factors in single cells, the distribution of staining suggests that most transcription factors are expressed in only a subset of type I cells and these subsets appear to be different and only partly overlapping for different transcription factors. This suggests that type I sensory cells are molecularly heterogeneous. A recent study has, for example, shown that different subpopulations of the ventral type I cells express Tlx and Soxb1c (Zieger, Garbarino et al. 2018). This heterogeneity may in part reflect different stages of differentiation (with Tlx–/SoxB1c+, and Tlx+/SoxB1c– cells representing early and late stages of differentiation, respectively), but may also indicate the existence of different subtypes of sensory cells. Nevertheless, the overall similar expression pattern of POU4, Islet, and Tlx suggests that these transcription factors are likely co-expressed in many sensory cells. They may, thus, confer a transcription factor expression profile to these cells resembling the profile of many of the sensory cells in tunicates (for which, however, Tlx has not been described) and of the hair cells (which may have lost Tlx secondarily), SSNs, GSNs and VSNs of vertebrates.

The putatively chemosensory type I neurons located in Pax4/6- and Six3/6-positive rostral head ectoderm that express vertebrate-type ORs and send their axons into the rostral most brain seem to share expression of POU4, Islet, and Tlx with other mechano- and/or chemosensory neurons, while lacking expression of transcription factors (Lhx2/9 and FoxG; Sp6-9 and DMRT4/5 have not been analyzed) that are specifically required for olfactory receptor neurons in vertebrates (see Chapter 6 in Schlosser 2021; Toresson et al. 1998; Albuixech-Crespo et al. 2017). This suggests that specialized rostral chemosensory neurons employing OR-type receptors may already have originated in the last common ancestor of chordates as a sister cell type of other POU4, Islet, and Tlx expressing mechano- and/or chemosensory neurons. However, the independent expansion of *OR* genes in amphioxus and vertebrates, the loss of *OR* genes in tunicates and the dispensability of POU4, Islet, and Tlx for the specification of olfactory receptor neurons in vertebrates indicates that these rostral chemosensory neurons then specialized and diversified further independently from each other in the different chordate lineages.

Whether the axonless secondary sensory cells of amphioxus (type II neurons, oral spine cells) correspond to the secondary sensory cells of tunicates and vertebrates or have evolved independently is currently difficult to determine. In vertebrates and tunicates, the specification of secondary sensory cells (or primary sensory cells whose axons don't reach the brain) probably depends on Atonal, while the specification of sensory neurons, which transmit sensory information to the brain, appears to depend on Neurogenin (see Section 3.1.2 above). There are probably other important differences in transcription factor expression, but these have not been well characterized. In amphioxus, Atonal expression has not yet been described, while data on Neurogenin expression are only available for embryonic stages and early larvae (Holland et al. 2000). The latter data suggest that neurons located in the dorsal neural tube (and presumably neural tube-derived) express Neurogenin. These neurons may include neurons innervating the oral spine cells. However, data for the oral spine cells themselves as well as the type II neurons and the sensory neurons, by which they are innervated, are not available. Therefore, it is currently unclear, whether the duplication and divergence of primary sensory cells into Atonal-dependent secondary sensory cells and Neurogenin-dependent sensory neurons already happened in the last common ancestor of chordates or only in the tunicate-vertebrate ancestor.

3.2.3   THE LAST COMMON CHORDATE ANCESTOR

Based on the many similarities between amphioxus and vertebrates, the last common ancestor of chordates was probably a quite amphioxus-like filter feeding animal. Its body plan remained conserved in the amphioxus lineage and probably also in the stem lineage of tunicates and vertebrates. After the divergence of tunicates and vertebrates, tunicates then evolved a highly modified body plan, whereas vertebrates maintained but embellished a more amphioxus like organization. Judged from the sensory cell types shared between amphioxus and other chordates, the last common ancestor of chordates probably was equipped with scattered POU4-, Islet-, and Tlx-dependent mechano- and/or chemosensory cells distributed throughout the non-neural ectoderm (Fig. 3.6). Some of these also co-expressed Six1/2 and Eya1 and possibly required these factors for their differentiation. However, there were no special domains of Six1/2 and Eya expression in the dorsal-most non-neural ectoderm adjacent to the neural plate, which may have evolved only in the tunicate-vertebrate lineage.

Like extant amphioxus and tunicates, the last common chordate ancestor probably had a group of specialized POU4-, Islet-, and Tlx-dependent chemosensory neurons in the rostral ectoderm whose axons entered the brain via a rostral nerve. Differentiation of this rostral subtype of sensory neurons may have dependent on additional transcription factors with expression in anterior non-neural ectoderm such as Pax4/6 and Six3/6. From this rostral subtype of sensory neurons some of the rostral sensory neurons in tunicates (e.g. PSNs, RTENs, aATENs) and the olfactory receptor neurons of vertebrates may have evolved.

Because only limited data are available from amphioxus, it currently remains unclear, whether a functional unit comprising an Atonal-dependent sensory cell (either primary sensory cell with a short axon or secondary sensory cell) and a Neurogenin-dependent sensory neuron had already evolved in the last common ancestor of chordates or whether this only happened in the tunicate-vertebrate lineage (Fig. 3.6), with secondary sensory cells evolving independently in amphioxus (as their innervation by a CNS-derived neuron suggests).

3.3 MECHANO- AND CHEMOSENSORY CELLS IN THE LAST COMMON DEUTEROSTOME ANCESTOR

3.3.1 MECHANO- AND CHEMOSENSORY CELL TYPES IN HEMICHORDATES AND ECHINODERMS

To gain insight into the mechano- and chemosensory cells present in the more distant last common ancestor of deuterostomes, we next have to consider the ambulacrarians (hemichordates and echinoderms), the sister group of the chordates within the deuterostomes. Among the ambulacrarians, the echinoderms display a radically modified body plan, while the pterobranch hemichordates have secondarily become sessile, leaving the enteropneust hemichordates as the most chordate-like forms, which presumably retained more of the ancestral deuterostome characters than pterobranchs and extant echinoderms (see Chapter 1). In spite of their different body plans, all ambulacrarians have relatively diffuse nervous systems, with neurons or sensory cells widely distributed in the ectoderm and with no complex sense organs.

Not much is currently known about mechano- and chemosensory cells in ambulacrarians. Several candidates have been identified in ultrastructural studies, but whether they serve as mechano- or chemoreceptors remains to be determined. It is currently also unclear, how hemichordates and echinoderms sense chemical stimuli. Genomic studies have found members of insect type chemoreceptors in both echinoderms and hemichordates but not in chordates (Robertson 2015), whereas vertebrate-type ORs have been identified in echinoderm but not hemichordate genomes (Raible et al. 2006; Krishnan et al. 2013). However, we do not know, in which cells these receptors are expressed, whether they are involved in chemosensation and, if so, which signal transduction cascades they activate.

In the ciliary bands and apical organs of hemichordate and echinoderm larvae, several types of sensory cells have been described including monociliated cells without microvilli, (Lacalli, Gilmour, and West 1990; Byrne et al. 2007; Kaul-Strehlow et al. 2015) as well as cells without cilia but with branched apical protrusions (Burke 1978; Lacalli et al. 1990; Lacalli and West 1993).

Ciliated cells with a collar of microvilli have been described in adult hemichordates and echinoderms (Fig. 3.9) and for some of these cells axonal processes have been found suggesting that they are primary sensory cells (Cobb 1968; Nørrevang and Wingstrand 1970; Dilly 1972; Burke 1980; Jørgensen 1989; Benito and Pardos 1997). In echinoderms, such putatively mechano- and/or chemosensory cells are mostly associated with the tube feet and pedicellariae (Cobb 1968; Burke 1980). Another type of mechano- or chemosensory cell, found in some echinoderms, is a bipolar cell with a dendrite reaching the surface of the epidermis, where it carries a cilium surrounded by microvilli (Whitfield and Emson 1983). The nerve nets of hemichordates also contain bipolar cells with dendrites projecting to the surface, but whether these are ciliated or not, and which sensory modality they mediate, has not been determined (Bullock 1945; Knight-Jones 1952).

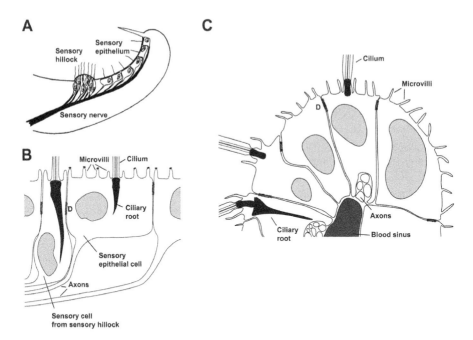

FIGURE 3.9   Sensory cells in ambulacrarians. (**A**) Diagram of a single jaw of a pedicellarium of a sea urchin (*Echinus*) showing the sensory hillock and the general sensory epithelium. (**B**) Detail of sensory cells from the sensory hillock and the general sensory epithelium. Both types of cells are primary sensory cells (with an axon) and carry microvilli and a single cilium with long ciliary roots (bundles of microtubules). Some microvilli end in a cap-like structure (black ovals). (**C**) Diagram of sensory cells in the tentacle of the pterobranch *Rhabdopleura*. Each tentacle has three rows of ciliated cells (one row of single cells with short cilia shown above; two rows of paired ciliated cells with long cilia; one of these rows is shown to the left). These are associated with bundles of axons on their basal side, suggesting that they are sensory cells. Most of the epithelial cells as well as the sensory cell have microvilli. ([A,B] Redrawn from Dilly 1972. D, desmosome. [C] Reprinted and redrawn with permission from Cobb (1968).)

3.3.2   TRANSCRIPTION FACTORS INVOLVED IN SPECIFYING HEMICHORDATE AND ECHINODERM MECHANO- AND CHEMOSENSORY CELL TYPES

Which transcription factors are involved in specifying the identity of sensory cells in ambulacrarians is also largely unknown. Only limited expression data and no single-cell RNA-Seq data of sufficiently advanced embryos or larval stages are currently available. In enteropneust hemichordates, transcription factors promoting neural progenitors such as SoxB1 as well as general neuronal markers like Hu/Elav are expressed in wide areas of the ectoderm, indicating that the neurons of its diffuse nerve net arise throughout the ectoderm (Lowe et al. 2003; Cunningham and Casey 2014). However, the expression of many transcription factors that define the identity of mechano- and chemosensory neurons of the PNS in chordates such as

POU4, Islet, or Tlx1/3, Hmx2/3, Gfi, Phox2 or of the bHLH factors Atonal, Ascl, and Neurogenin has not been described. Prox and BarH are expressed in partly overlapping domains of the mesosome, while DRG11 is mostly expressed in the collar cord (CNS) and in rare neurons in the prosome (proboscis), rostral to the mesosome (Lowe et al. 2003; Nomaksteinsky et al. 2009; Cunningham and Casey 2014). Six1/2 and Eya are mostly expressed in the developing gill slits but also have a transient expression domain in the anterior mesosome (Gillis, Fritzenwanker, and Lowe 2012).

In echinoderm larvae, neural progenitors expressing SoxB1 or SoxB2 are also distributed widely throughout the ectoderm, but neurons subsequently differentiate only in the regions of the developing ciliary bands and a few other well-defined domains (Yankura et al. 2013; Garner et al. 2016). The ciliary bands and/or apical organ were reported to express Islet, Prox, Pax2/5/8, Gfi as well as Ascl and Neurogenin, while POU4, Atonal, Six1/2, and Eya were undetectable in the ectoderm by in situ hybridization (Burke et al. 2006; Saudemont et al. 2010; Yankura et al. 2010; Materna et al. 2013; Slota, Miranda, and McClay 2019). After metamorphosis, POU4, Islet, Pax2/5/8, BarH1, Phox2, Atonal, and Eya were found to be expressed in the tube feet (Czerny et al. 1997; Burke et al. 2006; Agca et al. 2011). However, it is not clear whether these transcription factors are expressed in mechano-, chemo-, or photosensory cells, all of which are known to be enriched in tube feet and whether they are co-expressed in individual cells (Burke 1980; Ullrich-Lüter et al. 2011; Lesser et al. 2011; Agca et al. 2011). Other transcription factors involved in specifying mechano- or chemosensory cells in chordates were not analyzed.

3.3.3 THE LAST COMMON DEUTEROSTOME ANCESTOR

Taken together, the sparse information available from ambulacrarians do not allow us to reach any firm conclusions, on which mechano- or chemosensory cell types were likely present in the last common ancestor of deuterostomes. The available data suggest that this ancestor was equipped with one or several types of primary sensory cells expressing some of the transcription factors involved in the specification of mechano-and/or chemosensory cells in chordates. As further comparisons with other metazoans will show (see Sections 3.4 and 3.5), several of these transcription factors including POU4, Islet, Pax2/5/8, Gfi, Atonal, and Ascl appear to have functions for sensory identity already prior to the origin of deuterostomes. However, it is currently impossible to infer, in which combination these transcription factors were expressed in the last common deuterostome ancestor and how many discrete subtypes of sensory cells they defined. Interestingly, Six1/2 and Eya, which appear to be not expressed in any sensory cells in ambulacrarians, are expressed in subsets of sensory cells (including mechanosensory cells and rhabdomeric photoreceptors) not only in chordates but also in many protostomes and cnidarians (see below). This suggests that ambulacrarians have either lost a subset of Six1/2 and Eya-dependent sensory cells or express these transcription factors at very low levels.

3.4   MECHANO- AND CHEMOSENSORY CELLS IN THE LAST COMMON BILATERIAN ANCESTOR

The sister group of the deuterostomes are the protostomes, which comprise two major groups, the lophotrochozoans (including flatworms, mollusks, annelids among other groups) and the ecdysozoans (including arthropods, nematodes, and a number of smaller taxa) (Fig. 2.4). Protostomes and deuterostomes together form a mono-phyletic clade. The position of another group, the xenacoelomorphs (acoel flatworms and related taxa) within the bilateria is still controversial. While they are considered a basal group of deuterostomes by some studies (Philippe et al. 2011, 2019), other studies suggest that they are actually the sister group of the combined protostome-deuterostome clade and, thus, the most basal group of bilaterians (Cannon et al. 2016) (Fig. 1.8). Since little is known about the sensory cells of xenacoelomorphs, we can leave this question open for the following discussion. Comparisons of sensory cells between deuterostomes, protostomes, and outgroups of the bilaterians such as cnidarians should allow us some insights into the sensory cells present in the last common ancestor of bilaterians – sometimes called the "Urbilaterian".

However, this task is hampered by the extraordinary diversification of the proto-stomes into a multitude of highly divergent taxa, each with its own specializations and adaptations. This also resulted in many lineage-specific modifications of sen-sory cells and organs, which make it very difficult to identify commonalities and reconstruct the ancestral pattern. In addition, we know almost nothing about the genetic basis of sensory development for most of these groups. Only in one insect, the fruitfly *Drosophila melanogaster* and in the nematode *Caenorhabditis elegans* has the genetic basis of sensory development been studied in any detail. Both spe-cies belong to the ecdysozoans (Fig. 1.8), which have evolved a secreted cuticle as a barrier between the epidermis and the surrounding medium. Interpolated between sensory cells and the environment, the cuticle also necessitated changes in the mechanisms of sensory transduction leading to specialized types of sensory cells in both arthropods and nematodes, which are probably highly modified from the ancestral protostomian condition (reviewed in Hartenstein 2005; Bargmann 2006; Goodman 2006; Kernan 2007; Merritt 2007; Goodman and Sengupta 2019). The sensory cells found in annelids, mollusks, and other lophotrochozoans (Fig. 1.8), appear less specialized and more similar to the sensory cells known from deutero-stomes suggesting that they are much better guides for reconstructing ancestral bilaterian cell types (reviewed in Dorsett 1986; Budelmann 1989; Jørgensen 1989; Purschke 2005; Marlow et al. 2014; Williams and Jékely 2019). Unfortunately, we currently have only very limited knowledge from a few species on the genetic basis of sensory development in lophotrochozoans, in particular the polychaete annelid *Platynereis dumerilii*.

3.4.1   MECHANO- AND CHEMOSENSORY CELL TYPES IN PROTOSTOMES

It is impossible to do justice to the diversity of sensory cells and organs in proto-stomes in a few pages. The main purpose of the following brief overview, there-fore, is to highlight some of the recurrent features of sensory cells in protostomes and to briefly introduce types of sensory cells that are particularly well studied. Like in deuterostomes, sensory cells in protostomes often have a cilium surrounded by microvilli with many variations on this common theme (Figs. 3.10–3.13). These include changes in numbers of cilia, microvilli or both; loss of cilia or microvilli or both; motile or non-motile cilia; recessed or non-recessed cilia etc. (reviewed in Laverack 1988; Budelmann 1989; Jørgensen 1989; Schlosser 2018).

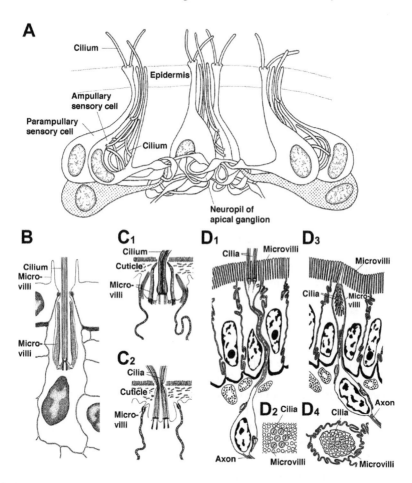

FIGURE 3.10 Mechano- and chemosensory cells in lophotrochozoans. (**A**) Apical organ in a larval mollusk (gastropods, prosobranchs). Non-sensory cells are stippled. (**B–D**) Mechano- or chemosensory cells in adult flatworms (**B**), annelids (**C**), and mollusks (**D**). (**B**) Primary sensory cell with single cilium surrounded by microvilli in the flatworm (turbellarian) *Parotoplana capitata*. (**C**) Apical region of monociliated (C_1) and multiciliated (C_2) sensory cells in the annelid (oligochaete) *Hirudo medicinalis*. (**D**) Type 1 (D_1, D_2) and type 2 (D_3, D_4) multiciliated primary sensory cells in the gastropod (prosobranch) *Nassarius reticulatus*. Cross-sections through cilia and microvilli are shown in D_3 and D_4. ([A] Reprinted with permission from Page and Parries 2000; [B] reprinted and modified with permission from Ehlers and Ehlers 1977; [C] reprinted and modified with permission from Phillips and Friesen 1982; [D] reprinted with modification with permission from Crisp 1971.)

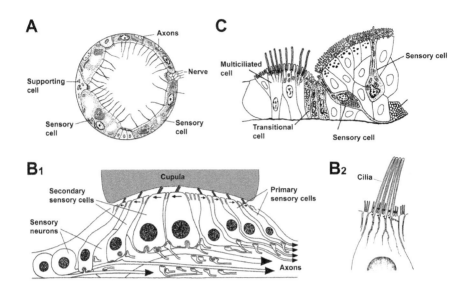

FIGURE 3.11   Mechano- and chemosensory organs in lophotrochozoans. (**A**) Diagram of a gastropod statocyst, lacking the statoconia. (**B**) Diagram of a cephalopod statocyst (*Octopus vulgaris*) with primary and secondary sensory cells and sensory neurons. Overview in **B₁**; detail of a secondary sensory cell with multiple cilia shown in **B₂**. (**B₁**) and Budelmann 1989 (**B₂**). (**C**) Diagram of an osphradium in gastropods (prosobranchs). ([A] Reprinted with permission from Dorsett 1986; [B] reprinted with modification with permission from Budelmann, Sachse, and Staudigl 1987; [C] reprinted with permission from Dorsett 1986.)

Sensory cells originate from the embryonic ectoderm and often remain part of the epidermis, although in other cases their cell body migrates to a subepithelial position and extends a dendrite to the surface of the epidermis, where cilia and/or microvilli may be found on the dendrite (Fig. 3.10). Most sensory cells are primary sensory cells, which send their axon to the CNS. However, secondary sensory cells without an axon also have been described in several taxa (Fig. 3.11B) (e.g. in cephalopod statocysts; Budelmann, Sachse, and Staudigl 1987). Sensory cells may occur as scattered cells distributed in the ectoderm or may form complex sense organs. Whether sensory cells serve as mechano- or chemoreceptors or mediate some other sensory modality (e.g. thermo- or hygroreceptors) is unknown for many protostomes or has been only tentatively proposed based on ultrastructural studies. Exceptions are the various sensory receptor cells and organs of nematodes and arthropods, for most of which the sensory modality has been firmly established in physiological experiments.

3.4.1.1   Lophotrochozoans

Many lophotrochozoans have a specialized larval stage, the trochophora larva, which has sensory cells together with neurosecretory cells concentrated in the apical organ (Fig. 3.10A). Apart from photoreceptors (see Chapter 4), the apical organ contains several types of ciliated receptor cells, which most likely serve as chemo- and/or mechanoreceptors. In mollusks, these include so-called parampullary cells with

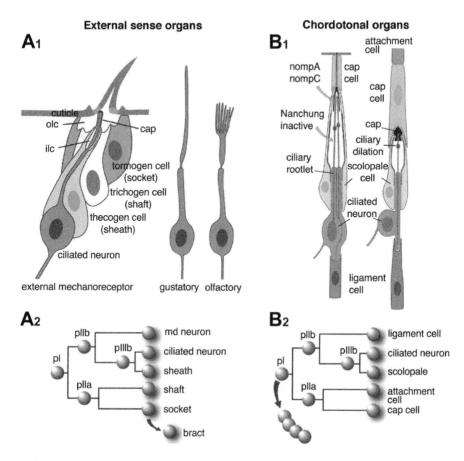

FIGURE 3.12 *Drosophila* external sense organs and chordotonal organs with cell lineage. (A) Diagram of external sense organs (**A₁**) and their cell lineage (**A₂**). Ilc, inner lymphatic cavity; olc, outer lymphatic cavity. Different types of sensory neurons are illustrated. (**B**) Diagrams of scolopidia of two types of chordotonal organs (**B₁**) and their cell lineage (**B₂**). On the left is a scolopidium with two neurons as typically found in the antenna. On the right is a scolopidium with one neuron as found in chordotonal organs of the body wall. (Reprinted with modification with permission from Merritt 2007.)

one or two cilia that protrude into the surrounding medium and ampullary cells with an invaginated apical surface carrying many cilia, which do not protrude beyond the surface (Fig. 3.10A) (Marois and Carew 1997; Croll and Dickinson 2004) and similar cell types have been identified in polychaetes (marine annelids with a larval stage) (Lacalli 1981; Marlow et al. 2014). In some species, the apical organ has been shown to be required for larvae to respond to settlement cues, which induce metamorphosis (Hadfield, Meleshkevitch, and Boudko 2000; Conzelmann et al. 2013). However, the apical organ regresses before metamorphosis in other species and, therefore, most likely serves additional functions (Lacalli 1981). In addition to the apical organ, isolated ciliated sensory cells have also been described in larval lophotrochozoans (e.g. Bezares-Calderón et al. 2018).

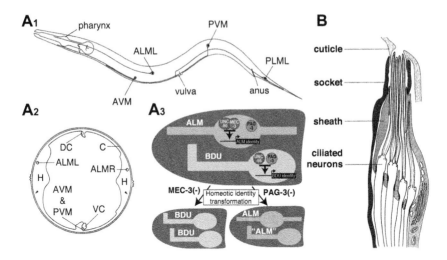

FIGURE 3.13   Sensory cells and organs in *Caenorhabditis elegans*. (**A**) Unciliated mecha-noreceptors (ALML, ALMR, AVM, PLML, PLMR, PVM) with long processes shown in a lateral view (A_1) and cross-section (A_2). A_3: Specification of sensory neuronal identity of left and right ALMs (ALML, ALMR) by POU4 (Unc-86) and Lhx1/5 (Mec-3). A combination of POU4 (Unc-86) and Gfi (Pag-3) specify a different type of sensory neuron (BDU) instead. Mutation of Lhx1/5 (Mec-3) and Gfi (Pag-3) causes a switch in neuronal identity. C, cuticle; DC, dorsal nerve cord; H, hypodermis; VC, ventral nerve cord. (**B**) Amphid with ciliated che-mosensory neurons. ([A] A_1, A_2: Reprinted with permission from Chalfie and Sulston 1981. A_3: Reprinted with permission from Gordon and Hobert 2015; [B] Reprinted with permission from Ward et al. 1975.)

In many adult lophotrochozoans including flatworms, mollusks, and annelids, ciliated sensory cells are found scattered throughout the epidermis or in a subepi-dermal position with a ciliated dendrite embedded in the epidermis (Fig. 3.10B–D). There can be one or more cilia and these can be either flush with the epidermis or recessed (Crisp 1971; Ehlers and Ehlers 1977; Phillips and Friesen 1982; Dorsett 1986; Wyeth and Croll 2011). Depending on the species, cilia may be surrounded by microvilli or not, and microvilli may appear as an unorganized array or may encircle the cilium as a collar. In flatworms, cells located adjacent to these ciliated sensory cells sometimes display elongated microvilli flanking the cilium (Fig. 3.10B) (e.g. Ehlers and Ehlers 1977; Budelmann 1989), suggesting that they may contribute to sensory transduction.

Apart from such isolated sensory cells, several types of mechano- and/or chemo-sensory organs are known from lophotrochozoans (Fig. 3.11). These include stato-cysts, which are usually fluid filled vesicles with ciliated receptor cells containing calcareous concretions. Gastropods and some other lophotrochozoans have a simple statocyst with multiciliated primary sensory cells (Fig. 3.11A) (reviewed in Dorsett 1986). Cephalopods have evolved a more complex and highly specialized statocyst with multiciliated sensory cells attached to a gelatinous cupula (Fig. 3.11B). Some of these are primary sensory cells, which send their axon into the CNS; others, however, are secondary sensory cells resembling vertebrate hair cells (Budelmann et al. 1987; Budelmann 1989). The secondary sensory cells synapse onto sensory neurons located nearby, which transmit stimuli to the CNS. The sensory cells of the cephalopod statocyst do not have microvilli and are activated by shearing forces between their multiple cilia (Budelmann 1989). This is in contrast to vertebrate hair cells, which contain many microvilli (stereovilli) in addition to a single cilium and are activated by shearing forces between the microvilli. Cephalopods also carry multiciliated primary sensory cells on their epidermis. They resemble those in the statocyst, but are arranged in lines and respond to water movements in striking convergence to the vertebrate lateral line (Budelmann and Bleckmann 1988).

Chemosensory cells are also found concentrated in specialized sense organs such as the osphradium (Fig. 3.11C), a lamellated structure found in the mantle cavity of many mollusks, or the rhinophores and nuchal organs located on the dorsal surface of the head in gastropods and polychaetes, respectively (Dorsett 1986; Emery 1992; Purschke 2005; Cummins et al. 2009). The primary sensory cells of the osphradium can be non-ciliated or ciliated often carrying one or two cilia with an intra- or subepithelial cell body embedded into microvillous supporting cells (Haszprunar 1985, 1992; Emery 1992). Similar cells are also found in the rhinophores and nuchal organs. The chemoreceptor proteins used in lophotrochozoans are still mostly unknown, except for mollusks, where they were first characterized in gastropods but later also found in cephalopods (Cummins et al. 2009; Albertin et al. 2015). They are not closely related to the chemoreceptors used in other phyla (e.g. vertebrate ORs or insect chemoreceptors) but rather comprise a distinct family of GPCRs resulting from a lineage specific expansion in mollusks (Cummins et al. 2009).

3.4.1.2   Ecdysozoans

In the other protostomian lineage, the ecdysozoans, sensory cells have been particularly well studied in *Drosophila* and *C. elegans* (Figs. 3.12 and 3.13). I will here only provide a brief overview of the sensory cells known from these developmental model organisms ignoring the staggering diversity of sensory receptors and organs found in other arthropods and nematodes (reviewed in Wright 1980; Hallberg and Hansson 1999; Hartenstein 2005; Merritt 2007; Keil 2012). In *Drosophila*, two types of sensory organs can be distinguished (reviewed in Hartenstein 2005; Kernan 2007; Merritt 2007). Type I sensory organs contain sensory cells embedded into a group of supporting cells; the sensory cells carry a single cilium but no microvilli (Fig. 3.12). Type II sensory organs comprise a single, unciliated multidendritic cell without supporting cells. All sensory cells are primary sensory cells that send their axon into the CNS.

The type I sensory organs include the external sense organs and the chordotonal organs. The external sense (ES) organs form bristles, which either serve mechano- or chemosensory (gustatory or olfactory) functions. They are typically made up of four cells: the sensory cell, a closely associated sheath (thecogen) cell and the shaft (trichogen) and socket (tormogen) cells (Fig. 3.12A). The sheath cell secretes extracellular matrix (encoded by the *NompA* gene), which forms the dendritic sheath encapsulating the cilium of the sensory cell (Hartenstein 2005). The shaft and socket cells, in turn, secrete the cuticle of the bristle shaft and socket, respectively. The mechanosensory bristles are activated by the displacement of the bristle and require TRPN channels (encoded by the *NompC* gene) for mechano-transduction. The campaniform (bell-shaped) sensilla found near the limb joints and along the wings and halteres are modified mechanosensory sense organs with a similar set of cell types, which respond to bending of the cuticle (Hartenstein 2005; Merritt 2007).

Chemosensory bristles, which are found on the antenna and mouthparts among other sites (Gupta and Rodrigues 1997; Clyne, Warr et al. 1999; Goulding, zur Lage, and Jarman 2000), are overall similar to mechanosensory bristles except that they have pores in their cuticle, are innervated by multiple neurons and express olfactory (OR) and gustatory (GR) receptor proteins (Fig. 3.12A) (review in Benton 2006; Merritt 2007; Spehr and Munger 2009; Kaupp 2010; Suh, Bohbot, and Zwiebel 2014; Gómez-Díaz et al. 2018; Robertson 2019). The insect OR and GR proteins are 7-transmembrane receptors that do not belong to the GPCRs, in distinction to the receptors that mediate olfactory stimuli in vertebrates or lophotrochozoans. While members of the insect GR family of unknown function have been found in many animals, they have expanded to form a family of gustatory receptors in arthropods (Robertson 2015, 2019). The ORs of insects have evolved from within the GR family at the base of the insect lineage. Insect ORs form heterodimeric receptors comprising the highly conserved co-receptor OR83b (also called Orco) and a ligand-specific receptor. Typically, only one ligand-specific receptor is expressed in each olfactory sensory neuron. The OR receptor dimers act as ligand gated ion channels modulated by cyclic nucleotides (Kaupp 2010; Robertson 2019). In addition, insects also use ionotropic receptors (IR) for chemoreception; these form a subfamily within the larger family of ionotropic glutamate receptors, which originated in the protostomes (Robertson 2019).

The chordotonal organs are not associated with bristles but are instead attached to the cuticle internally, where they serve as stretch receptors (Fig. 3.12B) (reviewed in Hartenstein 2005; Merritt 2007; Boekhoff-Falk and Eberl 2014). A specialized chordotonal organ is Johnston's organ in the insect antenna, which detects movements of the distal antennal segments in response to sound pressure thereby serving as an auditory organ. Chordotonal organs may be composed of multiple individual units called scolopidia. Each scolopidium in turn is made up of four cells, which correspond to the four cells of the ES organs and are generated in development by a similar cell lineage (Lai and Orgogozo 2004) (Fig. 3.12B).

The attachment and cap cells of the scolopidium correspond to the shaft and socket cells, respectively, of the ES organs, while the scolopale cell and neuron of the scolopidium correspond to the sheath cell and neuron of the ES organ, respectively. The scolopale cell forms a "dendritic cap" (encoded by the *NompA* gene) as well as the scolopale, a sheath enclosing a fluid-filled space surrounding the sensory cilia. The sensory cilia express TRPN and TRPV channels (encoded by *NompC* and the *Nanchung* and *Inactive* genes, respectively), which are probably activated by shearing forces between the cilium and the adjacent scolopale. Several homologs of proteins involved in mechanotransduction in the vertebrate hair cell also are involved in chordotonal organ function in *Drosophila*. These include homologs of the tip link protein Protocadherin-15 (Cad99c, which is located in the scolopale cell) and

Myosin VIIa, which is involved in tip link tethering in vertebrate hair cells (Todi et al. 2005; Li et al. 2016). Cad87A, a *Drosophila* homolog of Cadherin-23, the other component of vertebrate tip links, is also highly expressed in mechanosensory cells (Fung et al. 2008). Furthermore, Cadherin-23 was recently shown to be required for mechanotransduction in the cnidarian *Nematostella* (Tang and Watson 2014). Taken together, this suggests a possibly deep phylogenetic conservation of some aspects of mechanotransduction mechanisms in at least a subset of mechanoreceptors.

This evolutionary conservation of mechanotransduction mechanisms extends to TRP channels (even though their function has been replaced by other types of channels in the vertebrate hair cell). These evolved already in the last common ancestor of metazoans and choanoflagellates (Cai et al. 2015; Peng, Shi, and Kadowaki 2015). Interestingly, many TRP-channels including TRPC and TRPN are intrinsically mechanosensitive (Kang et al. 2010; Quick et al. 2012; Venkatachalam, Luo, and Montell 2014; Sexton et al. 2016). TRP channels have, thus, been proposed to play an ancient role in direct mechanotransduction before becoming secondarily responsive to a variety of additional stimuli that modulate phospholipase C (PLC) activity (see fig. 6.8B in Schlosser 2021 and Fig. 4.3B) (Liu and Montell 2015). PLC removes the bulky inositol (IP3) group from phospholipid PIP2 in the membrane (resulting in the production of DAG), thereby reducing membrane area locally. This causes membrane stretching, which activates nearby TRP channels (Hardie and Franze 2012; Liu and Montell 2015).

It has been argued that TRP channel function requires them to be localized together with their activating proteins in small cellular compartments (e.g. cilia or microvilli) (Henderson, Reuss, and Hardie 2000; Venkatachalam and Montell 2007; Fain, Hardie, and Laughlin 2010; Huang et al. 2010; Lange 2011; Hardie and Juusola 2015). Such local confinement is probably required for PLC-mediated PIP2 depletion, membrane stretching, and acidification to be effective, which contributes to the activation of multiple TRP channels (e.g. TRPV, TRPM, and TRPC). It also allows the local sequestration of Ca^{2+} entering the cell through these channels and enables positive feedback of TRP channel opening by Ca^{2+} ions. Indeed, several TRP channels show an ancient association with cilia predating the origin of animals, whereas TRPC channels are often associated with microvilli (Fain et al. 2010; Lange 2011; Sigg et al. 2017). It is currently poorly understood, how the different TRP channels in the mechanoreceptors of protostomes are activated and whether this is mediated by PLC (as in rhabdomeric photoreceptors; see Chapter 4) or by a different mechanism. However, the rapid activation kinetics of TRPN channels in *Drosophila* and *C. elegans* suggest that they may be directly mechanosensitive (Kang et al. 2010; Yan et al. 2013).

Apart from *Drosophila*, the nematode *C. elegans* is the second ecdysozoan model organism, for which mechano- and chemosensory cells have been extensively characterized (Fig. 3.13). Two types of mechanoreceptor neurons have been described in this and related nematode species. First, unciliated neurons, which send long processes filled with microtubules along the cuticle and respond to light touch (Fig. 3.13A) (Chalfie and Sulston 1981). Second, sensory cells with ciliated dendrites, which have their cilia embedded in the cuticle or exposed to the environment via pores in the cuticle (Ward et al. 1975). The transduction of mechanical stimuli is mediated by TRPV and TRPN channels in ciliated neurons and by DEG/ENAC channels in unciliated sensory neurons (Goodman 2006; Goodman and Sengupta 2019). Several types of chemosensory neurons have also been identified, which resemble the ciliated mechanoreceptors. These include the amphids associated with the pharynx (Fig. 3.13B) and the phasmids located near the anus. They employ chemoreceptor proteins, which have evolved by lineage specific expansion of various GPCRs and are unrelated to the chemoreceptors of arthropods, vertebrates, or lophotrochozoans (Robertson and Thomas 2006). Downstream of these receptors either TRPV or CNG channels are used for chemotransduction (Bargmann 2006).

In summary, this survey suggests that mechanosensory cells in the different taxa of protostomes have adopted a large variety of functional specializations. Nevertheless, general structural features (e.g. ciliated cells with or without microvilli) as well as molecular aspects of mechanotransduction pathways appear to be relatively widely conserved. This is also reflected in similar CoRNs as the next section will show. In contrast, chemosensory cells are structurally similar to mechanosensory cells but use independently evolved chemosensory cells molecules in different taxa. This suggests that chemosensory cells have evolved multiple times independently, probably by diversification of various types of cell-surface receptors and their expression in mechanoreceptor cells. This allowed pre-existing mechanoreceptor cells, already equipped with all the necessary machinery for electric and synaptic transmission and connected to the CNS, to be re-purposed into receptor cells conveying a different sensory modality.

3.4.2 Transcription Factors Involved in Specifying Mechano- and Chemosensory Cell Types in Protostomes

In spite of the diversity of mechano- and chemosensory cell types in protostomes, many of these cell types express proneural proteins (Ascl, Atonal) and subsets of other transcription factors that are thought to promote sensory cell identity in vertebrates and other deuterostomes including POU4, Islet, Pax2/5/8, Gfi, Prox, and BarH (see Fig. 3.4) as I will review in this section. This has been characterized in greatest detail for the ecdysozoan model organisms *D. melanogaster* and *C. elegans*. In these species, the role of some of these factors in specifying mechano- and chemosensory identity has also been experimentally confirmed. Furthermore, recent single-cell RNA sequencing data obtained for most cell types of *C. elegans* and for olfactory cells in *Drosophila* allow to assess whether these various transcription factors are indeed co-expressed in individual cells (Cao et al. 2017; Li et al. 2017; Packer et al. 2019). While the resolution of a similar dataset for the lophotrochozoan *Platynereis* (Achim et al. 2018) is still too low and does not allow the characterization of individual cells, co-expression of transcription factors in *Platynereis* can be determined based on a spatial expression atlas at single-cell resolution (Vergara et al. 2017).

3.4.2.1 Proneural Transcription Factors

As reviewed in Chapter 5 of the first volume (Schlosser 2021), the ability of the genes encoding the bHLH transcription factors Atonal and Ascl to initiate neuronal or sensory differentiation as "proneural" genes was first elucidated in *Drosophila*. These early studies also established that different bHLH factors promote different types of sensory cells, thereby contributing to the specification of sensory subtype identity (reviewed in Bertrand, Castro, and Guillemot 2002; Huang, Chan, and Schuurmans 2014; Allan and Thor 2015). Atonal is essential for development of chordotonal organs and – together with the atonal-related factor Amos – of olfactory sensory neurons (Jarman et al. 1993; Gupta and Rodrigues 1997; Goulding et al. 2000), whereas Ascl is required for the differentiation of mechano- and chemosensory ES organs (Ghysen and Dambly-Chaudière 1989). Similarly, either Atonal or Ascl serve as proneural factors for various subsets of mechano- or chemoreceptor neurons in the nematode *C. elegans* (Zhao and Emmons 1995; Frank, Baum, and Garriga 2003; Poole et al. 2011). The expression patterns of Atonal and Ascl in the polychaete *Platynereis* suggest similar roles for these bHLH transcription factors in lophotrochozoans (Denes et al. 2007; Simionato et al. 2008).

Since two distinct subsets of mechanoreceptors, which rely on Atonal and Ascl as proneural factors, respectively, have been found in many bilaterians, it has been previously suggested (Schlosser 2005, 2018) that these may comprise two evolutionarily distinct lineages of mechanoreceptors, one Atonal- and the other one Ascl-dependent, derived from two distinct cell types in the last common ancestor of bilaterians. However, several observations suggest that Atonal and Ascl may initially have played partly redundant roles and that predominance of one or the other as major proneural factor for a particular cell type may have evolved several times independently. First, co-expression of these factors is observed in most sensory cells of some millipedes and crustaceans (Dove and Stollewerk 2003, Klann and Stollewerk 2017) as well as in other bilaterian taxa (e.g. tunicates as discussed in Section 3.1.2 above) in contrast to the situation in *Drosophila*, where Atonal is expressed in the sensory cells of chordotonal organs and Ascl in the sensory cells of external sensillae. Furthermore, the ES organs are also morphologically more similar to chordotonal organs in crustaceans than in insects, and their sheath cell forms a scolopale resembling those of chordotonal organs (Hartenstein 2005). Taken together with the molecular similarities of mechanotransduction mechanisms (dendritic sheath/cap; TRP channels) and the striking similarities in developmental lineage between ES organs and chordotonal organs (Fig. 3.12), this strongly suggests that these two types of sense organs and their constituent cell types are homologous (Lai and Orgogozo 2004; Hartenstein 2005) but have become dependent on different proneural genes in insects. The ease of transforming ES organs into chordotonal organs and vice versa by overexpression or mutation of another transcription factor encoded by the *Cut* gene, further supports this interpretation (Bodmer et al. 1987; Blochlinger, Jan, and Jan 1991). Hence, Atonal- and Ascl-dependence is evolutionarily flexible and not mutually exclusive, making these bHLH factors poor indicators of cell type identity.

3.4.2.2   Other Transcription Factors

The Pax2/5/8 transcription factor (*Drosophila*, shaven and sparkling; *C. elegans*, Egl-38 and Pax-2) is also widely expressed in most protostomian mechano- and chemosensory cells. In the polychaete *Platynereis* Pax2/5/8 is co-expressed with Atonal and TRP channels in many putatively mechano- or chemosensory cells (Denes et al. 2007) and it is also expressed in the gastropod statocyst and other sensory organs in mollusks (O'Brien and Degnan 2003; Wollesen et al. 2015). It is not widely expressed in mechano- or chemosensory neurons of *C. elegans* (Wang, Greenberg, and Chamberlin 2004; Park et al. 2006; Packer et al. 2019). However, in *Drosophila* it is expressed in both the developing ES organs and chordotonal organs and is required for the specification of shaft versus socket cells during sensory bristle development (Czerny et al. 1997; Fu et al. 1998; Kavaler et al. 1999). It is reasonable to assume that the various cell types composing the ES and chordotonal organs do not only share the same developmental lineage (Fig. 3.12), but also evolved as sister cells by duplication and divergence of a common ancestral sensory cell type (Lai and Orgogozo 2004). Under this hypothesis, Pax2/5/8 may have had an ancient bilaterian function in the specification of mechanosensory cells. After the diversification of this cell type into the various components of the ES and chordotonal organs in arthropods, it may have remained essential for the specification of only one of these sister cells, but may have been functionally replaced by other factors in other sister cells.

The POU4 transcription factor (*Drosophila*, Acj6; *C. elegans*, Unc-86), which appears to play a central role for the specification of sensory cells in chordates, is also widely but not universally expressed in mechano- and chemosensory cells of protostomes (reviewed in Gold et al. 2014). In lophotrochozoans, it has been shown to be expressed in the statocyst of gastropods and in various sensory cells in *Platynereis*, where it overlaps with Islet, Prox, Phox2, BarH, COE, GATA1/2/3, Neurog, and Six4/5 but not Eya or Pax2/5/8 (O'Brien and Degnan 2002; Vergara et al. 2017). In *C. elegans* the POU4 homolog Unc-86 is required for the specification of touch neurons (Zhang et al. 2014, 2018; Serrano-Saiz et al. 2018) and some chemosensory (oxygen sensing) neurons (Qin and Powell-Coffman 2004). In *Drosophila*, the POU4 homolog Acj6 is not expressed in mechanosensory neurons but controls the specification of a subset of olfactory neurons (Clyne, Certel et al. 1999; Certel et al. 2000; Lee and Salvaterra 2002) and in combination with other transcription factors it specifies the subtype identity of olfactory sensory cells (Jafari et al. 2012; Li et al. 2017). Taken together, this suggests that POU4 may have played an important role in specifying at least some mechano- and/or chemosensory cells in the last common ancestor of protostomes although it may subsequently have become dispensable and/or functionally replaced with other transcription factors in some derivative sensory cell types.

Like in chordates, POU4 in protostomes is often co-expressed with Islet or other LIM-homeodomain proteins in developing mechano- or chemosensory cells (e.g. Vergara et al. 2017; Packer et al. 2019). LIM proteins appear to play an important role in the specification of different neuronal subtypes in these sensory cells as well as in the CNS of both protostomes and deuterostomes (reviewed in Hobert and Westphal 2000; Shirasaki and Pfaff 2002; Allan and Thor 2003). POU4 and LIM-proteins have been shown to cooperatively interact during sensory specification. This has been extensively studied in *C. elegans*, where a protein complex of POU4 (Unc-86) and Lhx1/5 (Mec-3) is required for the specification of unciliated touch receptors (Fig. 3.13A) (Xue, Tu, and Chalfie 1993; Gordon and Hobert 2015). *C. elegans* homologs of the LIM homeodomain proteins Lhx2/9 (Ttx-3), Lhx1/5 (Mec-3, Lin-11), Lhx3/4 (Ceh-14), and Lmx (Lim-6) are expressed in various subsets of its unciliated touch receptors and of ciliated mechano- and chemosensory neurons (Cao et al. 2017; Packer et al. 2019), suggesting that they may contribute to subtype specification. However, Islet (Lim-7), which plays a central role for mechano- and chemosensory specification in vertebrates and possibly some lophotrochozoans, has so far not been implicated in sensory development in nematodes (Voutev et al. 2009; Bhati et al. 2017) although it is expressed in some of its ciliated mechanosensory neurons (Cao et al. 2017; Packer et al. 2019). In *Drosophila*, the Islet homolog Tailup (Tup) promotes sensory bristle (ES) formation in some body parts but inhibits bristle formation in others (Biryukova and Heitzler 2005; de Navascues and Modolell 2010). Taken together, this suggests that Islet does not play the same central role for specification of mechano- and chemosensory cells in some protostomes as in chordates or cnidarians but rather may be involved in the specification of sensory subtypes. Whether the Islet-dependent subtypes in different taxa are evolutionarily related or whether Islet has acquired its role in subtype specification several times independently remains to be determined.

Homologs of the vertebrate transcription factors Gfi1, Prox1/2, and BarH are also expressed in various subsets of sensory cells in different protostomes (including lophotrochozoans and ecdysozoans) and were shown to play central roles for the specification of mechano- and chemosensory cells in *Drosophila* (Jafar-Nejad and Bellen 2004; Vergara et al. 2017). The *Drosophila* Gfi homolog, Senseless, is expressed in the sensory organ precursors of its ES and chordotonal organs (including Johnston's organ), and is required for their proper specification in synergy with proneural bHLH transcription factors (Nolo, Abbott, and Bellen 2000; Jafar-Nejad et al. 2003; Jafar-Nejad and Bellen 2004; Acar et al. 2006; Kirjavainen et al. 2008; Jarman 2014). During ES development, Prospero, the *Drosophila* Prox homolog is distributed by an asymmetric cell division to the pIIb cell (Fig. 3.12) giving rise to sensory neuron and thecogen cell (Manning and Doe 1999; Reddy and Rodrigues 1999). In the absence of prospero, the fate of pIIb is converted into its sibling pIIa, which forms the trichogen and tormogen cells, resulting in the loss of sensory neuron and thecogen cell. The BarH homologs BarH1 and BarH2 are also expressed in the pIIb cell and play a role in the specification of different ES subtypes (Higashijima et al. 1992).

In *C. elegans*, the BarH homolog Ceh-30 has been implicated in the specification of a male-specific subpopulation of sensory neurons (Schwartz and Horvitz 2007). While the *C. elegans* homolog of Gfi, Pag-3, is also expressed in cell lineages giving rise to sensory cells, it ultimately promotes non-sensory over sensory cell identity (Gordon and Hobert 2015). Similarly, the *C. elegans* Prox homolog (Pros-1) is expressed in glial cells of the amphid sensory organ, rather than in its sensory neurons (Kage-Nakadai et al. 2016). This suggests that Gfi, Prox, and BarH may have an ancient role for the development of at least some sensory lineages in all protostomes, but may then have been recruited for the specification of different cell types in different groups.

Other transcription factors with important roles in the specification of chordate sensory cells or neurons (Fig. 3.4) have also been implicated in the specification of some sensory cell types in protostomes. This includes the closely related NK5 (=Hmx2/3) and Tlx transcription factors, which together with Prox and BarH belong to the Natural Killer (NK) homeodomain family proteins (Saudemont et al. 2008). Together with some other NK subfamilies, NK5 and Tlx are expressed in the developing sensory cells of *Platynereis* (Saudemont et al. 2008). While NK5 transcription factors are also expressed in the developing PNS of *Drosophila*, no function of NK5 or Tlx (Clawless/C15) for sensory development has been reported (Wang et al. 2000; Stathopoulos et al. 2002; Reim, Lee, and Frasch 2003; Kojima, Tsuji, and Saigo 2005). However, in *C. elegans* NK5 (= Mls-2) specifies the identity of one ciliated mechanosensory cell type (Kim, Kim, and Sengupta 2010). Based on this limited information, it is currently difficult to draw any firm conclusions on the role of NK family transcription factors in ancestral protostomes. We also know relatively little about the role of paired type homeodomain proteins DRG11 and Phox2. DRG11 is broadly expressed in sensory neurons in mollusks, while Phox2 is largely confined to motor neurons in both mollusks and nematodes (Van Buskirk and Sternberg 2010; Nomaksteinsky et al. 2013) suggesting that a differential role of DRG11 and Phox2 in specifying GSNs versus VSNs, respectively may have originated only in vertebrates.

Finally, Six1/2 (*Drosophila*, Sine oculis; *C. elegans*, Ceh-32, Ceh-33, Ceh-34) and Six4/5 (*Drosophila*, Six4; *C. elegans*, Ceh-35) transcription factors and their Eya cofactors were shown to be expressed in the developing rhabdomeric photoreceptors of different protostomes (insects, planarians, and polychaetes) and were shown to be required for their proper specification in *Drosophila* and planarians (see Chapter 4). There is currently no evidence for any important role of these factors for the specification of other sensory cell types in protostomes. However, Six1/2, Six4/5 and/or Eya are also expressed in other ectodermal cells in protostomes, which may include mechano- and or chemosensory cells (Kumar and Moses 2001; Posnien et al. 2011; Vergara et al. 2017; Packer et al. 2019). They were also reported to be expressed in mechano- as well as photoreceptive cells in cnidarians (see Section 3.5.2 below) (Nakanishi et al. 2010; Hroudova et al. 2012). This suggests that these factors have an ancient association with development of at least a subset of mechano- and photosensory cells predating the origin of bilaterians but possibly lost in many mechanoreceptors of protostomes.

3.4.3 THE LAST COMMON BILATERIAN ANCESTOR

The great diversity of sensory cells and organs in protostomes makes it very difficult to reconstruct with any degree of certainty the sensory cell types present in the last common ancestor of protostomes and to infer, which of these were shared with the deuterostome ancestor. We can, therefore, draw only some tentative conclusions on the sensory cell types of the last common ancestor of bilaterians. These very likely included primary mechanosensory cells with one or several apical cilia surrounded by microvilli, since most mechanosensory cell types in protostomes and deuterostomes represent variations on this common theme and similar cell types have been described in bilaterian outgroups, such as the cnidarians discussed below. The shared expression of some microRNAs (e.g. miR183) specifically in sensory cells throughout bilaterians also suggests a common evolutionary origin of these cells (Pierce et al. 2008; Christodoulou et al. 2010).

Some components of the mechanotransduction mechanism of this ancestral mechanosensory cell type, including Protocadherin-15, Cadherin-23, Myosin VII, and TRP channels, still play roles for mechanotransduction in the mechanosensory cells of modern protostomes and deuterostomes. However, chemosensory cells probably evolved many times independently in different lineages by provisioning mechanosensory cells with freshly expanded families of cell surface receptors enabling the distinction of different chemical stimuli. Sensory receptor cells mediating other sensory modalities (e.g. thermo- or hygroreceptors) may also have evolved repeatedly from this cell type.

The shared functions of transcription factors between subsets of sensory cells in deuterostomes and protostomes further suggest that these transcription factors may already have played a role in specifying the identity of mechanosensory cells in the last common bilaterian ancestor. Thus, Atonal and Ascl most likely served as proneural factors that initiated their sensory differentiation program, probably in a partly redundant fashion. Pax2/5/8, Gfi, Prox and BarH and probably POU4 and Islet may have been involved in the CoRN conferring mechanosensory cell type identity. Other LIM homeodomain proteins, NK5, Tlx, Six1/2, Six4/5, and Eya probably also played a role in specifying at least some subtypes of mechanosensory cells. Somewhat surprisingly, many of the core regulators of mechanosensory cells (e.g. Atonal, POU4, Gfi, Prox, BarH, Six1/2) are also shared with a subset of photoreceptors (rhabdomeric photoreceptors), which do however depend on Pax4/6 rather than Pax2/5/8. This points to a deep evolutionary relationship between photoreceptors and mechano- or chemoreceptors, which will be discussed in detail in the next chapter.

When mechanosensory cells underwent independent diversification in the different protostomian lineages or produced sister accessory cells (such as the shaft, sheath, and socket cells of ES organs), some of these CoRN members may have maintained a central regulatory role for only one of the sister cell types, while new transcription factors were probably recruited for the specification of others. Alternatively, there may have been several subtypes of mechanoreceptors already in the last common ancestor of bilaterians, which may have differed from each other regarding the expression of POU4, Six1/2, NK5, LIM homeodomain proteins and other transcription factors. These subtypes may have subsequently persisted in some of the descendant taxa or may have given rise to a new assortment of different subtypes in the descendant taxa by recombination and rewiring of transcription factors.

3.5   MECHANO- AND CHEMOSENSORY CELL TYPES IN THE LAST COMMON ANCESTOR OF EUMETAZOANS AND METAZOANS

While some recent phylogenies place the xenacoelomorphs, simple flatworm-like animals as the sister taxon of bilaterians, other phylogenies consider them basal deuterostomes (Philippe et al. 2011; Cannon et al. 2016; see Chapter 1). Due to these uncertainties, I will not consider these animals further here, which carry monociliated mechano- and/or chemosensory receptors (either with or without a microvillar collar) on their surface that resemble those of bilaterians (Martínez, Hartenstein, and Sprecher 2017). Leaving the enigmatic xenacoelomorphs aside, the sister group of the bilaterians are the cnidarians comprising the medusozoans (hydrozoans, scyphozoans, and cubozoans), which have a life cycle with alternating polyp and medusa stages, and the anthozoans (sea anemones, corals, and relatives), which lack a medusa stage (Fig. 1.8). Traditionally, another non-bilaterian group, the comb jellies or ctenophores, were considered to be closely related to the cnidarians, with the combined cnidarian-ctenophore lineage – the coelenterates – forming the sister group of the bilaterians. Together, coelenterates and bilaterians were thought to comprise the eumetazoans, animals with a multitude of specialized cell types (Fig. 1.8). Furthermore, the simple placozoans were considered to be the sister group of eumetazoans (Srivastava et al. 2008), while sponges (Porifera) were suggested to be the most basally branching lineage of metazoans (Fig. 1.8) (Srivastava, Simakov et al. 2010; Dohrmann and Wörheide 2013). In this scenario, neurons and sensory cells originated only in the eumetazoans as specialized cell types, since these cell types are found in cnidarians but neither in sponges nor in placozoans (Jacobs et al. 2007; Galliot et al. 2009; Watanabe, Fujisawa, and Holstein 2009).

However, recent phylogenomic studies have called this view into question and instead suggest that ctenophores rather than sponges are the most basal metazoan clade (Ryan et al. 2013; Moroz et al. 2014; Whelan et al. 2017). This suggestion has in turn be criticized by other studies that support the traditional view (Telford, Budd, and Philippe 2015; Feuda et al. 2017; Simion et al. 2017), leaving the position of ctenophores currently in doubt. Moreover, while ctenophores have neurons and sensory cells, they lack many of the neuron-specific genes found in eumetazoans (Ryan et al. 2013; Moroz et al. 2014) or do not co-express them in the same cell types (Sebé-Pedrós, Saudemont et al. 2018). Taken together with the proposed basal phylogenetic position of ctenophores, this has led to suggestions that either sponges and placozoans have lost a primitive type of neuron or ctenophores evolved neurons convergently to eumetazoans (Ryan et al. 2013; Moroz et al. 2014). Should, alternatively, ctenophores be more closely related to cnidarians, these peculiarities of their neuronal differentiation program would suggest that they evolved highly divergent lineage-specific neuronal cell types. Because of the uncertain status of ctenophores and their sensory cells, I will consider them only briefly here and base inferences on the sensory cell types present in the last common ancestor of eumetazoans mostly on comparisons between cnidarians and bilaterians.

3.5.1 MECHANO- AND CHEMOSENSORY CELL TYPES IN CNIDARIANS, CTENOPHORES, PLACOZOANS, AND SPONGES

3.5.1.1 Cnidarians

Cnidarians have a diffuse nerve net with neurons scattered throughout the surface of the body, some of which are equipped with a cilium (often surrounded by microvilli) and may serve as photo-, mechano-, and/or chemosensory cells (e.g. Lentz and Barrnett 1965; Westfall and Kinnamon 1978; Westfall, Sayyar, and Elliott 1998; Westfall, Elliott, and Carlin 2002; Nakanishi et al. 2008; Galliot et al. 2009; Nakanishi et al. 2012; Rentzsch, Layden, and Manuel 2017). The planula larvae of some cnidarian species also have a specialized apical tuft comprising the elongated cilia of a cluster of ciliated cells (Chia and Koss 1979). These apical tuft cells have been shown in some taxa to promote metamorphosis presumably in response to settlement cues and express TRP channels; they, thus, may play some sensory role (Rentzsch et al. 2008; Sinigaglia et al. 2015). However, apical tuft cells do not have axons, do not form synapses, and do not express any neuronal genes (Rentzsch et al. 2017; Sebé-Pedrós, Saudemont et al. 2018). Therefore, they are probably not related to neuron-like sensory cells unlike the sensory cells of the apical organ in the trochophora larvae of lophotrochozoans.

Only found in cnidarians, the stinging cells or cnidocytes are highly specialized sensory-effector cells equipped with a modified cilium, the cnidocil, surrounded by microvilli (Fig. 3.14A). Stimulation of the cnidocil triggers exocytosis of a venom filled capsule – the cnidocyst – equipped with a protrusible tubule for predation (reviewed in Tardent 1995; Holstein 2012). As already mentioned in Chapter 2, the cnidocyst is a novel structure that evolved in the cnidarian lineage and combines several cnidaria-specific proteins (minicollagens, NOWA, spinalin), an enzyme acquired by horizontal gene transfer (poly-γ-glutamate synthase) and other proteins (Denker et al. 2008; Babonis and Martindale 2014).

Several lines of evidence suggest that cnidocytes are specialized neuron-like sensory cells, which probably evolved from sensorineural cells in the last common metazoan ancestor (reviewed in Galliot et al. 2009; Babonis and Martindale 2014; Rentzsch, Juliano, and Galliot 2019). First, cnidocytes and neurons were shown to

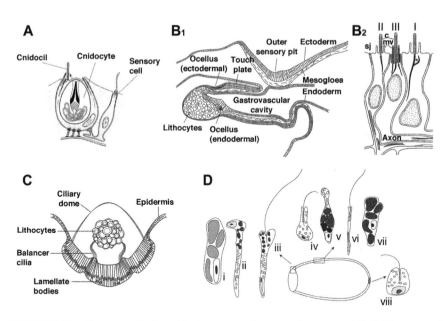

FIGURE 3.14 Sensory cells in cnidarians, ctenophores, and sponges. (**A**) Cnidocytes and associated sensory cell in the cnidarian *Hydra*. (**B**) Rhopalium in the scyphozoan medusa *Aurelia aurita* with ocelli, statocyst with lithocytes and touch plate (**B₁**). Detail of three types of mechanosensory cells of touch plate shown in **B₂**. c, cilia; mv, microvilli; sj, septate junction. (**C**) Apical organ in ctenophores. The lamellate bodies may serve in photoreception. (**D**) Putative sensory cells in the posterior (i–iii), middle (iv–vii) and anterior (viii) region of the larva of the sponge *Amphimedon queenslandica*. ([A] Reprinted with modification with permission from Holstein 2012, after Hobmayer et al. 1990; [B] B₁: redrawn from Chapman 1999; B₂: reprinted with permission from Hündgen and Biela 1982; [C] redrawn from Hernandez-Nicaise 1984; [D] reprinted with permission from Mah and Leys 2017.)

develop from a common progenitor in both hydrozoans and anthozoans (Holstein and David 1990; Miljkovic-Licina et al. 2007; Richards and Rentzsch 2015). Second, there is physiological evidence that stimulation of one cnidocyte can activate other cnidocytes. This suggests that they have neuron like functions even though presynaptic terminals have not been unequivocally identified (Thurm et al. 2004; Anderson and Bouchard 2009). Third, cnidocytes and sensory neurons cluster together as related cell types in some single-cell RNA-Seq analyses (Siebert et al. 2019). Fourth and finally, their differentiation appears to be regulated by partially shared sets of transcription factors (including Ascl) as will be discussed in the next section.

Although cnidocytes are widely considered to be mechanoreceptors, which are activated by mechanical stimulation of the cnidocil, they in fact respond optimally only when stimulated by mechanical and chemical cues (Thurm et al. 2004; Anderson and Bouchard 2009). Moreover, some contact-dependent responses to chemical stimuli were shown to be directly mediated by the cnidocil but the chemical cell-surface receptors involved have not yet been characterized (Thurm et al. 2004).

In addition, cnidocytes receive direct or indirect synaptic input from proper sensory cells, which respond to mechanical, chemical, and/or light stimuli and modulate cnidocyte discharge in response to these various stimuli (Fig. 3.14A) (Hobmayer, Holstein, and David 1990; Kass-Simon and Hufnagel 1992; Westfall et al. 1998, 2002; Holstein 2012; Plachetzki, Fong, and Oakley 2012). In hydrozoans, these sensory cells and various types of cnidocytes are found clustered together in so-called battery complexes. The sensory cells usually carry a single apical cilium surrounded by a collar of stereovilli (microvilli with an actin rich core). At least for some of these sensory cells it has been shown that their responsiveness to mechanical stimuli is modulated by chemicals (N-acetylated sugars and proline), which alter the length of stereovilli (Watson and Hessinger 1992; Mire-Thibodeaux and Watson 1994; Watson and Hudson 1994). This indicates that these sensory cells are multimodal and respond to both chemical and mechanical stimuli. The widespread expression of opsin and other phototransduction genes in cells of the battery complex except the cnidocytes suggests that at least some of these cells may also respond to light (Plachetzki et al. 2012; see Chapter 4). Whether additional sensory cell types exist that act as dedicated mechano-, chemo-, or photoreceptors, still needs to be clarified.

Mechanotransduction in cnidarians appears to depend on Cadherin-23 (a component of the tip link of vertebrate hair cells) and TRP-channels (TRPA, TRPN, and TRPP), while phototransduction depends on opsins and CNG channels. Hence, cnidarians share at least some of the molecular components of sensory transduction machineries with bilaterians (Chapter 4) (Mahoney et al. 2011; Plachetzki et al. 2012; Tang and Watson 2014; Schüler et al. 2015; McLaughlin 2017). However, it is currently unclear how chemical stimuli are transduced in cnidarians. Apart from receptors for the sugars and amino acids mentioned above, it is long known that *Hydra* and other cnidarians have a receptor for the peptide glutathione (GSH), which elicits a feeding response (Loomis 1955; Lenhoff 1961; Morita and Hanai 1987). However, the molecular nature of these various chemoreceptors is still unknown. Although genes related to vertebrate ORs and insect chemoreceptors have been found in cnidarian genomes, they do not appear to play a role in chemoreception in these animals (Churcher and Taylor 2011; Saina et al. 2015).

In addition to cnidocytes and their associated sensory cells, some cnidarians also have specialized and more complex sense organs (Singla 1975; Hündgen and Biela 1982; Nakanishi, Hartenstein, and Jacobs 2009). These are found mostly in the medusa stage of the medusozoa. In hydrozoan medusae, there are statocysts containing a single cell filled with a crystalline concretion – a so-called lithocyte – that weighs down on monociliated receptor cells (with or without microvilli) nearby (Singla 1975).

In scyphozoans and cubozoans there are complex, club-shaped structures – so-called rhopalia – which are attached to the rim of their bell and combine multiple photo-, mechano-, and possibly chemoreceptive sensory organs including a statocyst and several ocelli or eyes (Fig. 3.14B) (Hündgen and Biela 1982; Nakanishi et al. 2009). In cubozoans, each rhopalium contains six different eyes, some of which are equipped with a lens and a complex multicellular retina and are probably capable of image-formation (see Chapter 4; Yamasu and Yoshida 1976; Piatigorsky and Kozmik 2004; Nilsson et al. 2005). In the scyphozoan *Aurelia*, the gravistatic part of the rhopalium comprises several sensory epithelia including one known as the "touch plate" containing mechanosensory cells with a single cilium surrounded by a ring of microvilli (Fig. 3.14B) (Hündgen and Biela 1982; Nakanishi et al. 2009). These mechanosensory cells appear to be stimulated by a multicellular lithocyst (sometimes misleadingly called a "statocyst", a term that should be reserved for the complete gravistatic organ comprising lithocyst plus sensory epithelia). The lithocyst is composed of endodermal lithocytes, which contain crystalline concretions and are surrounded by an ectodermal epithelium. Depending on the position of the medusa, the lithocyst will stimulate different mechanosensory cells allowing the detection of the direction of gravity.

3.5.1.2   Ctenophores

Like cnidarians, ctenophores have a diffuse nervous system, although the homology of their neurons and sensory cells with those of cnidarians and bilaterians has been questioned (Ryan et al. 2013; Moroz et al. 2014). Several types of receptor cells can be found scattered throughout the outer epithelium, which carry either cilia or microvilli or both and may mediate mechanical and/or chemical stimuli (Tamm 2014; Norekian and Moroz 2019a, 2019b). The cilia of ctenophore sensory cells are anchored in a peculiar onion shaped root, while their microvilli are often short and peg-like.

In addition to these scattered sensory cells, ctenophores possess one complex apical sense organ, which has been shown to act as a mechano- and photosensory organ. The apical organ comprises a conglomeration of lithocytes with intracellular concretions covered by a cupula and supported on long and motile balancer cilia extending from the presumed mechanosensory cells (Fig. 3.14C) (Krisch 1973; Tamm 2014). Stimulation of different balancer cilia by shifting positions of the lithocytes allows the detection of gravity and the apical organ, thus, serves as a statocyst. In the floor of the apical organ there are cells containing lamellate bodies formed by ingrown immotile cilia and expressing opsin that function as photoreceptors (Horridge 1964; Schnitzler et al. 2012).

Unlike cnidarians, ctenophores have no cnidocytes. Instead they capture prey with another type of specialized cells, the colloblasts (Franc 1978; Eeckhaut et al. 1997; Babonis et al. 2018). Colloblasts consist of a basal part containing extensible filaments and an apical part filled with vesicles containing an adhesive. Contact with prey leads to the release of the vesicles and extension of the basal filaments, thereby ensnaring the prey in sticky protrusions. It is still not clear, how mechanical stimuli are transduced during colloblast stimulation. However, since these cells have no cilia or microvilli and display no other structures that could mediate mechanical stimuli, colloblasts themselves may not be sensory cells and their stimulation may rather depend on signals from adjacent sensory cells.

Based on lineage tracing experiments, which showed that colloblasts and neurons arise from common developmental progenitors, it has recently been suggested that colloblasts and neurons in ctenophores evolved, respectively, from secretory cells and neurons sharing a common developmental progenitor in the last common ancestor of ctenophores and eumetazoans (Babonis et al. 2018). However, there is currently no evidence for transcription factors shared between colloblasts and neurons that would support this scenario.

3.5.1.3   Placozoans and Sponges

In contrast to cnidarians and ctenophores, placozoans and sponges have no true neurons or sensory cells. While many molecular components involved in synaptic transmission (e.g. proteins of the postsynaptic density) and sensory transduction (e.g. TRP and CNG ion channels) are present in these animals, they are not all co-expressed in a single cell-type (Sakarya et al. 2007; Srivastava et al. 2008; Srivastava, Larroux et al. 2010; Plachetzki et al. 2012; Ludeman et al. 2014; Mah and Leys 2017; Sebé-Pedrós, Chomsky et al. 2018; Wong et al. 2019). Other components of sensory transduction are lacking (e.g. opsins in sponges) (Mah and Leys 2017; Sebé-Pedrós, Chomsky et al. 2018; Wong et al. 2019). Nevertheless, placozoans and sponges (both larvae and adults) are able to respond to mechanical, chemical, and light stimuli indicating that at least some of their cell types also can transduce and transmit sensory stimuli, even though they may not be dedicated sensory cells and possibly may serve other functions as well (Renard et al. 2009; Ryan and Chiodin 2015; Mah and Leys 2017; Mayorova et al. 2018; Leys et al. 2019).

In larvae of different sponge species, various unciliated or ciliated cells known as flask cells, globular cells, or cruciform cells have been suggested to play a sensory role (Fig. 3.14D) (Leys and Degnan 2001; Richards et al. 2008; Renard et al. 2009; Nakanishi et al. 2015; Mah and Leys 2017). Whether these various cell types are evolutionarily related is still unclear since little is known about their transcription factor profile. Many of these cell types contain large numbers of small vesicles and, thus, also may play a secretory role. Furthermore, the flask cells of *Amphimedon* were recently shown to be required for the initiation of metamorphosis in response to metamorphic cues (Nakanishi et al. 2015).

Adult sponges are able to respond with slow contractions to a variety of stimuli but it is largely unknown how these stimuli are perceived or transmitted (reviewed in Renard et al. 2009; Mah and Leys 2017; Leys et al. 2019). Recently, cells with non-motile cilia have been described in the osculum of some sponges (the opening, through which water is expelled), which could play a role in the monitoring of water flow (Ludeman et al. 2014). More generally, the choanocytes of sponges, which line its internal chambers and are mainly involved in feeding, have been proposed to be potential evolutionary precursors of sensory cells, mainly because of their resemblance to many eumetazoan sensory cells that bear a cilium surrounded by a microvillar collar (Fritzsch et al. 2007; Jacobs et al. 2007; Renard et al. 2009).

In line with this suggestion, recent RNA-Seq studies show that some neuronal proteins such as those involved in postsynaptic scaffolding and vesicle trafficking are co-expressed at high levels in choanocytes (Musser et al. 2019; Wong et al. 2019). However, the same proteins are also expressed to some extent in other sponge cells, while other classes of neuronal proteins, such as presynaptic proteins are not enriched in choanocytes (Musser et al. 2019; Wong et al. 2019). A recent single cell RNA-Seq study found that some presynaptic genes are instead enriched in so-called amoeboid-neuroid cells of sponges (Musser et al. 2019). These cells were found in choanocyte chambers and extend protrusions that contact the microvilli of choanocytes (Musser et al. 2019). Whether these contacts are involved in transmitting signals across a synapse-like structure between amoeboid-neuroid cells and choanocytes, potentially allowing the modulation of the beating of choanocytes in response to food particles as the authors of this study suggest, remains to be confirmed. It also remains unclear, whether and how amoeboid-neuroid cells respond to environmental stimuli as would be expected from this hypothesis.

3.5.2 TRANSCRIPTION FACTORS INVOLVED IN SPECIFYING MECHANO- AND CHEMOSENSORY CELL TYPES IN CNIDARIANS, CTENOPHORES, PLACOZOANS, AND SPONGES

3.5.2.1 Cnidarians

In cnidarians, like in bilaterians, sensory neurons as well as cnidocytes appear to express at least one of the bHLH transcription factors Ascl and Atonal (Grens et al. 1995; Müller et al. 2003; Hayakawa, Fujisawa, and Fujisawa 2004; Seipel, Yanze, and Schmid 2004; Nakanishi et al. 2010; Hroudova et al. 2012; Layden, Boekhout, and Martindale 2012; Richards and Rentzsch 2014; Watanabe et al. 2014; Sebé-Pedrós, Saudemont et al. 2018). The majority of sensory cells or neurons including putative mechano- and chemoreceptors, may in fact express both Atonal and Ascl consecutively with Atonal acting as a proneural factor in proliferating cells that initiates neuronal/sensory differentiation and Ascl regulating subsequent stages of differentiation (Richards and Rentzsch 2015).

Other transcription factors that play important roles for the specification of mechano- and chemosensory neurons in chordates already appear to play similar roles in cnidarians. For example, POU4 is expressed in the mechanoreceptors of statocysts and rhopalia in medusozoans (Nakanishi et al. 2010; Hroudova et al. 2012). POU4 is also expressed in scattered sensory and neuronal cells including cnidocytes in the anthozoan *Nematostella* and was shown to be required for terminal differentiation of these cells (Tournière et al. 2020). A recent single-cell RNA-Seq study in *Nematostella* found POU4 to be highly expressed in a large subset of neuronal/sensory cell types (Sebé-Pedrós, Saudemont et al. 2018). Genes with enriched expression in neuronal/sensory cell types and cnidocytes also frequently displayed POU4 binding motifs in their enhancers (Sebé-Pedrós, Saudemont et al. 2018). Many of the neuronal/sensory cell types also co-expressed Islet at low levels, while high levels

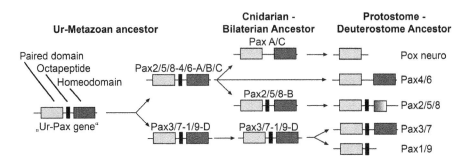

FIGURE 3.15 Evolution of Pax transcription factors. See text for details. (Reprinted with modification with permission from Matus et al. 2007.)

of Islet (together with GATA, OtxC, and ETS transcription factors) were present in most of the neuronal/sensory cell types that did not express POU4 (Sebé-Pedrós, Saudemont et al. 2018).

The *PaxB* gene of cnidarians is closely related to the *Pax2/5/8* genes of bilaterians, while a gene related to *Pax4/6* genes has not be found in cnidarians (Miller et al. 2000; Piatigorsky and Kozmik 2004). However, PaxB has a domain structure (paired domain, octapeptide, homeodomain) that combines features of the Pax2/5/8 (paired domain, octapeptide) and Pax4/6 (paired domain, homeodomain) lineages suggesting that the ancestral gene that gave rise to the *Pax2/5/8* and *Pax4/6* lineages was *PaxB*-like (Fig. 3.15) (Kozmik et al. 2003). Cnidarian PaxB is expressed in mechano- and photo receptors of the cubozoan rhopalia and in scattered ectodermal cells, which likely include sensory and neuronal cells in hydrozoans and anthozoans (Gröger et al. 2000; Kozmik et al. 2003; Piatigorsky and Kozmik 2004; Matus et al. 2007). This suggests that the ancestral *PaxB*-like gene in the eumetazoan ancestor may also have had a role in specification of both photoreceptors and mechanoreceptors or of multimodal cells and that *Pax2/5/8* and *Pax4/6* adopted specific functions in the specification of mechano- and photoreceptors, respectively, only in bilaterians (Kozmik et al. 2003; Piatigorsky and Kozmik 2004).

The recent single-cell RNA-Seq study in *Nematostella* confirmed low-level PaxB expression in many neuronal/sensory cells, but also revealed that PaxB is expressed at even higher levels in many epidermal cells (Sebé-Pedrós, Saudemont et al. 2018). Although cnidocytes share developmental and probably also evolutionary precursors with cnidarian neurons, they express PaxA rather than PaxB (Sebé-Pedrós, Saudemont et al. 2018; Siebert et al. 2019). Furthermore, PaxA together with Mef2 was shown to be required for cnidocyte specification (Babonis and Martindale 2017), suggesting that these transcription factors are members of the CoRN defining this novel cnidarian cell type.

Similar to PaxB, Six1/2, and Eya were found to be expressed in putative mechanosensory and photosensory cells of the statocysts and rhopalia in medusozoans (Bebenek et al. 2004; Stierwald et al. 2004; Graziussi et al. 2012; Hroudova et al. 2012). In anthozoans, the expression of Six1/2 is confined to a small subpopulation of sensory cells or neurons, while the expression of Eya has not been described

(Sebé-Pedrós, Saudemont et al. 2018). Other transcription factor with roles in chordate sensory identity, including DMRT4/5 (DMRTa2), Runx, and Rx were also confined to small subsets of neurons and sensory cells in *Nematostella* (Sebé-Pedrós, Saudemont et al. 2018) while expression of many other transcription factors (e.g. Gfi, Prox, BarH, NK5, Tlx) has not been described. Whereas no clear Phox2 ortholog could be identified in cnidarians (Nomaksteinsky et al. 2013), related paired like homeodomain transcription factors were found to be expressed in nerve cells and cnidocytes of *Hydra* (Gauchat et al. 1998).

The recent analyses of sensory and neuronal cell-types following single-cell RNA-Seq in *Nematostella* and *Hydra* confirmed that many transcription factors expressed in sensory and neuronal cell types in chordates and protostomes, are also enriched in subsets of sensory/neuronal cell types in cnidarians (Sebé-Pedrós, Saudemont et al. 2018; Siebert et al. 2019). However, they also found that combinations of transcription factors that specify particular sensory or neuronal cell types are not conserved between these different groups. This suggests that sensory cells/neurons are evolutionarily related but have diversified independently in the different lineages.

3.5.2.2   Ctenophores

Another single-cell RNA Seq study recently has provided some insights into the transcription factor expression profile in ctenophores although our knowledge of gene expression in this group remains very limited (Sebé-Pedrós, Chomsky et al. 2018). Interestingly, none of the cell types identified in this study displayed a distinct neuronal signature and it remains currently unclear, which transcription factor expression profile characterizes ctenophore neurons or sensory cells. Instead sensory and neuronal genes including those encoding transcription factors were expressed in different cell types. For example, Islet and Runx were found in non-sensory cells carrying motile cilia (comb cells), while NK5 was expressed in muscle cells (Sebé-Pedrós, Chomsky et al. 2018). However, a previous report suggested that Islet together with Tlx is also expressed in the mechanosensory cells of the polar fields of the apical organ (Derelle and Manuel 2007; Simmons, Pang, and Martindale 2012). Another NK-related transcription factor, BarH is expressed broadly in the ectoderm but not in the apical organ (Pang and Martindale 2008). A POU4 ortholog has not been identified in ctenophores (Gold et al. 2014). Taken together, these observations suggest that ctenophore neurons and sensory cells are indeed quite different from those of other metazoans. However, they do not allow us to determine, whether ctenophore neurons have evolved convergently to cnidarian and bilaterian neurons or share a common ancestor with them but subsequently diverged significantly in the ctenophore lineage.

3.5.2.3   Placozoans and Sponges

Although sponges do not have sensory cells or neurons, most of the transcription factors involved in specifying sensory and neuronal cell types in eumetazoans were identified in sponge genomes (Srivastava, Simakov et al. 2010; Mah and Leys 2017). These transcription factors include bHLH factors related to Atonal and Ascl (Simionato et al. 2007; Fortunato et al. 2016) that serve as proneural genes in cnidarians and bilaterians as well as many of transcription factors involved in specifying the identity of their sensory cell types (POU4, Islet, PaxB, Six1/2, Eya, the NK-family members Prox, BarH, NK5, Tlx, and paired-like homeodomain transcription factors related to Phox2 and DRG11).

In some sponge larvae, the Atonal-related transcription factor is expressed in globular cells, which have a potential sensory function although they are not neurons (Richards et al. 2008). The expression of Ascl related transcription factors in larval or adult sponges has not been described. In the cruciform cells of other sponge larvae, which also have been proposed to be non-neuronal cells serving a sensory function, PaxB, Six1/2, Eya, and NK5 were expressed (Fortunato et al. 2014; Fortunato, Leininger, and Adamska 2014). However, which gene batteries are regulated by these transcription factors in larval sensory cells is currently unclear.

In adult sponges, Atonal is enriched together with Islet, PaxB, Six1/2, and other transcription factors in pinacocytes, which thereby have a transcription factor signature that is most similar to sensorineural cells in eumetazoans (Hill et al. 2010; Rivera et al. 2013; Musser et al. 2019). However, Islet is also enriched (together with KLF5, COE, and other transcription factors) in amoeboid/neuroid cells and PaxB, Six1/2 and Eya in choanocytes, suggesting that many transcription factors that adopt specific sensorineural functions in eumetazoans play broader roles for multiple cell types in sponges (Srivastava, Simakov et al. 2010; Fortunato et al. 2014; Musser et al. 2019). Whether any of the transcription factors expressed in amoeboid-neuroid cells and choanocytes, is involved in coordinating the expression of pre- and post-synaptic proteins enriched in these cells, respectively (Musser et al. 2019), remains to be investigated.

Although *POU4* has been identified in the sponge genome, its expression has not been described (Gold et al. 2014). Similarly, several *NK* related genes, which play important roles in conferring neuronal identity in eumetazoans (e.g. *Prox*, *BarH*, *NK5*, *Tlx*) are present in the sponge genome but apart from *NK6/7* in choanocytes their expression is unclear (Seimiya et al. 1994; Larroux et al. 2007; Gazave et al. 2008). In placozoans, POU4 has recently been found to be enriched in epithelial cells (Sebé-Pedrós, Chomsky et al. 2018), a cell type that placozoans share with eumetazoans but that is not present in sponges, raising the possibility that it was initially recruited into the CoRN for epithelial cells before becoming specifically required for neurons and sensory cells in eumetazoans. Other neuronal transcription factors, such as PaxB, NK5, and COE are enriched in various peptidergic cell types in placozoans, which do, however, not co-express other synaptic or neuronal genes (Sebé-Pedrós, Chomsky et al. 2018).

In summary, while pre- and postsynaptic proteins were found enriched in amoeboid-neuroid cells and choanocytes, respectively, only few of the transcription factors expressed in these cells are shared with sensory cells of eumetazoans. Whether any of these transcription factors control the assembly of pre- or postsynaptic proteins in amoeboid-neuroid cells and choanocytes, respectively, is currently unknown. Instead, many transcription factors suggested to be core regulators of sensorineural identity in eumetazoans (e.g. Atonal, PaxB, Six, Islet) were found to be co-expressed in pinacocytes, which provide the external cover of sponges and include contractile cells (Musser et al. 2019) (see below). Taken together (and under the assumption that sponges and placozoans have not secondarily lost neurons) this suggests that sensory cells in the eumetazoans may have evolved by recombination and integration of several modules originally activated in different cells of the last common ancestor of all metazoans, probably by establishing a new CoRN between transcription factors regulating these various modules.

3.5.3 The Last Common Eumetazoan/Metazoan Ancestor

The many similarities between the mechanosensory cells in cnidarians and bilaterians suggest that they are derived from primary sensory cells capable of mechanotransduction in the last common ancestor of eumetazoans (cnidarians, bilaterians, and possibly ctenophores). This primary sensory cell was probably equipped with one or several apical cilia surrounded by microvilli and may have employed some of the same molecular components (including Cadherin-23 and TRP channels) as bilaterian mechanoreceptors for mechanotransduction. The cnidocytes of cnidarians and their various chemosensory and/or multimodal sensory cell types probably also evolved from this cell type. However, this does not imply that the eumetazoan sensory cells in question were necessarily dedicated mechanoreceptors. They may have instead been multimodal cells capable of transducing multiple sensory stimuli. Indeed, after providing a phylogenetic survey of photoreceptors in Chapter 4, I will argue that specialized mechanoreceptors and several lineages of photoreceptors originated from a multimodal mechano- and photoreceptive cell type in early eumetazoans.

The importance of Ascl and/or Atonal-related transcription factors for the early specification of neurons and sensory cells in cnidarians and bilaterians, suggests that a central role of these bHLH transcription factors for the regulation of generic neuronal differentiation programs already had evolved in this eumetazoan ancestor. Furthermore, POU4, Islet, and a PaxB-like transcription factor may have played roles for large and partly overlapping subsets of neuronal and sensory cell types in the eumetazoan ancestor similar to extant cnidarians and may have adopted more specific roles for neurons of the peripheral nervous system (sensory neurons for POU4 and PaxB; sensory and motor neurons for Islet) only in bilaterians. Other transcription factors showing expression in only a small subset of neuronal/sensory cell types such as Six1/2 and its cofactor Eya may have had a much more restricted function in only some neuronal/sensory cell types of the eumetazoan ancestor. We currently have not enough information about Gfi as well as NK-related (Prox, BarH, NK5, Tlx) and Paired-related (Phox2, DRG11) homeodomain transcription factors to infer their potential role in the eumetazoan ancestor.

Taken together, this suggests that a sensory cell type specified by a CoRN involving at least some of the transcription factors required for specification of bilaterian mechanosensory cells already existed in ancestral eumetazoans. However, as we will see in Chapter 4, many of these transcription factors are also shared with (some lineages of) photoreceptors. Moreover, transcription factors that are differentially required for specification of mechano- (Pax2/5/8) or photoreceptors (Pax4/6, Otx) in bilaterians, either have not yet arisen by gene duplication or do not show such differential expression in cnidarians. This suggests that sensory cell types conveying different cell modalities may not be properly genetically individualized in cnidarians, possibly reflecting the situation in the last common ancestor of eumetazoans. The ancestral mechanosensory cell-type may, thus, have been a multimodal mechano- and photoreceptive cell with individualized mechanoreceptors and photoreceptors only evolving in the bilaterian lineage.

In spite of new data on the cell type-specific expression of transcription factor expression in sponges and placozoans, we know even less about the cell types present in the last common ancestor of sponges, placozoans and eumetazoans. Due to the uncertain phylogenetic position of ctenophores, it is not even clear whether ctenophores were also derived from this ancestor (making it the last common ancestor of all metazoans) or whether ctenophores are its sister taxon. Be this as it may, the neuronal cell types of ctenophores appear to be highly peculiar either because they diverged substantially from the cell types of an eumetazoan ancestor shared with bilaterians and cnidarians or because their neurons evolved convergently and are not homologous to the neurons of bilaterians and cnidarians.

Although it has been proposed that sponges and placozoans may be secondarily simplified and have lost neurons (Ryan et al. 2013), there is currently little evidence to support this scenario. If we assume instead that sponges and placozoans maintain, to some extent, the primitive situation found in the common ancestor with cnidarians and bilaterians, comparisons between these taxa are likely to shed light on the evolutionary origin of neurons. These comparisons suggest that neurons and sensory cell types evolved as novelties in the eumetazoan lineage, probably by establishing a new CoRN. This may have involved the forging of new regulatory links between transcription factors regulating different functional modules (e.g. post- and presynaptic proteins) and initially expressed in different cell types, thereby bringing them together in a common cell type and establishing a new CoRN.

4 Evolution of Photosensory Cell Types

In the last chapter, I used a comparative approach to reconstruct the evolutionary history of mechanosensory cells by proceeding backwards, step by step, from vertebrates to the last common ancestor of metazoans. This survey revealed that the various mechanosensory cells and sensory neurons of vertebrates originated by diversification of one mechanosensory cell type (or several related mechanosensory cell types) that can be traced back all the way to the ancestors of bilaterians. Chemoreceptors probably evolved many times independently from this lineage of cells. In the present chapter, I will attempt to reconstruct the evolution of photoreceptors in a similar fashion. This will reveal that two lineages of photoreceptors, known as rhabdomeric and ciliary photoreceptors, respectively, have an equally deep pedigree that can be followed back to stem bilaterians. Although we lack sufficient data to fully illuminate the origin of these distinct cell types in early animals, several lines of evidence suggest that rhabdomeric and ciliary photoreceptors as well as mechanoreceptors originated from a common sensorineural precursor cell type and that this happened in the stem lineage of bilaterians.

4.1 CILIARY AND RHABDOMERIC TYPES OF PHOTORECEPTORS

As already briefly discussed in Chapter 7 of the first volume (Schlosser 2021), the distinction of photoreceptors into ciliary and rhabdomeric types was first proposed by Richard M. Eakin (e.g. Eakin 1963, 1979) on morphological grounds (Fig. 4.1). In the ciliary type of photoreceptor, photopigments are localized to expanded parts of the ciliary membrane, whereas in the rhabdomeric type such membrane expansions are derived from microvilli. It was originally proposed that within the bilaterians ciliary photoreceptors are confined to deuterostomes and rhabdomeric photoreceptors to protostomes (Eakin 1963). However, recent evidence suggests that both types of photoreceptors are present in both lineages. This has already been discussed for vertebrates, where rods and cones are ciliary, and intrinsically light-sensitive retinal ganglion cells are rhabdomeric photoreceptors (see Chapter 7 in Schlosser 2021). Additional examples from other deuterostomes and protostomes will be discussed in more detail below. In the decades since Eakin's proposal, we have also learned that the differences between the two types of photoreceptors run much deeper than membrane morphology. Ciliary and rhabdomeric photoreceptors not only use photopigments that belong to different clades of opsins (c-opsins vs. r-opsins, respectively) and rely on different phototransduction mechanisms, they are also specified by distinct transcription factors (Arendt and Wittbrodt 2001; Arendt 2003) (Fig. 4.1). I will briefly summarize these differences in the following section before providing a phylogenetic survey of photoreceptors.

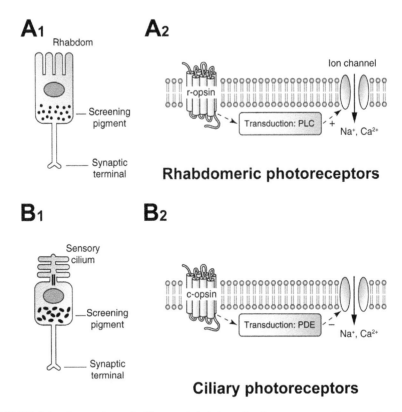

Rhabdomeric photoreceptors

Ciliary photoreceptors

FIGURE 4.1   Comparison of ciliary and rhabdomeric photoreceptors. See text for details. (Reprinted with permission from Nilsson and Arendt 2008.)

4.1.1 PHOTOPIGMENTS

The classical ciliary (c-opsins) and rhabdomeric opsins (r-opsins) were first identified from photoreceptors of vertebrates and arthropods, respectively, and are employed in most ciliary and rhabdomeric photoreceptors known from bilaterians. In addition, however, a group of different opsins, the cnidopsins (originally called "cnidops") were found in cnidarians (Plachetzki, Degnan, and Oakley 2007). There is also a fourth lineage of opsins, known as "group 4" opsins (or "tetropsins"), which is found in many bilaterians (Porter et al. 2012). Group 4 opsins include neuropsins, G_0 opsins (because they couple with $G\alpha_0$), and retinal G protein coupled receptors (RGR); the latter do not serve as visual pigments but act as photoisomerases that convert all-trans-retinal back to 11-cis retinal, thereby reconstituting the light-responsive form of the photopigment. Recently, additional opsins have been found in protostomes that are closely related to cnidopsins and have been termed xenopsins (Ramirez et al. 2016; Vöcking et al. 2017; Rawlinson et al. 2019).

The existence of these four lineages of opsins is well supported by most analyses (Fig. 4.2) (Feuda et al. 2012, 2014; Porter et al. 2012; Hering and Mayer 2014; Liegertova et al. 2015; Rawlinson et al. 2019; Fleming et al. 2020). However, the

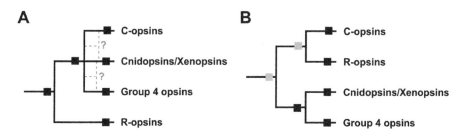

FIGURE 4.2 Opsin evolution. (**A**) Consensus tree of several recent studies. Strongly supported branches indicated with black squares. It is still unresolved, whether cnidopsins/xenopsins are more closely related to c-opsins or to group 4 opsins (grey hatched lines). (**B**) Opsin evolution. Strongly supported branches indicated with black squares, weakly supported branches with grey squares. ([A] Feuda et al. 2012, 2014; Hering and Mayer 2014; Liegertova et al. 2015; Rawlinson et al. 2019; Fleming et al. 2020; [B] Ramirez et al. 2016.)

relation between them is controversial. A recent study, which also posits the existence of several additional clades of opsins, suggests that c- and r-opsins are most closely related and jointly form the sister group of a clade comprising cnidopsins/xenopsins and group 4 opsins (Fig. 4.2B) (Ramirez et al. 2016). However, the deep nodes of the opsin tree in this study are not well supported, raising doubts about the proposed branching order. Most studies instead find strong support for r-opsins being the sister clade to a clade comprising c-opsins, cnidopsins/xenopsins, and group 4 opsins, even though it remains disputed whether cnidopsins/xenopsins are more closely related to c-opsins or to group 4 opsins (Fig. 4.2A) (Feuda et al. 2012, 2014; Hering and Mayer 2014; Liegertova et al. 2015; Rawlinson et al. 2019; Fleming et al. 2020). Moreover, photoreceptors expressing cnidopsins/xenopsins or group 4 opsins are morphologically of the ciliary type (Kojima et al. 1997; Martin 2002; Passamaneck et al. 2011; Vöcking et al. 2017) and these opsins are co-expressed or known to couple with $G\alpha_i$ or $G\alpha_0$ subunits in bilaterians, similar to c-opsins (Kojima et al. 1997; Rawlinson et al. 2019).

Taken together, this suggests the existence of a clade of "ciliary" opsins in the wider sense which include c-opsins, cnidopsins/xenopsins, and group 4 opsins. These are mostly employed in photoreceptors of the morphologically ciliary type, in which the photopigments reside in the cilium. To avoid confusion, I will refer to this larger clade of opsins as "ciliary" opsins (in quotes), reserving c-opsins for one of its subclades. In contrast, r-opsins which form the sister clade to "ciliary opsins" are typically found in photoreceptors of the rhabdomeric type, in which the photopigments reside in microvilli.

4.1.2 Phototransduction Mechanisms

In vertebrate ciliary photoreceptors, light-activated c-opsins typically activate G-proteins containing a $G\alpha_t$ subunit; $G\alpha_t$ then activates phosphodiesterase (PDE) leading to hydrolysis of cGMP; consequently, cyclic nucleotide-gated (CNG) cation channels are inhibited resulting in hyperpolarization (Fig. 4.3A) (reviewed in

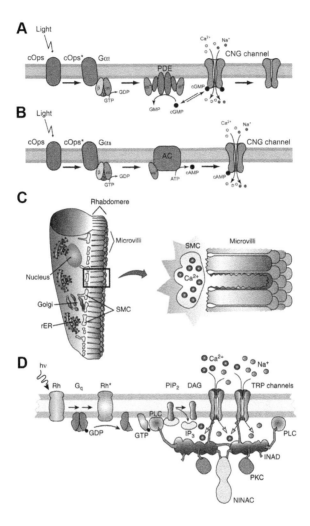

FIGURE 4.3 Phototransduction in ciliary and rhabdomeric photoreceptors. (**A, B**) Phototransduction cascade in ciliary photoreceptors of vertebrates (**A**) and cnidarians (**B**). α, β, γ, subunits of G-protein; α_t^*/α_s^*, activated (GTP-conjugated) α-subunits of $G\alpha_t/G\alpha_s$; AC, adenylate cyclase; ATP, adenine triphosphate; cAMP, cyclic AMP; CNG channel, cyclic nucleotide gated channels; cOps, "ciliary" opsin; cOps*, photoactivated cOps; $G\alpha_t/G\alpha_s$, G protein containing α_t/α_s subunit; PDE, phosphodiesterase. (**C**) Rhabdomeric photoreceptor (left) and detail of microvilli with adjacent submicrovillar cisternae (SMCs) containing Ca^{2+} (right). (**D**) Phototransduction cascade in rhabdomeric photoreceptors of *Drosophila*. See text for explanation. In other species, IP3-induced Ca^{2+} release from the SMCs (not shown) contributes to phototransduction. hv, light; Rh*, activated form of the photopigment rhodopsin; G_q, G protein containing α_q subunit; GDP, guanosine diphosphate; GTP, guanosine triphosphate; PLC, phospholipase C; PIP_2, phosphatidylinositol 4,5-bisphosphate; IP_3, inositol 1,4,5-triphosphate; DAG, diacylglycerol; PKC, protein kinase C; rER, rough endoplasmic reticulum; NINAC, class III myosin; INAD, a protein containing PDZ binding domains responsible for forming the signaling complex in a fly microvillus. ([C,D] Reprinted with permission from Fain, Hardie, and Laughlin 2010.)

Lamb, Collin, and Pugh 2007; Lamb 2009, 2013; Shichida and Matsuyama 2009; Arshavsky and Burns 2012; see also Chapter 7 in Schlosser 2021). However, $G\alpha_t$ (transducin) evolved only in vertebrates from a duplicate $G\alpha_i$ (Lamb and Hunt 2017; Lokits et al. 2018). Ciliary photoreceptors in invertebrates use a $G\alpha_i$-dependent phototransduction mechanism, in which c-opsin activates $G\alpha_i$. The latter then probably acts by inhibiting adenylylcyclase (AC), thereby reducing cAMP production which in turn inhibits CNG channels (Lamb and Hunt 2017; Lamb 2020). Additional phototransduction mechanisms have been described for other ciliary photoreceptors. In scallops, an unknown opsin (possibly of the G_0 class) activates $G\alpha_0$ which has been shown to activate a guanylylcyclase; the resulting increase in cGMP opens a potassium channel leading to hyperpolarization of the cell (Kojima et al. 1997; Gómez and Nasi 2000). Whether and how this mechanism evolved from a canonical ciliary phototransduction pathway is currently unclear.

Finally, some of the cnidarian-specific opsins (cnidopsins; see below) signal via a cascade that resembles canonical ciliary phototransduction in using CNG channels but relies on the different $G\alpha$ subunit $G\alpha_s$, while the mechanism is still unresolved for others (Koyanagi et al. 2008; Kozmik et al. 2008; Plachetzki, Fong, and Oakley 2010; Liegertova et al. 2015). Phototransduction mediated by $G\alpha_s$ differs from the canonical ciliary phototransduction mechanism in leading to depolarization rather than hyperpolarization in light (Fig. 4.3B). This is due to the light mediated increase in cAMP resulting in CNG activation, while PDE appears not to be involved (Koyanagi et al. 2008). However, another member of the cnidopsin/xenopsin clade, a flatworm xenopsin was recently shown to link to $G\alpha_i$, indicating that it may signal via the canonical ciliary phototransduction pathway (Rawlinson et al. 2019). This suggests that cnidopsins/xenopsins and c-opsins may have originally relied on the same $G\alpha_i$-dependent phototransduction pathway, but that divergent phototransduction pathways evolved for at least some of the cnidopsins in cnidarians.

In contrast, rhabdomeric photoreceptors use a different phototransduction mechanism. Light-activated r-opsins activate G-proteins containing a $G\alpha_q$ subunit; $G\alpha_q$ then activates phospholipase C (PLC) which cleaves the phospholipid PIP_2 into IP_3 (inositol 1,4,5-tris-phosphate) and DAG (diacylglycerol) and activates calcium channels of the transient receptor potential (TRP) type resulting in depolarization (Fig. 4.3C, D) (reviewed in Montell 2012; Angueyra et al. 2012; Ferrer et al. 2012). Inactivation of photoactivated opsin requires phosphorylation by rhodopsin kinases and arrestin binding for both ciliary and rhabdomeric photoreceptors; however, phosphorylation is mediated by members of two different lineages of rhodopsin kinases in ciliary (GRK1/7/4/5/6) and rhabdomeric (GRK2/3) photoreceptors (Kikkawa et al. 1998; Lee, Xu, and Montell 2004; Alvarez 2008; Lamb et al. 2018; Lamb 2020).

4.1.3 Transcription Factors Involved in Specification of Photoreceptors

Detlev Arendt first emphasized that apart from their reliance on different phototransduction pathways, ciliary and rhabdomeric photoreceptors are also specified by different transcription factors, suggesting that they are distinct cell types (Arendt 2003). Whereas Pax4/6 and Otx transcription factors appear to be essential for both ciliary and rhabdomeric photoreceptors, Atonal, POU4, BarH, and Prox as well as persistent Pax4/6 expression have been proposed to be specifically required for the specification of rhabdomeric and Rx for the specification of c-opsin-dependent ciliary photoreceptors (Arendt 2003). Indeed, we have seen in Volume 1 (Chapter 7 of Schlosser 2021) that the ciliary photoreceptors in vertebrates (e.g. rods and cones of the retina; photoreceptors in brain) depend on a core regulatory network (CoRN) involving persistent Rx/Rax, Otx2, and Otx5/Crx, while the intrinsically photosensitive retinal ganglion cells, which are modified rhabdomeric photoreceptors, rely on a CoRN involving persistent Pax6, Atoh7, POU4f2, and Islet1 expression. As discussed below, Six1/2, Six4/5, and Eya may also be required for rhabdomeric, but not ciliary photoreceptors in invertebrates (but not vertebrates).

However, phototransduction cascades in ciliary and rhabdomeric photoreceptors use components encoded by paralogous genes arising from gene duplication (e.g. c-opsin vs. r-opsin; $G\alpha_i$ vs. $G\alpha_q$; GRK1/7/4/5/6 vs. GRK2/3). Taken together with the shared dependence of ciliary and rhabdomeric photoreceptors on Pax4/6 and Otx transcription factors, this suggests that they evolved as sister cell types by duplication and divergence from a common ancestral photoreceptive cell type (Arendt 2003). Given that ciliary and rhabdomeric photoreceptors have been identified in both protostomes and deuterostomes, this must have happened already in the ancestors of bilaterians (Fig. 2.8).

This scenario for the evolutionary origin of photoreceptors has been elaborated without considering mechanoreceptors. However, several observations suggest that the evolutionary origins of photoreceptors and mechanoreceptors may be linked (Kozmik et al. 2003; Fritzsch et al. 2005; Schlosser 2018). First, there are striking similarities between the CoRNs of mechanoreceptors and rhabdomeric photoreceptors, both of which rely on Atonal, POU4, BarH, and Prox and other transcription factors such as Gfi and Six1/2 or Six4/5 (see below and Chapter 3). Second, while *Pax4/6* plays a central role for specification of photoreceptors, the related *Pax2/5/8* genes are crucial for many mechanosensory cells; both lineages of *Pax* genes originated from a common precursor resembling the *PaxB* gene of cnidarians. And third, the presence of sensory cells in cnidarians that respond to both light and mechanical stimuli suggests that similar multimodal cells may have been present in the eumetazoan ancestor. Dedicated photo- and mechanoreceptors may then have evolved from such multimodal cells in the ancestors of bilaterians.

The following survey of photoreceptor evolution will allow us to explore these connections further and to develop a more inclusive scenario for the evolutionary origin of mechano- and photoreceptors.

4.2 PHOTOSENSORY CELLS IN THE LAST COMMON TUNICATE-VERTEBRATE ANCESTOR

4.2.1 PHOTOSENSORY CELL TYPES IN TUNICATES

The best studied photoreceptors in tunicates are the photoreceptors of the pigmented ocellus located in the sensory vesicle of the ascidian larva (review in Kusakabe and Tsuda 2007; Esposito et al. 2015) (Fig. 4.4A, B). These are ciliary photoreceptors with branched cilia that express c-opsins and $G\alpha_i$ and respond to light with

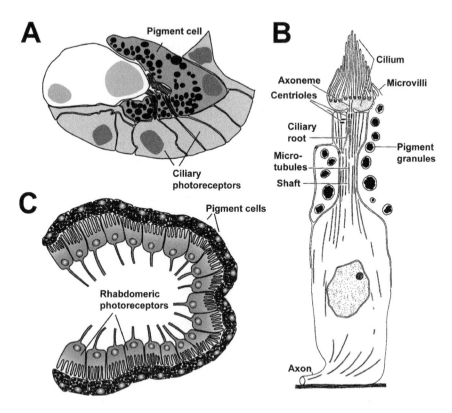

FIGURE 4.4   Tunicate photoreceptors. (**A**) Ciliary photoreceptors and pigment cells in larva of ascidian *Ascidia nigra*. (**B**) Detail of ascidian ciliary photoreceptor. (**C**) Pigment-cup eye with rhabdomeric photoreceptors in oozooid of salp *Thalia democratica*. ([A] Reprinted with permission from Arendt and Wittbrodt 2001 (after Jefferies 1986); [B] reprinted and modified with permission from Eakin and Kuda 1971; [C] reprinted and modified with permission from Braun and Stach 2017.)

hyperpolarization similar to the photoreceptors of the vertebrate retina (Eakin and Kuda 1971; Gorman, McReynolds, and Barnes 1971; Kusakabe et al. 2001; Yoshida et al. 2002). However, $G\alpha_t$ (transducin) is not present in the tunicate (or amphioxus) genome suggesting that $G\alpha_i$ mediates phototransduction in these ciliary photorecep- tors. Two groups of ciliary photoreceptors are associated with the ocellus in *Ciona* on the right side of the larval brain that may have different functions in mediat- ing negative phototaxis (group I) or a looming-object escape response (group II), respectively. They have been proposed to be homologous to photoreceptors of the vertebrate retina and pineal organ, respectively (Horie et al. 2008; Salas et al. 2018; Kourakis et al. 2019). A third group of ciliary photoreceptors is associated with the pigmented statocyte on the left side (see Chapter 3). Their function is poorly under- stood, but the late larval onset of c-opsin expression in these cells suggests that they may be involved in the initiation of metamorphosis (Horie et al. 2008).

In addition to these ciliary photoreceptors, rhabdomeric photoreceptors have been described from the adult siphons of ascidians (Dilly 1973) and morphologically similar photoreceptors have been found in salps (Fig. 4.4C) (Gorman et al. 1971; Braun and Stach 2017). However, salp photoreceptors respond with hyperpolarization to light, suggesting that they may not employ a typical rhabdomeric phototransduction cascade (Gorman et al. 1971). While r-opsins and $G\alpha_q$ are found along with c-opsins and $G\alpha_i$ in tunicate genomes (Yoshida et al. 2002; Porter et al. 2012; Cronin and Porter 2014), it is currently not clear whether they are employed in any tunicate photoreceptors. So far, $G\alpha_q$ expression has only been reported in the trunk and tail of the ascidian larva, where no known photoreceptors are located (Yoshida et al. 2002).

4.2.2 TRANSCRIPTION FACTORS INVOLVED IN SPECIFYING TUNICATE PHOTOSENSORY CELL TYPES

Whereas nothing is known about the transcription factors expressed in the candidate rhabdomeric photoreceptors of salps and adult ascidians, the transcription factors expressed in the ciliary photoreceptors of larval ascidians have been well characterized. They include both Otx and Pax4/6 which are widely expressed in the sensory vesicle of ascidian larvae including their photoreceptors (Wada, Holland, and Satoh 1996; Mazet et al. 2003; Irvine et al. 2008). However, their role for photoreceptor specification in tunicates has not been analyzed. In addition, the transcription factor Rx is expressed more narrowly in photoreceptors, pigment cells and a subset of neurons of the sensory vesicle and was shown to be essential for the specification of these photoreceptors (D'Aniello et al. 2006; Oonuma and Kusakabe 2019).

A recent single-cell RNA-seq study of *Ciona* larvae has confirmed that Otx, Pax4/6, and Rx but not Atonal are expressed in a subset of c-opsin-positive cells (Cao et al. 2019). This suggests that these cells express a profile of transcription factors implicated in the CoRN of ciliary but not rhabdomeric photoreceptors. The failure of this study to detect these transcription factors in other subsets of c-opsin expressing cells may well be an artefact due to methodological limitations in detecting rare transcripts; however this remains to be verified in further studies. Surprisingly, this study also reported that subsets of c-opsin-positive cells express transcription factors Pax2/5/8 and POU4, previously implicated in the CoRNs of mechanoreceptors or rhabdomeric photoreceptors but not ciliary photoreceptors (Cao et al. 2019). However, neither Pax2/5/8 nor POU4 expression could be confirmed in candidate cells by in situ hybridization (Candiani et al. 2005; Mazet et al. 2005) suggesting that these transcription factors are only expressed at low levels and possibly without a major regulatory function in these cells.

4.2.3 THE LAST COMMON TUNICATE-VERTEBRATE ANCESTOR

Taken together with the evidence for the existence of ciliary photoreceptors in vertebrates (see Chapter 7 in Schlosser 2021), these findings suggest that ciliary photoreceptors were present in the last common ancestor of vertebrates and tunicates and subsequently gave rise to the rods and cones of vertebrates and the photoreceptor cells in the sensory vesicle of ascidian larva. However, a new phototransduction mechanism evolved in the ciliary photoreceptors of vertebrates, in which reduction of cyclic nucleotide levels was achieved by $G\alpha_t$ mediated activation of PDE rather than by $G\alpha_i$ mediated inhibition of AC. In the absence of transcription factor data, evidence for the existence of rhabdomeric photoreceptors in tunicates is currently incomplete. However, the fact that the intrinsically photosensitive retinal ganglion cells of vertebrates express components of the rhabdomeric phototransduction cascade and are specified by similar transcription factors (including Atonal, POU4, BarH, Prox) than the rhabdomeric photoreceptors of protostomes (see below and Chapter 7 in Schlosser 2021), indicates that rhabdomeric photoreceptors were also present in the tunicate-vertebrate ancestor but were possibly lost or modified in some clades or life stages of tunicates.

4.3 PHOTOSENSORY CELLS IN THE LAST COMMON CHORDATE ANCESTOR

4.3.1 PHOTOSENSORY CELL TYPES IN AMPHIOXUS

In amphioxus, both ciliary and rhabdomeric types of photoreceptor cells have been reported and c-opsins, r-opsins, group 4 opsins, and opsins of uncertain affiliation were identified in the genome (Pantzartzi et al. 2017; review in Lacalli 2004; Pergner and Kozmik 2017). Ciliary photoreceptors are found in the frontal eye at the rostral tip of the neural tube and in the so-called lamellar body, while two types of rhabdomeric photoreceptors – the Joseph cells and the photoreceptors of the Hesse organs – are distributed along the neural tube (Fig. 4.5).

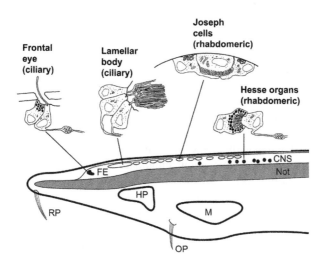

FIGURE 4.5 Amphioxus photoreceptors. The four types of photoreceptors (two ciliary, two rhabdomeric) in an amphioxus larva are depicted. CNS, central nervous system; HP, Hatschek's pit; M, mouth; Not, notochord; OP, oral papillae; RP, rostral papillae. (Overview redrawn after Lacalli 2004. Details of sensory cells reprinted with permission from Lacalli 2004.)

The unpaired frontal eye is located at the rostral end of the neural tube comprising several rows of cells (Lacalli, Holland, and West 1994; Lacalli 1996; Vopalensky et al. 2012). Based on its location and its association with pigment cells, it has been proposed to be the homolog of the vertebrate paired eyes (which also develop from an unpaired precursor region, the eye field, in the embryonic forebrain) (Lacalli et al. 1994). The frontal eye has been shown to function in maintaining vertical orientation of the body during larval feeding and has been implicated in the startle response of adults to sudden illumination (Guthrie 1975; Stokes and Holland 1995). The row 1 cells, which are sandwiched between the more anterior pigment cells and the more posterior row 2–4 cells, serve as ciliary photoreceptors that express c-opsin and $G\alpha_i$, but carry relatively unspecialized cilia (Figs. 4.5, 4.6A) (Lacalli et al. 1994; Lacalli 1996; Vopalensky et al. 2012).

The unpaired lamellar body of amphioxus is located more posterior in a region that has been proposed to correspond to the posterior forebrain of vertebrates or a combined diencephalic/mesencephalic region (Lacalli 1996; Albuixech-Crespo et al. 2017). The cells of the lamellar body have highly branched cilia that are lined up in parallel to form the name-giving lamellae, similar to the cilia of ascidian photoreceptors or the pineal photoreceptors in vertebrates (Fig. 4.5) (Eakin and Westfall 1962; Ruiz and Anadón 1991b). Thus, based on its location and the morphology of its receptors, the lamellar body is widely considered a homolog of the vertebrate pineal gland (Lacalli et al. 1994; Wicht and Lacalli 2005). Since the lamellar body is very well developed in larvae, but disaggregates in adult amphioxus, it has been suggested to be involved in regulating circadian rhythms such as diurnal migrations in the larval stage (Lacalli et al. 1994). However, no c-opsin was yet found to be expressed in the lamellar body and direct evidence for its role in photoreception is still missing (Vopalensky et al. 2012).

The organs of Hesse (also known as dorsal organs) are located in the ventral spinal cord all along the body axis but excluding the cerebral vesicle (Fig. 4.5) (Nakao 1964; Ruiz and Anadón 1991a; Pergner and Kozmik 2017). Each Hesse organ comprises a single rhabdomeric photoreceptor with an axon surrounded by one cup-shaped pigment cell (Fig. 4.5) (Nakao 1964; Ruiz and Anadón 1991a; Castro et al. 2006; Pergner and Kozmik 2017). The only exception appears to be the anterior-most Hesse organ which contains two photoreceptors and one pigment cell (Lacalli 2002a). There are around 1500 Hesse organs on each side in adult amphioxus making them the most abundant photoreceptors in this animal (Nakao 1964). It has been suggested that Hesse organs may allow amphioxus to detect how deeply it is buried in sediments (Lacalli 2004), but their association with a pigment cup suggests that they play additional roles in detecting light direction.

The second type of rhabdomeric photoreceptors in amphioxus, the Joseph cells, are located in the dorsal neural tube posterior to the lamellar body and persist in the adult stage (Figs. 4.5, 4.6A) (Welsch 1968; Ruiz and Anadón 1991a; Pergner and Kozmik 2017). Joseph cells are single photoreceptor cells that are unpigmented and are not associated with pigment cells (Fig. 4.5) (Welsch 1968; Watanabe and Yoshida 1986; Ruiz and Anadón 1991a). They probably have an axon; however, this could still not be shown unequivocally (Welsch 1968; Castro et al. 2015). Their function is poorly understood, but they have been proposed to mediate a shadow response and/or circadian rhythms in adult amphioxus (Pergner and Kozmik 2017; Lacalli 2018).

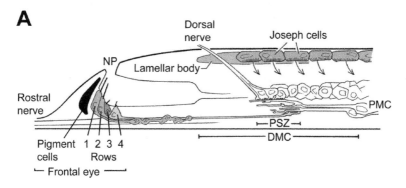

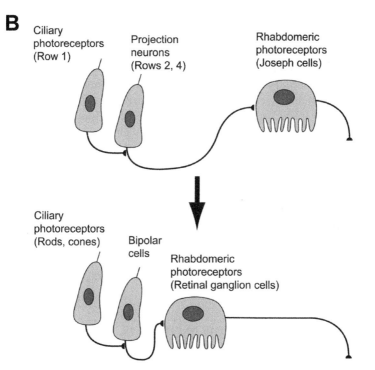

FIGURE 4.6 Model of vertebrate retina evolution. (**A**) Lateral view of anterior nerve cord of amphioxus. A row 4 projection neuron from the frontal eye is shown to make synaptic contact to neurons of the primary motor centre (PMC). There may also be indirect inputs to the Joseph cells (direct inputs have not been described). The (unknown) outputs of the Joseph cells are indicated by small arrows. DN, dorsal nerve; DMC, di-mesencephalic region; NP, neuropore; PMC, primary motor center; PSZ, primary synaptic zone; RN, rostral nerve. (**B**) Model of evolution of cell types in the vertebrate retina (below) from an amphioxus like chordate ancestor (above). See text for details. ([A] Reprinted and modified with permission from Lacalli 2018.)

Recently, phototransduction in both Hesse organs and Joseph cells has been studied extensively. This revealed that both types of photoreceptors are not only morphologically of the rhabdomeric type, but also use the r-opsin melanopsin as a visual pigment and employ a classical rhabdomeric phototransduction cascade with a $G\alpha_q$ type G protein, PLC, and TRP-type calcium channels for signal transduction (Koyanagi et al. 2005; Gómez, Angueyra, and Nasi 2009; Angueyra et al. 2012; Pulido et al. 2012; Ferrer et al. 2012; Peinado et al. 2015).

4.3.2 TRANSCRIPTION FACTORS INVOLVED IN SPECIFYING AMPHIOXUS PHOTOSENSORY CELL TYPES

The row 1 cells of the frontal eye are not only morphologically of the ciliary type and express components of ciliary phototransduction (c-opsin, $G\alpha_i$), they also express Otx, Pax4/6, and possibly Rx transcription factors. Pax4/6 and Rx are also being found in rows 3 and 4 (Williams and Holland 1996; Glardon et al. 1998; Kozmik et al. 2007; Vopalensky et al. 2012). Taken together, this suggests that row 1 cells are ciliary photoreceptors and potential homologs of the cones and rods of the vertebrate retina (Fig. 4.6A). However, expression of neither Otx nor Rx was found in the lamellar body, raising further doubts on whether the ciliated cells of the lamellar body are indeed ciliary photoreceptors (Vopalensky et al. 2012).

Very little is known about the transcription factor profiles of amphioxus' rhabdomeric photoreceptors, the Hesse organs and Joseph cells. However, these photoreceptors are known to express Pax2/5/8 and Hesse organs also express Six4/5 and Eya but not Pax4/6 or POU4, the latter being a core transcription factor of most rhabdomeric photoreceptors (Kozmik et al. 1999; Candiani et al. 2006; Kozmik et al. 2007). Therefore, while morphology, phototransduction and Six4/5 and Eya expression suggest that at least some of these cells are typical rhabdomeric photoreceptors, we still lack further evidence from transcription factor expression to fully confirm this hypothesis.

4.3.3 THE LAST COMMON CHORDATE ANCESTOR

In summary, the presence of putative ciliary and rhabdomeric photoreceptors in amphioxus as well as in tunicates and vertebrates (see Chapter 7 in Schlosser 2021) indicates that both types of photoreceptor existed in the last common ancestor of chordates. This animal may have even been equipped with an anterior and a posterior unpaired eye, which may have given rise to frontal eye and lamellar body in amphioxus and to the paired eyes and unpaired pineal eye/gland in vertebrates, respectively. Alternatively, the anterior eye may have been paired already in the chordate ancestor, but became simplified to an unpaired structure in amphioxus (due to the loss of mechanisms splitting the initially unpaired eye field in the embryonic forebrain). We currently lack data that would allow us to decide between these two scenarios. In any case, the anterior-most eye (or pair of eyes) of this ancestor probably contained ciliary photoreceptors that gave rise to the cones and rods of the vertebrate retina. But did it also contain cells that gave rise to other cells in the vertebrate retina, such as the retinal ganglion cells (RGCs) and can we find such cells still in present day amphioxus?

One candidate for such RGC homologs is the cells of row 2–4 in the amphioxus frontal eye (Fig. 4.6A). Row 1 cells of amphioxus form synapses with axons of the adjacent row 2–4 cells and the latter send axon to regions of its brain with diencephalic or di-mesencephalic characteristics (Lacalli 1996, 2002b; Suzuki et al. 2015; Albuixech-Crespo et al. 2017; Lacalli 2018). This has led to suggestions that row 2–4 cells may be homologs of RGCs, which already form retinal-diencephalic (pretectal) or possibly retinal-mesencephalic (tectal) connections in amphioxus (Lacalli 1996; Vopalensky et al. 2012; Suzuki et al. 2015; Pergner and Kozmik 2017; Lacalli 2018). However, this suggestion conflicts with the proposed homology of RGCs with rhabdomeric photoreceptors (Arendt 2003).

An alternative and functionally plausible model for the evolution of the vertebrate retina was put forward by Lamb (Lamb 2009). He proposed (see Chapter 7 in Schlosser 2021) that the vertebrate retina evolved in ancestral chordates, when ciliary photoreceptors (the precursors of cones and rods) formed synaptic contacts with rhabdomeric photoreceptors (the precursors of RGCs) located nearby. This may have allowed ciliary photoreceptors, which are more efficient in dark environments (like the deep ocean), to take over a major role in photoreception, while retaining pre-established connections of rhabdomeric photoreceptors to the brain's visual centers.

Although none of the rhabdomeric photoreceptors of amphioxus are located near the row 1 cells, the anterior-most Joseph cells are found just behind the lamellar body in a region that may correspond to the vertebrate tectum or pretectum and receives axons from rows 2 and/or 4 of the frontal eye (Fig. 4.6A) (although direct synaptic inputs from the frontal eye to Joseph cells have not been described) (Lacalli 2018). Since Joseph cells are widely distributed along the anteroposterior axis, it is reasonable to assume that their distribution may have extended further anteriorly, possibly into the vicinity of the frontal eye in the last common chordate ancestor. This makes them interesting candidates for RGC homologs according to Lamb's model, while row 2–4 cells may be homologs to bipolar cells, which connect ciliary and rhabdomeric photoreceptors (Fig. 4.6B). As discussed in Chapter 7 of Volume 1 (Schlosser 2021), the bipolar cells and photoreceptors of the vertebrate retina may share a common origin from ciliary photoreceptors based on similarities in their CoRNs (with Rx and Otx2 being required for specification of both cell types) and the use of ribbon synapses. Conversely, the amacrine and horizontal cells of the vertebrate retina may have a common origin with the RGCs from rhabdomeric photoreceptors (see Chapter 7 of Schlosser 2021; Arendt 2003; Lamb et al. 2007; Lamb 2009).

4.4   PHOTOSENSORY CELLS IN THE LAST COMMON DEUTEROSTOME ANCESTOR

4.4.1   PHOTOSENSORY CELL TYPES IN AMBULACRARIANS

Many members of the ambulacrarians (hemichordates and echinoderms), the sister group of chordates within the deuterostomes, are able to respond to light by dispersed dermal light sensitivity. Eyes consisting of aggregations of photoreceptors

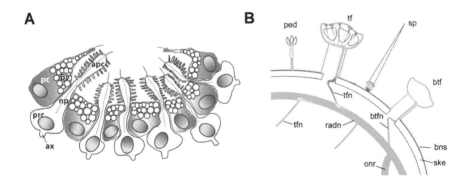

FIGURE 4.7 Photoreceptors in ambulacrarians. (**A**) Diagram of a longitudinal section through the eyecup of the tornaria larva of the enteropneust *Glossobalanus marginatus* with rhabdomeric photoreceptors. Apc, apical process of photoreceptor cell; ax, axon emerging from base of photoreceptor cell; np, neck region of photoreceptor cell; pc, pigment cell; prc, photoreceptor cell; pv, pigment vesicle. (**B**) Diagram showing the distribution of photoreceptors expressing c-opsin (yellow) and r-opsin (red) in the sea urchin *Strongylocentrotus purpuratus* and their connection to the nervous system (green). Bns, basiepithelial nervous system; btf, buccal tube foot; btfn, btf nerve; onr, oral nerve ring; ped, pedicellaria; radn, radial nerve; ske, skeleton; sp, spine; tf, tube foot; tfn: tubefoot nerve. ([A] Reprinted with permission from Braun et al. 2015; [B] reprinted with permission from Ullrich-Lüter et al. 2013.)

and pigment cells are found only in a few groups. However, all major groups of opsins are present in ambulacrarian genomes (except for G_0 opsins in hemichordates) and both ciliary and rhabdomeric photoreceptors have now been identified (D'Aniello et al. 2015; Ramirez et al. 2016).

A pair of cup-shaped eyes with interspersed photoreceptors and pigment cells develop on the apical pole in the tornaria larvae of some hemichordates (Fig. 4.7A) (Brandenburger, Woollacott, and Eakin 1973; Nezlin and Yushin 2004; Braun et al. 2015). The photoreceptors in these eyes carry a single cilium that is surrounded by many microvilli. Cup-shaped ocelli with similar photoreceptors have also been found in some sea cucumbers and in the optic cushion located at the tip of the arms in sea stars (Takasu and Yoshida 1983; Yoshida, Takasu, and Tamotsu 1984; Johnsen 1997). The photoreceptors in these ocelli were initially proposed to be of the ciliary type (Eakin 1963). However, the elaborate structure of their microvilli, which undergo dynamic changes in response to light in echinoderms and express r-opsins in hemichordate larvae, suggest that they are rhabdomeric photoreceptors (Takasu and Yoshida 1983; Braun et al. 2015). Unfortunately, we currently know nothing about transcription factor expression to support this hypothesis. Recently, a different type of eye using ciliary photoreceptors and expressing a G_0-type opsin and a "ciliary" signature of transcription factors (including Rx; see below) has been described in sea urchin larvae (Valero-Gracia et al. 2016; Valencia et al. 2019).

Even in echinoderms that lack eyes, the skin and tube feet are long known to be photosensitive (Yoshida et al. 1984; Ramirez et al. 2011) and some of the underlying receptors have recently been identified (Fig. 4.7B). Rhabdomeric photoreceptors with a single cilium surrounded by many microvilli that express an r-opsin (opsin4) have been described in the tube feet of sea urchins (Agca et al. 2011; Lesser et al. 2011; Ullrich-Lüter et al. 2011). In addition, slender monociliated ciliary photoreceptors that express c-opsin (opsin1) have been found in tube feet, although most of them are associated with skeletal elements (e.g. pedicellariae and spines) in sea urchins and sea stars (Ullrich-Lüter, D'Aniello, and Arnone 2013).

4.4.2 Transcription Factors Involved in Specifying Ambulacrarian Photosensory Cell Types

The unequivocal assignment of ambulacrarian photoreceptors to ciliary or rhabdomeric types is currently hindered by our limited knowledge of the expression and function of transcription factors in these cells (see also Chapter 3). In both directly and indirectly developing hemichordates, Rx and Otx are involved in patterning of the anteroposterior axis and are expressed in anterior and intermediate ring-like domains, respectively, that do not extend to the anterior/apical pole (Lowe et al. 2003; Gonzalez, Uhlinger, and Lowe 2017). Thus, Rx and Otx as well as Pax4/6 are excluded from the apical ectoderm where the ocelli with putatively rhabdomeric photoreceptors develop (Gonzalez et al. 2017). The absence of Rx in these cells lends some support to the proposal, that they are not ciliary photoreceptors. However, BarH is not expressed in these cells and it is currently unknown whether any of the other transcription factors, proposed to define rhabdomeric photoreceptor identity (e.g. Atonal, POU4, Prox, Gfi) are expressed in these cells (Gonzalez et al. 2017).

In contrast, Rx and Otx together with some other transcription factors (Six3, Tbx2/3) and a G_0 opsin were found to be expressed in the photoreceptors of the paired apical eye spots of sea urchin larvae, supporting their identification as ciliary photoreceptors (Valencia et al. 2019). However, Pax4/6, Six1/2, Eya, and Dach were not expressed in these cells (but rather were co-expressed in the hydrocoel pore). In adult sea urchins, the expression of POU4 and the Atonal homolog Ato6 (but not of Pax4/6 or Six3/6) were enriched in the disks of tube feet, where r-opsin is also enriched providing circumstantial evidence that the r-opsin expressing cells in the tube feet are rhabdomeric photoreceptors (Agca et al. 2011).

4.4.3 The Last Common Deuterostome Ancestor

Although further support is needed, these data taken together with the diversity of opsins expressed in various sensory cells of ambulacrarians suggest that ciliary and rhabdomeric photoreceptors are present in ambulacrarians like in many chordates. It is therefore safe to assume that both types of photoreceptors were present in the last common ancestor of deuterostomes.

Interestingly, Pax4/6 has so far not been shown to be expressed in either ciliary or rhabdomeric photoreceptors in ambulacrarians adding further examples to the growing list of photoreceptors – found in several taxa scattered among bilaterians (see below) – that develop independently of Pax4/6. However, apart from these scattered exceptions, Pax4/6 is widely expressed in ciliary and rhabdomeric photoreceptors of most chordates (except the rhabdomeric photoreceptors in amphioxus; see above) and protostomes (see below) and persistent Pax4/6 expression is essential for specification of rhabdomeric photoreceptors in protostomes and of their putative homologs, the RGCs in vertebrates (e.g. Halder, Callaerts, and Gehring 1995; Sheng et al. 1997; Marquardt et al. 2001; Oron-Karni et al. 2008). This suggests that Pax4/6 also played a role for development of both types of photoreceptor in the last common ancestor of deuterostomes and that sustained Pax4/6 expression was required for the specification of rhabdomeric photoreceptors. This implies that the ciliary and rhabdomeric photoreceptor in the larvae of hemichordates and echinoderms probably became independent of Pax4/6 secondarily, possibly because the function of Pax4/6 has been taken over by some other transcription factor. Clearly, this hypothesis needs to be confirmed in further studies.

4.5 PHOTOSENSORY CELLS IN THE LAST COMMON BILATERIAN ANCESTOR

4.5.1 PHOTOSENSORY CELL TYPES IN PROTOSTOMES

Both rhabdomeric and ciliary photoreceptors are present in protostomes, the sister group of deuterostomes within the bilateria (Figs 4.8, 4.9), where they form a large variety of eyes but also occur as extraocular photoreceptors (reviewed in Eakin 1963; Eakin 1972; Vanfleteren and Coomans 1976; Salvini-Plawen and Mayr 1977; Arendt and Wittbrodt 2001; Randel and Jékely 2016).

4.5.1.1 Rhabdomeric Photoreceptors

The eyes of most protostomes contain predominantly rhabdomeric photoreceptors, which are equipped with numerous microvilli (and often carry a single cilium as well), express r-opsin and use a $G\alpha_q$, PLC, and TRP (TRPC) dependent phototransduction (review in Arendt and Wittbrodt 2001; Fain, Hardie, and Laughlin 2010). In the lophotrochozoans (Fig. 4.8), rhabdomeric photoreceptors have been described, for example, in the adult eyes of *Platynereis* and other polychaetes and in eyes of many mollusks (Land 1984; Arendt et al. 2002; Purschke 2005; Wilkens and Purschke 2009; Vöcking, Kourtesis, and Hausen 2015). Rhabdomeric photoreceptors are also found in eyes adjacent to but outside of the apical organ in trochophora larvae and in the apical ectoderm of other larvae (Arendt et al. 2002; Randel et al. 2013; Vöcking et al. 2015). In the ecdysozoans (Fig. 4.9), well studied eyes with rhabdomeric photoreceptors include the larval and adult eyes of arthropods such as Bolwig's organ in the *Drosophila* larva and the compound eye and ocelli in the adult (Paulus 1979; Land and Fernald 1992; Hardie and Postma 2008; Land and Nilsson 2012; Montell 2012; Friedrich 2013). The compound eyes consist of numerous identical units (ommatidia), each with multiple photoreceptors (e.g. eight photoreceptors in *Drosophila*) (Fig. 4.9).

It has been proposed that rhabdomeric photoreceptors became the dominant type of photoreceptor in the eyes of protostomes because of intrinsic properties of

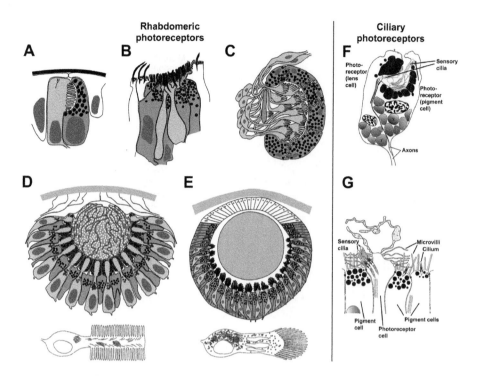

FIGURE 4.8 Rhabdomeric and ciliary photoreceptor cells in lophotrochozoans. (**A–E**) Rhabdomeric photoreceptors (light grey) and associated pigment cells (dark grey) in (**A**) Trochophora larva of *Platynereis dumerilii* (polychaetes, annelids); (**B**) Trochophora larva of *Lepidochiton cinereus* (polyplacophores, mollusks); (**C**) Pigment cup eye of *Dugesia japonica* (turbellaria, flatworms); (**D**) Eye of adult *Platynereis dumerilii* (polychaetes, annelids) with single photoreceptor cell shown below; (**E**) Eye of adult *Helix pomatia* (gastropods, mollusks) with single photoreceptor cell shown below. (**F**) Ciliary photoreceptors expressing c-opsin in larva of *Terebratalia transversa* (brachiopods). One of the two photoreceptors contains an apical intracellular lens-like structure (lens cell), while the other one contains pigment granules (pigment cell). (**G**) Photoreceptor in larva of *Leptochiton asellus* (polyplacophores, mollusks). This photoreceptor expresses both xenopsin and r-opsin and displays both sensory cilia and multiple slender horizontal microvilli. ([A-E]: Reprinted with permission from Arendt and Wittbrodt 2001. Details of photoreceptors in [D,E] reprinted with permission from Whittle 1976; [F] reprinted with permission from Passamaneck et al. 2011; [G] reprinted with permission from Vöcking et al. 2017.)

rhabdomeric phototransduction (Fain et al. 2010). These include the sequestration of all molecular components (r-opsin, PLC, and TRPC channels) into the small space of a microvillus, signal amplification due to a positive feedback of calcium ions on TRP channels and the photoconvertability of r-opsin (Fain et al. 2010). These properties underlie the higher photosensitivity, temporal resolution and dynamic range of rhabdomeric photoreceptors compared to ciliary photoreceptors and their ability to photoregenerate pigment and, thus, may help to explain why they came to be

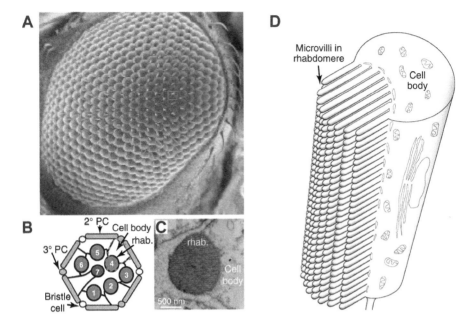

FIGURE 4.9 Rhabdomeric photoreceptor cells in the compound eye of arthropods. (**A**) Scanning electron microscope (SEM) image of an adult compound eye of *Drosophila* with approximately 750–800 ommatidia. The red box indicates one ommatidium. (**B**) Diagram illustrating the various cell types of one ommatidium (distal region) in a cross section. Seven photoreceptor cells, each containing a rhabdomere, are shown (the eighth photoreceptor cell is positioned below the seventh photoreceptor cell and is not visible in this plane of section). The broken blue line indicates a single photoreceptor cell. Rhab., rhabdomere; 2° PC, secondary retinal pigment cells; 3° PC, tertiary retinal pigment cells. (**C**) Transmission EM cross-section through one photoreceptor cell. (**D**) Diagram showing a longitudinal view of one photoreceptor cell with rhabdomere. (Reprinted with permission from Montell 2012.)

prevalent in the eyes of most bilaterians. In contrast, ciliary photoreceptors were maintained extraocularly for specialized functions such as detecting light levels or coordinating circadian rhythms (see below). Only when rods evolved in vertebrates, did ciliary photoreceptors acquire a similar sensitivity than rhabdomeric receptors allowing the evolution of a more energy efficient duplex retina with slow, high sensitivity rods for vision in the dark and fast, low sensitivity cones for vision in bright light (Fain et al. 2010).

Rhabdomeric photoreceptors expressing r-opsin also mediate extraocular photosensitivity in protostomes. Many annelids, mollusks, and other lophotrocozoans have extraocular cells with microvilli and/or cilia, sometimes with microvilli clustered in an intracellular cavity (phaosomes), that presumably serve as photoreceptors for extraocular photosensitivity (Light 1930; Röhlich, Viragh, and Aros 1970; Ramirez et al. 2011; Porter 2016). Similar photoreceptive cells are also found in the genital region of butterflies (Miyako, Arikawa, and Eguchi 1993).

4.5.1.2 Ciliary Photoreceptors

Whereas rhabdomeric photoreceptors predominate in most protostomes, ciliary photoreceptors expressing c-opsin have now been identified in several taxa. They were initially found outside of the cerebral eyes and deeply inside the brain. First reported in the polychaete *Platynereis* (Arendt et al. 2004), c-opsin expressing cells in the brain have recently also been described for several arthropods (Velarde et al. 2005; Eriksson et al. 2013; Beckmann et al. 2015). In addition, extraocular photoreceptors resembling the ciliary type have also been identified in the general ectoderm of mollusks. In the snail *Lymnaea*, for example, such cells occur in the mantle edge, siphons, and tentacles. The photoresponsiveness of these organs can be blocked by CNG inhibitors but not TRPC inhibitors suggesting use of a ciliary phototransduction pathway (Zylstra 1971; Pankey et al. 2010; Ramirez et al. 2011).

Recent evidence suggests, however, that c-opsins may be absent from the genomes of protostomes other than annelids and arthropods and that other "ciliary" opsins in the wider sense – viz. xenopsins and potentially G_0 opsins – may mediate phototransduction in some photoreceptors that are morphologically of the ciliary type (Vöcking et al. 2017; Rawlinson et al. 2019; Wollesen, McDougall, and Arendt 2019). Xenopsins (initially classified as c-opsins) have first been identified in the eyes as well as in apical extraocular photoreceptors of brachiopod larvae with enlarged ciliary membranes (Fig. 4.8F) (Passamaneck et al. 2011). Xenopsin is also expressed in photoreceptors of the larval eyes of the mollusk *Leptochiton* (Fig. 4.8G) (Vöcking et al. 2017). While initially classified as rhabdomeric photoreceptors based on their elaborate microvilli and the co-expression of r-opsin, $G\alpha_q$, and TRPC channels, these cells have subsequently been shown to also carry cilia and were proposed to combine rhabdomeric and ciliary identity (Vöcking et al. 2015, 2017). In flatworms, the larval and adult eyes are composed mostly of rhabdomeric photoreceptors expressing r-opsin. However, xenopsin was recently been shown to be expressed in a single ciliary photoreceptors in the larval eye as well as in the ciliary photoreceptor of the larval epidermal eye and in extraocular ciliary photoreceptors (with hundreds of cilia in an intra-cellular vacuole or phaosome) of the adult (Rawlinson et al. 2019). Co-expression of flatworm xenopsin with $G\alpha_i$ and its loss of activity after application of $G\alpha_i$ inhibitors suggest that it signals by a canonical ciliary phototransduction cascade (Rawlinson et al. 2019).

In contrast, ciliary photoreceptors in the eyes of scallops have been shown to use a divergent, $G\alpha_0$-dependent phototransduction pathway as described above (Kojima et al. 1997; Gómez and Nasi 2000). G_0 opsins and other group 4 opsins (peropsin, neuropsin) are also expressed in the larval and adult eyes of some polychaetes and mollusks as well as in some extraocular photoreceptors (Tosches et al. 2014; Gühmann et al. 2015; Ayers et al. 2018; Wollesen et al. 2019). Furthermore, extraocular photoreceptors with G_0 expression have been shown to mediate some responses to light such as the shadow reflex in adult *Platynereis* (Ayers et al. 2018). However, it is currently unclear, which phototransduction pathway is used by the various photoreceptors expressing group 4 opsins.

Ciliary photoreceptors have also been described adjacent to the apical organ of trochophora and other larvae in lophotrochozoans (Sensenbaugh and Franzén 1987;

Wilkens and Purschke 2009; Passamaneck et al. 2011; Marlow et al. 2014; Tosches et al. 2014; Veraszto et al. 2018). In the polychaete *Platynereis*, where this has been particularly well studied, a few scattered c-opsin expressing photoreceptors develop immediately adjacent to the apical organ, while the rhabdomeric photoreceptors of the larval eyes develop further laterally (Randel et al. 2013; Marlow et al. 2014; Tosches et al. 2014). While the rhabdomeric photoreceptors of the larval eyes function in phototaxis (Jékely et al. 2008; Randel et al. 2013), the ciliary photoreceptors control circadian behaviors via regulation of melatonin production and mediate downward swimming in response to non-directional UV light (Tosches et al. 2014; Veraszto et al. 2018). In addition to c-opsins, several group 4 opsins (neuropsin, peropsins, and G_0 opsins) are expressed in many cells of the apical organ and the surrounding ectoderm in *Platynereis* (Marlow et al. 2014; Tosches et al. 2014; Williams et al. 2017), but the functional significance of this is currently not understood.

Taken together, these findings suggest that ciliary photoreceptors are present in addition to rhabdomeric photoreceptors in many protostomes. Furthermore, ciliary photoreceptors in protostomes are a heterogeneous group of cells, which may express any of the "ciliary" opsins in the wider sense (c-opsins, xenopsins, group 4 opsins) and employ different phototransduction mechanisms. While often found extraocularly, they appear to be interspersed with rhabdomeric photoreceptors in the eyes of some protostomes such as flatworms and – based on ultrastructural evidence – in some polychaetes (Purschke 2005; Wilkens and Purschke 2009). In the eyes of scallops, there are two separate retinae, one with depolarizing, $G\alpha_q$ dependent rhabdomeric photoreceptors and another one with hyperpolarizing, $G\alpha_0$ dependent ciliary photoreceptors (Barber, Evans, and Land 1967; McReynolds and Gorman 1970; Kojima et al. 1997). Because ciliary photoreceptors in the eyes of extant protostomes express xenopsin, whereas c-opsin expressing photoreceptors appear restricted to the brain, it has been proposed that the last common ancestor of all bilaterians may have had eyes composed of ciliary photoreceptors expressing xenopsin intermingled with rhabdomeric photoreceptors, while ciliary photoreceptors expressing c-opsin were confined to the brain (Vöcking et al. 2017). However, so far this scenario is supported only by sparse data from few protostomians and more information is needed to corroborate its validity.

4.5.1.3 Photoreceptors of Uncertain Identity

Ciliary and rhabdomeric photoreceptors are typically recognizable as clearly distinct cell types. However, "ciliary" opsins are sometimes co-expressed with r-opsins in single photoreceptor cells as described for xenopsin in the larval eyes of *Leptochiton* (Vöcking et al. 2017), for G_0 opsin in larval and adult eyes of *Platynereis* (Gühmann et al. 2015) and for several group 4 opsins (neuropsins, peropsins, G_0 opsins) and r-opsin in the apical organ of *Platynereis* (Williams et al. 2017). Co-expression of a c-opsin with an r-opsin (melanopsin) was also described for the retina of zebrafish embryos (Davies et al. 2011). It is currently unclear, whether these photoreceptor cells belong to the ciliary or rhabdomeric photoreceptor type, having merely recruited components of the rhabdomeric or ciliary phototransduction cascade, respectively, or whether they represent cell types of truly mixed ciliary-rhabdomeric identity. In the latter case they may represent either rare survivors of an ancestral lineage of unspecialized photoreceptors prevalent prior to the ciliary-rhabdomeric split or novel cell types arising by recombination of ciliary and rhabdomeric photoreceptor types. Further insights into the CoRNs among transcription factors guiding the specification of these cells will be needed, to decide among these alternative scenarios.

Finally, there is evidence for extracellular photosensitivity in protostomes, which is probably mediated by other cell types, not homologous to ciliary or rhabdomeric photoreceptors. For example, chromatophores also express the r-opsin melanopsin and are directly photosensitive in cephalopods similar to the chromatophores in many vertebrates (Provencio et al. 1998; Kasai and Oshima 2006; Mathger, Roberts, and Hanlon 2010; Ramirez et al. 2011). In the nematode *Caenorhabditis elegans*, ciliary neurons do not use opsins but rather a protein related to invertebrate gustatory receptors (GR) as photopigment, which activates CNG channels via G proteins and guanylylcyclase (Ward et al. 2008; Liu et al. 2010; Gong et al. 2016). Another GR related protein (GR28b) mediates photosensitivity of cells with dendritic arbors in larvae of *Drosophila*, but signaling through a TRPA channel (Xiang et al. 2010). These cell types most likely have acquired photosensitivity independently from ciliary and rhabdomeric photoreceptors and will not considered further here.

4.5.2 TRANSCRIPTION FACTORS INVOLVED IN SPECIFYING PHOTOSENSORY CELL TYPES IN PROTOSTOMES

Like in deuterostomes, Pax4/6 and Otx homologs are widely expressed in ciliary and rhabdomeric photoreceptors of protostomes and are required as upstream regulators for all photoreceptors. Additional transcription factors are then required specifically for the specification of rhabdomeric (e.g. Atonal, POU4, BarH, and Prox) versus ciliary (e.g. Rx) photoreceptors. The function of these transcription factors has been studied extensively in *Drosophila* (Fig. 4.10). Together with expression data, which are available for protostomes of many different taxa (both ecdysozoans and lophotrochozoans), they allow us to draw a relatively detailed picture of the role of transcription factors in the specification of photoreceptors in protostomes.

4.5.2.1 Role of Pax4/6 and Otx in Rhabdomeric and Ciliary Photoreceptors

The role of *Pax4/6* as a core regulator for eye development has first been demonstrated in a series of elegant experiments for the *Drosophila Pax4/6* (=*eyeless*) gene (e.g. Quiring et al. 1994; Halder et al. 1995; Sheng et al. 1997). These revealed that Pax4/6 is essential for development of the eye and all its cell types and can initiate the formation of extra eyes in other imaginal discs when expressed ectopically. Subsequently, Pax4/6 was found to be expressed in the eyes of many other arthropods and lophotrochozoans that are composed of rhabdomeric photoreceptors (flatworms, Callaerts et al. 1999; Salo et al. 2002; mollusks, Tomarev et al. 1997; Hartmann et al. 2003; Vöcking et al. 2015; annelids, Arendt et al. 2002; nemertines, Loosli, Kmita-Cunisse, and Gehring 1996; reviewed in Arendt 2003). More recently, Pax4/6 was also shown to be expressed in the ciliary and xenopsin expressing photoreceptors of brachiopods and mollusks (Passamaneck et al. 2011; Vöcking et al. 2015).

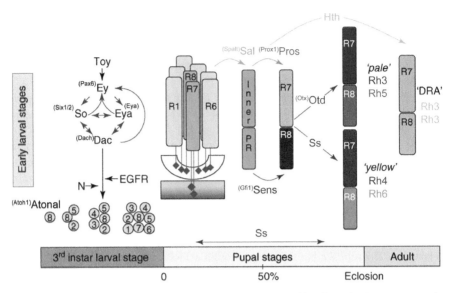

FIGURE 4.10 Transcription factors involved in the specification of photoreceptors in *Drosophila*. During larval stages the eye-antennal disc is specified by the eye determination factors Twin of eyeless (Toy), Eyeless (Ty), Sine oculis (So), and Eyes absent (Eya). Photoreceptors (R1–R8) are then recruited sequentially into an ommatidium. R8 is the first photoreceptor to be specified by expression of Atonal. Subsequently, R2 and R5, later R3 and R4, and finally R1 and R6 are induced by signals from adjacent photoreceptor cells. Decisions for different photoreceptor cell fates are then made in the late larval and early pupal stages. First, a generic color photoreceptor fate is established through the action of Spalt (Sal). Then R7 and R8 cells are distinguished from each other by the specific expression of Prospero (Pros) and Senseless (Sens) in R7 and R8, respectively. Different combinations of subtypes of R7 and R8 ("pale", "yellow" or "dorsal rim area (DRA)"), which express different opsins (Rh 3, 4, 5, or 6), are then specified by the action of Orthodenticle (Otd), Spineless (Ss) or Homothorax (Hth). Homologous transcription factors or cofactors in vertebrates are indicated in parentheses. EGFR, Epidermal growth factor receptor; N, Notch. (Reprinted with modification with permission from Morante, Desplan, and Celik 2007.)

However, in vertebrates, where *Pax4/6* has duplicated into *Pax4* and *Pax6* and Pax4 has lost a role in eye development, Pax6 is only transiently required in progenitors of the ciliary photoreceptors (see Chapter 7 in Schlosser 2021). It is subsequently downregulated in differentiating rods and cones and is dispensable for the activation of rod- or cone-specific differentiation genes such as *c-opsin* (Belecky-Adams et al. 1997; Perron et al. 1998; Vopalensky and Kozmik 2009). Whether this is also true for ciliary photoreceptors in protostomes is currently unknown.

In contrast, Pax6 continues to be expressed in the retinal ganglion cells of the vertebrate retina, which have been proposed to be highly modified rhabdomeric photoreceptors, as does Pax4/6 in the differentiating rhabdomeric photoreceptors of protostomes (Arendt 2003). In *Drosophila*, Pax4/6 is expressed in rhabdomeric photoreceptors of both the larval Bolwig organ and the adult complex eyes and directly regulates *r-opsin* gene transcription (Sheng et al. 1997). Persistent expression of

Pax4/6 was also observed in some rhabdomeric photoreceptors of lophotrochozoans (Callaerts et al. 1999; Arendt et al. 2002; Vöcking et al. 2015). This suggests that while Pax4/6 is required in progenitors of both ciliary and rhabdomeric photoreceptors, it may only be required in rhabdomeric photoreceptors for later stages of differentiation (Arendt 2003).

In spite of the central role of Pax4/6 for photoreceptor development in most bilaterians, examples of eyes with rhabdomeric photoreceptors that develop in the absence of Pax4/6 have been found in some arthropods (*Limulus*, Blackburn et al. 2008) and lophotrochozoans (adult eyes and extraocular photoreceptors in *Platynereis*, Arendt et al. 2002; Backfisch et al. 2013; posterior eyes of *Leptochiton* larvae, Vöcking et al. 2015) like in some deuterostomes (see above). Pax4/6 is also dispensable for the development of rhabdomeric photoreceptors in the Bolwig organ of *Drosophila* larvae (Suzuki and Saigo 2000) and in the regenerating eyes of turbellarian flatworms (Pineda et al. 2002). However, development of the Bolwig organ requires the joint activity of Six1/2 and Eya upstream of Atonal (Suzuki and Saigo 2000). These transcription factors (Six1/2, Atonal) or cofactors (Eya) are also required as CoRN members for the development of the adult complex eyes, where Six1/2 and Eya engage in a network of positive cross-regulatory interactions with Pax4/6 (Fig. 4.10) (Halder et al. 1998).

The existence of such a positively autoregulatory network may facilitate the loss or substitution of some of its members if expression of some of its members is sufficient to activate a set of target genes (Fig. 4.11). This may help to explain how Pax4/6 or any other transcription factor with an essential and evolutionarily widely conserved role for the differentiation of a particular cell type, can nevertheless be lost from the CoRN of this cell type or substituted by another transcription factor in

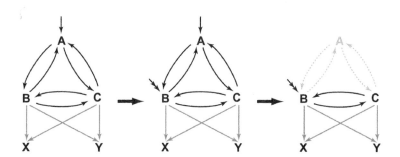

FIGURE 4.11 Model for evolutionary changes in core regulatory networks (CoRN). Left panel: The original CoRN involves initial activation of a transcription factor A by some upstream regulator (arrow). A then activates an autoregulatory network between transcription factors A, B, C, some of which regulate transcription of a battery of differentiation genes, here represented by X and Y. Middle panel: Additional ways of activating the autoregulatory network, e.g. by a new upstream activator of B (double arrow) may be established during evolution without affecting CoRN function. Right panel: Since B can now activate the autoregulatory network and expression of X and Y independently of A, this allows the evolutionary loss of A (or its substitution by other transcription factors) without compromising the regulatory function of the CoRN.

some lineages (an example for "developmental systems drift" or "genetic piracy" as discussed in Chapter 2).

Like Pax4/6, the Otx transcription factors (known as Otd or Orthodenticle in arthropods) is expressed in both rhabdomeric and ciliary photoreceptors in arthropods (Vandendries, Johnson, and Reinke 1996) and lophotrochozoans (Umesono, Watanabe, and Agata 1999; Arenas-Mena and Wong 2007; Passamaneck et al. 2011; Vöcking et al. 2015). However, unlike Pax4/6, Otx appears to be maintained in differentiating photoreceptors of both ciliary and rhabdomeric lineages and is required for their differentiation (Arendt 2003; Vopalensky and Kozmik 2009). In *Drosophila*, Otd is required for the differentiation of a subset of ommatidial photoreceptors (Fig. 4.10) and directly regulates r-opsin and presumably several other genes involved in photoreceptor differentiation and phototransduction (Tahayato et al. 2003; Ranade et al. 2008; Viets, Eldred, and Johnston 2016). This mirrors the role of the Otx family member Crx in the ciliary photoreceptors of vertebrates, which directly regulates transcription of c-opsin and other phototransduction genes (PDE, arrestin and guanylate cyclase) via binding to cis-regulatory elements (Chen et al. 1997; Furukawa, Morrow, and Cepko 1997; Qian et al. 2005). Otx, therefore, seems to be a core regulator and CoRN member for photoreceptor specification throughout bilaterians.

4.5.2.2   Transcription Factors in Rhabdomeric Photoreceptors

In addition to Pax4/6 and Otx, the differentiation of rhabdomeric photoreceptors appears to specifically require Six1/2, Atonal, and additional transcription factors (e.g. POU4, Gfi, BarH, Prox). Six1/2 (*Drosophila*: Sine oculis) and Six4/5 transcription factors and their Eya cofactors were shown to be expressed in the developing rhabdomeric photoreceptors of many different protostomes (insects, planarians, and polychaetes) and functional studies in *Drosophila* and planarians demonstrated that Six1/2 and Eya are essential for photoreceptor specification (Bonini, Leiserson, and Benzer 1993; Cheyette et al. 1994; Serikaku and O'Tousa 1994; Bonini et al. 1997; Pignoni et al. 1997; Halder et al. 1998; Pineda et al. 2000; Suzuki and Saigo 2000; Arendt et al. 2002; Mannini et al. 2004; Vöcking et al. 2015). Taken together with the expression of Six4/5 and Eya in the rhabdomeric Hesse organs and the absence of Six1/2, Six4/5, and Eya from ciliary photoreceptors of amphioxus (Kozmik et al. 2007), this suggests that Six1/2, Six4/5, and Eya may be specific components of the CoRN for rhabdomeric photoreceptors in all bilaterians.

Similarly, *Atonal* is required in *Drosophila* as a proneural gene for the specification of rhabdomeric photoreceptors in the larval eye (Bolwig's organ) as well as for the earliest forming photoreceptors R8 in the adult eye (Fig. 4.10) (Jarman et al. 1994; Daniel et al. 1999; Suzuki and Saigo 2000; Frankfort and Mardon 2002; Hsiung and Moses 2002). It is also expressed in the rhabdomeric photoreceptors of other protostomes including lophotrochozoans (e.g. *Platynereis* larval and adult eyes; planarian eyes) (Arendt et al. 2002; Martin-Durán, Monjo, and Romero 2012). Together with the importance of Atonal-related transcription factors for the development of retinal ganglion cells in vertebrates (see Chapter 7 in Schlosser 2021), this supports the proposal that Atonal may play a central role for specification of rhabdomeric photoreceptors in all bilaterians (Arendt 2003).

POU4 is also expressed in some rhabdomeric photoreceptors in the lophotrochozoans including the larval (but not adult) eyes of *Platynereis* and putative photoreceptors in some mollusks (O'Brien and Degnan 2000, 2002; Tessmar-Raible 2004;
Backfisch et al. 2013). However, the *Drosophila* POU4 ortholog Acj6 is not expressed
in any of its rhabdomeric photoreceptors (Certel et al. 2000). Instead Acj6 is co-
expressed with Atonal in neurons of the lobula complex neurons, which receive input
from the rhabdomeric photoreceptors of its compound eye via intermediary neurons
(Erclik et al. 2008). This raises the possibility that the POU4 expressing optic relay
neurons and rhabdomeric photoreceptors in arthropods may have evolved as sister
cells by duplication of ancestral rhabdomeric photoreceptors, which subsequently
lost POU4 dependence in the photoreceptor lineage but retained it in relay neurons
(Nilsson and Arendt 2008; Erclik et al. 2009). In cephalopods, which most likely
evolved complex eyes independently of arthropods, POU4 is also absent from photoreceptors but is expressed in the nearby neurons of the optic ganglia (Wollesen et al.
2014). This suggests that a similar sequence of events may have happened several
times independently, when complex eyes with elaborate optical circuits evolved from
simpler structures. In each instance, multiple rounds of duplication and divergence
of an ancestral type of rhabdomeric photoreceptor may have given rise to the various
new cell types making up these eyes or circuits. However, the alternative possibility
that POU4 was recruited to target cells of rhabdomeric photoreceptors several times
independently can currently not be ruled out.

As discussed in Chapter 3, Six1/2, Six4/5, Atonal, and POU4, which may be part
of the CoRN defining rhabdomeric photoreceptor identity, are also likely core regulators of mechanosensory cells. This striking overlap between the CoRNs of rhabdomeric photoreceptors and mechanoreceptors extends even further to transcription
factors Gfi, BarH, and Prox, which play essential roles for specification of mechanosensory cells in many bilaterians (see Chapter 3). All of these transcription factors
have been implicated in the specification of rhabdomeric photoreceptors in arthropods. The *Drosophila* ortholog of Gfi, Senseless, is required for the specification of
the earliest forming photoreceptors (R8), while the *Drosophila* ortholog of Prox,
Prospero, is required for specification of the adjacent R7 photoreceptors in the compound eye (Fig. 4.10) (Nolo, Abbott, and Bellen 2000; Frankfort et al. 2001; Cook
et al. 2003; Xie et al. 2007). BarH instead specifies the fates of outer photoreceptors,
which develop after R7 and R8 have formed, and of pigment cells (Higashijima
et al. 1992; Hayashi, Kojima, and Saigo 1998). Much less is known about the roles
of Gfi, BarH, and Prox in lophotrochozoans, although Prox was recently shown to be
expressed in the rhabdomeric photoreceptors of some mollusks (Vöcking et al. 2015).
With exception of Islet, which does not appear to be required for the specification
of rhabdomeric photoreceptors (Thor and Thomas 1997), we currently know nothing about potential roles in photoreceptors of other transcription factors involved in
specification of mechanosensory cells (e.g. the NK related factors NK5 and Tlx and
the paired like homeodomain factors Phox2 and DRG11).

4.5.2.3 Transcription Factors in Ciliary Photoreceptors

Whereas many transcription factors have been implicated in the specification of
rhabdomeric photoreceptors, so far only one candidate, Rx, has been identified that

appears to be specifically required for ciliary photoreceptors (Arendt 2003; Lamb et al. 2007). The well-characterized rhabdomeric photoreceptors in *Drosophila* or flatworms do not express Rx and develop in an Rx-independent way (Eggert et al. 1998; Salo et al. 2002; Davis et al. 2003; Mannini et al. 2008; Martin-Durán et al. 2012). In contrast, Rx plays a central role in the specification of ciliary photoreceptors in tunicates and vertebrates as discussed above and in Chapter 7 of Volume 1 (Schlosser 2021). The c-opsin expressing ciliary photoreceptors identified in the brain of *Platynereis* are also expressing Rx (Arendt et al. 2004; Tessmar-Raible et al. 2007). Furthermore, the c-opsin expressing photoreceptors of the *Platynereis* larva are possibly all located in the ring of Rx expression that encircles the apical organ (Marlow et al. 2014). However, many of the cells that express group 4 opsins (neuropsins, peropsins, G_0 opsins) in *Platynereis* clearly develop in a more apical position devoid of Rx, suggesting that they are specified in an Rx independent way (Marlow et al. 2014; Williams et al. 2017). Unfortunately, it is not known whether any of the recently described xenopsin expressing ciliary photoreceptors in protostomes co-express Rx (Passamaneck and Di Gregorio 2005; Vöcking et al. 2017; Rawlinson et al. 2019). Taken together, this suggests that Rx may play an evolutionarily conserved role for the specification of at least the subset of ciliary photoreceptors that expresses c-opsin. Whether Rx also plays a role in the specification of other ciliary photoreceptors that use xenopsins or group 4 opsins or whether these require other transcription factors (and which ones) remains to be investigated.

4.5.3 THE LAST COMMON BILATERIAN ANCESTOR

In summary, although there are still many gaps in our knowledge, the available data support the idea that the last common bilaterian ancestor was equipped with different types of photoreceptors, a rhabdomeric photoreceptor and one or several types of ciliary photoreceptors and that these cell types persist in many extant lineages of protostomes and deuterostomes. Apart from some exceptions with secondarily modified CoRNs, specification of both rhabdomeric and ciliary photoreceptors requires Otx and at least transitory Pax4/6 activity. In addition, specification of rhabdomeric photoreceptors requires sustained Pax4/6 expression as well as Atonal, Six1/2 or Six4/5, POU4 and possibly Gfi, BarH, and Prox. A CoRN comprising these transcription factors then activates genes encoding r-opsins and many other components of the rhabdomeric phototransduction cascade (e.g. $G\alpha_q$, PLC, TRPC). In contrast, specification of ciliary photoreceptors requires Rx in addition to Otx and transient Pax4/6 expression to activate genes encoding c-opsins and other components of ciliary phototransduction cascades (e.g. $G\alpha_i$ or $G\alpha_t$, PDE or AC, CNG channels). However, it is currently unknown, whether any other core transcription factors cooperate with Rx in the CoRN for ciliary photoreceptors. It is also unknown, whether photoreceptors that express other "ciliary" opsins, viz. xenopsins or group 4 opsins, are specified by the same or by a different CoRN.

These uncertainties surrounding the identity of ciliary photoreceptors raise the possibility, that rhabdomeric photoreceptors may not be the sister cell type of ciliary photoreceptors in general, but may have arisen as sister cells of only a sublineage of ciliary photoreceptors (making ciliary photoreceptors a paraphyletic cell type family), for example, a Rx-dependent and c-opsin expressing sublineage. Furthermore, the striking overlap between the CoRNs of rhabdomeric photoreceptors and mechanoreceptors suggests that these two cell types also originated from a common ancestral cell type existing prior to the origin of bilaterians. Thus, while the shared dependence on Pax4/6 and Otx and the use of related phototransduction mechanisms suggests that rhabdomeric photoreceptors are more closely related to (c-opsin expressing) ciliary photoreceptors than to mechanoreceptors, other similarities in CoRNs support the conflicting view, that rhabdomeric photoreceptors and mechanoreceptors are more closely related. Before we can consider possible scenarios that can resolve this apparent conflict below, we need to extend our survey of photoreceptors to the most basal metazoans, the cnidarians, ctenophores, and sponges.

4.6   PHOTOSENSORY CELLS IN THE LAST COMMON EUMETAZOAN AND METAZOAN ANCESTORS

4.6.1   Photosensory Cell Types in Cnidarians, Ctenophores, Placozoans, and Sponges

Some of the xenacoelomorphs possess simple eyes with a single pigment cell enclosing a small number of photoreceptors lacking both cilia and microvilli (Martínez, Hartenstein, and Sprecher 2017). However, their proposed phylogenetic position as sister group of the bilaterians is still contentious (see Chapters 1 and 3; Fig. 1.8) and I will, therefore, not consider them further here. The undisputedly basal metazoans – cnidarians, ctenophores, and sponges – are all able to perceive light, although only cnidarians and ctenophores have specialized photoreceptor cells. Very little is known about the photoreceptors in ctenophores (see below), whose systematic position is also still disputed (Fig. 1.8). I will, therefore, focus here mostly on the photoreceptors of cnidarians, which in comparison with bilaterian photoreceptors should provide some insight into the types of photoreceptors present in the last common ancestor of eumetazoans.

4.6.1.1   Cnidarians – Photoreceptors with "Ciliary" and "Rhabdomeric" Morphology

Extraocular photosensitivity (independent from eyes) is found in both anthozoans and medusozoans in all life stages including planula larvae, polyps, and medusae (in medusozoans). While several different cell types have been shown to be photoresponsive in cnidarians including gonads and muscle cells, for most species we do not know, which cells mediate responsiveness to light (reviewed in Martin 2002; Ramirez et al. 2011). However, ciliated sensory cells that are scattered through the diffuse nerve net are the most likely candidates. In hydromedusans, some sensory cells displaying lamellated ciliary membranes have been identified as putative photoreceptors (Bouillon and Nielsen 1974). In the polyp of *Hydra*, ciliated sensory cells of the battery complexes associated with cnidocytes express an opsin (the cnidopsin HmOps2), CNG channels and arrestin and regulate cnidocyte discharge in response to light (Fig. 4.14A) (Plachetzki, Fong, and Oakley 2012). This strongly suggests that these cells serve as photoreceptors even though they carry unspecialized cilia and microvilli and show no surface expansion of their apical membrane.

Similarly, sensory cells with a single, unspecialized cilium – often surrounded by a ring of microvilli – are widespread in cnidarians and have been described in planula larvae and adults of medusozoans and anthozoans (e.g. Lentz and Barrnett 1965; Westfall and Kinnamon 1978; Westfall, Sayyar, and Elliott 1998; Westfall, Elliott, and Carlin 2002; Nakanishi et al. 2008, 2012; Piraino et al. 2011; Mason et al. 2012). The simple structure of these cells gives no indication of their modality and it is, thus, typically not known whether they function as mechano-, chemo-, and/ or photoreceptors. However, some evidence from *Hydra* suggests that at least some of these cells are multimodal and respond to more than a single kind of sensory stimulus. In Chapter 3, I have reviewed evidence that at least some sensory cells of the battery complexes, which innervate the cnidocytes in *Hydra*, respond to both mechanical and chemical stimuli (Watson and Hessinger 1992; Mire-Thibodeaux and Watson 1994; Watson and Hudson 1994;). The widespread expression of opsins and CNG channels in cells of the battery complex (except cnidocytes) further-more indicates that at least some of these cells additionally act as photoreceptors (Plachetzki et al. 2012).

Whereas all cnidarians show extraocular photosensitivity, eyes consisting of groups of photoreceptor cells associated with shielding pigment are only found in one of the two major clades of cnidarians, the medusozoans (comprising the hydro-zoans, scyphozoans, and cubozoans), but not in their sister clade, the anthozoans (corals and their allies) (reviewed in Martin 2002). In medusozoans, eyes evolved several times independently (Picciani et al. 2018). With very few exceptions they are found only in the medusa stage, while polyps have no eyes. The gamut of eyes in medusozoans ranges from simple eye spots to pigment cup eyes to complex lens eyes (Martin 2002; Piatigorsky and Kozmik 2004; Kozmik et al. 2008) (Fig. 4.12). In hydromedusae, the ocelli are usually found in bulbs at the base of the tentacle, whereas the eyes of scyphozoans and cubozoans are part of the rhopalia, club-shaped sensory appendages attached to the bell margin. The eyes of cubozoans (box jel-lyfish) are particularly diverse and sophisticated. Six different eyes are present in each cubozoan rhopalium and two of those have a complex structure with lens and cornea (Fig. 4.12B, C) (Martin 2002; Piatigorsky and Kozmik 2004). The cubozoan photoreceptors also contain pigment granules and function as both photoreceptors and pigment cells (Fig. 4.12C). However, in many hydrozoans and scyphozoans, the photoreceptors are unpigmented and are surrounded by or interspersed with distinct pigment cells (Fig. 4.12A).

The photoreceptors of cnidarian eyes are morphologically of the ciliary type and typically carry a single cilium with a typical 9x2+2 arrangement of microtubules (Martin 2002). While in the eyes of some hydromedusae, the cilium is simple and unspecialized, in the photoreceptors of most cnidarian eyes, the membrane of the cilium is expanded or, in cubozoans, contains apically stacked discs of membranes resembling vertebrate rods (Fig. 4.12A–C) (Singla 1974; Martin 2002). Accordingly, the photoreceptors have been classified as "ciliary", ever since Eakin introduced his dichotomous classification (Eakin 1963). Recently, a rare case of photoreceptor cells

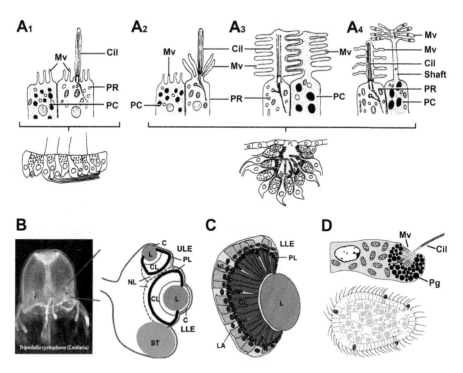

FIGURE 4.12 Cnidarian photoreceptors. (**A**) Ciliary photoreceptors in hydromedusae form ocelli with pigment cells (PC) and photoreceptors (PR). Upper row shows cellular details; lower row depicts overall shape of ocellus (**A₁**: flat ocellus; **A₂**-**A₄**: cup-shaped ocellus). **A₁**: *Leuckartiara*, **A₂**: *Asterias*, **A₃**: *Polyorchis* with interdigitation of the villous processes of the receptor and pigment cells; **A₄**: *Bougainvillia* with spatial differentiation of the villous processes of receptor and pigment cells. Cil, sensory cilium; Mv, microvilli. (**B**) Medusa of the cubozoan *Tripedalia cystophora* with sagittal section through rhopalium shown in detail. Upper (ULE) and lower (LLE) lens eyes contain a cornea (C), a lens (L), and a retina consisting of a ciliary layer (CL), a pigment layer (PL) and a neural layer (NL). S, stalk; St, statocyst. (**C**) Diagram of lower lens eyes (LLE) of *Tripedalia* with two types of photoreceptor cells expressing two different types of cnidopsins. The ciliary segments of type-B receptor cells (dark grey), which express Tcop13 dominate the ciliary layer (CL). Scattered among these are the cone-shaped projections of type-A photoreceptor cells (light grey), which express Tcop18. Cell bodies of both types of receptors occupy the neural layer. Axons of type-A photoreceptor cell bodies create a compact layer (LA) surrounding the whole retina. (**D**) Diagram of pigmented photoreceptor (above) and its distribution (below) in the larva of *Tripedalia cystophora*. Cil, sensory cilium; Mv, microvilli; Pg, pigment granules. ([A₁-A₄] Reprinted with permission from Singla 1974; [B] Photograph of *Tripedalia* reprinted with permission from Kozmik et al. 2003. Diagram of rhopalium reprinted with permission from Liegertova et al. 2015; [C] Reprinted with permission from Liegertova et al. 2015; [D] Reprinted with permission from Nordström et al. 2003.)

resembling the rhabdomeric type has been described in a cubozoan planula larva (Nordström et al. 2003). These cells form an apical cup filled with many microvilli and a single cilium and surrounded by pigment granules (Fig. 4.12D). Around a dozen of these axonless cells are sprinkled throughout the ectoderm of the larva without any apparent connection to neurons. It has, therefore, been proposed that they may be both sensory and effector cells, which may be able to modulate the movement of the cilium in response to light (Nordström et al. 2003).

However, I will argue in the remainder of this section that despite their morphological similarities, "ciliary" or "rhabdomeric" photoreceptors of cnidarians are probably not one-to-one homologous to the ciliary or rhabdomeric photoreceptors of bilaterians, respectively. At least some cnidarian "ciliary" photoreceptors use different photopigments (cnidopsins instead of c-opsins) and phototransduction mechanisms (using $G\alpha_s$ instead of $G\alpha_i$ or $G\alpha_t$) than typical bilaterian ciliary photoreceptors. In turn, no r-opsins have been found in cubozoans (Liegertova et al. 2015; see below) calling the classification of the photoreceptors in cubozoan larvae as "rhabdomeric" into question. Furthermore, there is so far little evidence that either of the "ciliary" or "rhabdomeric" photoreceptors of cnidarians share a CoRN of transcription factors specifically with ciliary or rhabdomeric photoreceptors of bilaterians, respectively. I will, therefore, propose below that cnidarian photoreceptors are independent offshoots from an ancestral multimodal eumetazoan cell type that gave rise to multiple distinct cell types in bilaterians including mechanoreceptors, ciliary, and rhabdomeric photoreceptors.

4.6.1.2 Cnidarians – Photopigments and Phototransduction

Cnidarians probably use several different photopigments including cryptochromes and opsins and a number of different phototransduction mechanisms to sense light. Cryptochromes are widely expressed in ectodermal cells of corals (Levy et al. 2007) and some opsins are expressed in multiple tissues in both medusozoans and anthozoans (Suga, Schmid, and Gehring 2008; Mason et al. 2012; Liegertova et al. 2015; Quiroga Artigas et al. 2018; Macias-Muñoz, Murad, and Mortazavi 2019). However, other opsins (of the cnidopsin clade) are specifically enriched in sensory cells, for example in the battery complex of *Hydra* and the photoreceptors in the eyes of hydro- or cubomedusae (Koyanagi et al. 2008; Kozmik et al. 2008; Suga et al. 2008; Plachetzki et al. 2012; Liegertova et al. 2015).

Several different clades of opsins have been described in cnidarians but their relationship with each other and with the opsins of bilaterians is still contentious. Some studies report a close relationship of cnidarian opsins with bilaterian c-opsins, r-opsins, and G_0-opsins (e.g. Feuda et al. 2012), while other studies indicate that cnidarian opsins are only distantly related to the visual opsins of bilaterians, with one major clade, the cnidopsins forming the sister group to the bilaterian xenopsins (e.g. Ramirez et al. 2016). The problem of clarifying the phylogeny of cnidarian opsins is exacerbated by many lineage specific expansions of different opsins in different cnidarian groups (Macias-Muñoz et al. 2019). However, while many questions remain, the majority of cnidarian opsin sequences can be assigned to the strongly supported clade of cnidopsins (Fig. 4.2) (e.g. Plachetzki et al. 2007; Feuda et al. 2012; Porter et al. 2012; Liegertova et al. 2015; Macias-Muñoz et al. 2019). The sister group relationship of cnidopsins and bilaterian xenopsins also is strongly supported by recent studies (Ramirez et al. 2016; Vöcking et al. 2017; Rawlinson et al. 2019). However, in addition to cnidopsins, several other opsins have been identified in cnidarians. Some of these (acropsin3, which is only found in anthozoans) have been assigned to the r-opsins with relatively high confidence, while the phylogenetic affiliations of other opsins is still unclear (Suga et al. 2008; Feuda et al. 2012; Feuda et al. 2014; Macias-Muñoz et al. 2019; Fleming et al. 2020).

Our knowledge of phototransduction in cnidarians is limited and is best studied in cubozoans, where candidate photoreceptors express multiple opsins of the cnidopsin clade together with the $G\alpha$ subunit $G\alpha_s$ (Koyanagi et al. 2008; Kozmik et al. 2008; Liegertova et al. 2015). Pharmacological inhibition of $G\alpha_s$ inhibits phototactic behavior in these animals, suggesting an essential role of $G\alpha_s$ in phototransduction (Liegertova et al. 2015). Upon illumination, cAMP levels increase (probably due to $G\alpha_s$ mediated activation of adenylylcyclase), presumably leading to the opening of CNG channels and depolarization of photoreceptors (Fig. 4.13A) (Koyanagi et al. 2008; Liegertova et al. 2015). Although not directly proven in cubozoans, CNG channels were shown to be required for phototransduction in *Hydra* (Plachetzki et al. 2010), while a depolarizing response of photoreceptors to light was demonstrated for photoreceptors in hydromedusans (Arkett and Spencer 1986). Thus, a similar $G\alpha_s$ based mechanism may be operating in photoreceptors of different cnidarian lineages although this needs to be supported by further evidence. This $G\alpha_s$ based mechanism differs significantly from the $G\alpha_i$ $G\alpha_t$ or $G\alpha_0$ based mechanisms described for ciliary photoreceptors in bilaterians, all of which lead to hyperpolarization of photoreceptors in response to light (see above; Fig. 4.3A, B).

However, in addition to the $G\alpha_s$ based mechanism just described, cnidarians probably use other phototransduction mechanisms as suggested by several observations. First, in cubozoans only a small subset of opsins (cnidopsins) was able to couple with $G\alpha_s$ suggesting that other opsins use different phototransduction mechanisms (Liegertova et al. 2015). Some of the bilaterian xenopsins, which are the bilaterian opsins most closely related to cnidopsins, are co-expressed with $G\alpha_i$ suggesting that they may employ a phototransduction mechanism similar to the well-characterized c-opsin expressing ciliary photoreceptors of bilaterians (Rawlinson et al. 2019). However, so far there is no evidence that $G\alpha_i$ (or $G\alpha_0$) is involved in phototransduction in any cnidarians and further studies are needed to elucidate how cnidopsin expressing, but $G\alpha_s$ independent photoreceptors perceive light. Second, an anthozoan r-opsin (acropsin3), which is expressed in planula larvae, has been shown to couple with the $G\alpha_q$ subunit (Mason et al. 2012). Taken together with the presence of *PLC* and *TRPC* genes in cnidarian genomes, this raises the possibility that some anthozoan photoreceptors perceive light by a mechanism resembling rhabdomeric phototransduction (Mason et al. 2012). If confirmed by further evidence this would lend credence to the scenario that rhabdomeric phototransduction may have originated prior to the cnidarian-bilaterian split, but may have been lost together with r-opsins in medusozoans (Mason et al. 2012).

The assumed presence of multimodal sensory cells in ancestral eumetazoans that were able to respond to both mechanical stimuli and light, similar to some of the sensory cells in extant cnidarians, may help to explain the evolution of the rhabdomeric

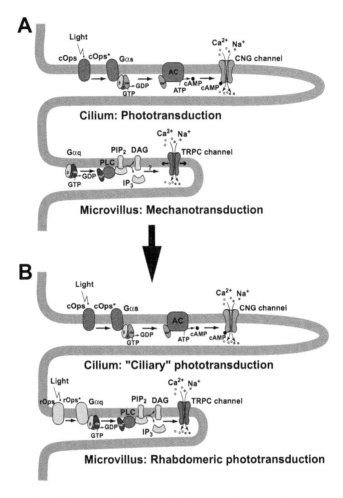

FIGURE 4.13 Model for evolution of the rhabdomeric phototransduction cascade. (**A**) In a multimodal eumetazoan receptor cell, photo- and mechanotransduction may have occured in parallel. Probably, phototransduction possibly using a "ciliary" opsin, its associated $G\alpha$ subunit (e.g. cnidopsin and $G\alpha_s$ as illustrated here) and CNG channels occurred predominantly in cilia, while mechanotransduction using TRPC channels occurred in microvilli. TRPC channels were intrinsically mechanosensitive (opposing arrows) but may also have responded to membrane stretching mediated by PLC activation. (**B**) After r-opsins originated by gene duplication from "ciliary" opsins and were co-expressed in the same cell, they may have evolved new protein interaction capacities, allowing them to interact with a different $G\alpha$ subunit ($G\alpha_q$), thereby linking them to PLC-mediated TRPC activation. α, β, γ, subunits of G-protein; α_s*/α_q*, activated (GTP-conjugated) α-subunits of $G\alpha_s$ and $G\alpha_q$, respectively; AC, adenylate cyclase; ATP, adenine triphosphate; cAMP, cyclic AMP; CNG channel, cyclic nucleotide gated channels; cOps, "ciliary" opsin (e.g. cnidopsin); cOps*, photoactivated cOps; DAG, diacylglycerol; $G\alpha_q$, G protein containing α_q subunit; $G\alpha_s$, G protein containing α_s subunit; GDP, guanosine diphosphate; GTP, guanosine triphosphate; IP_3, inositol 1,4,5-triphosphate; PLC, phospholipase C; PIP_2, phosphatidylinositol 4,5-bisphosphate; rOps, rhabdomeric opsin; rOps*, photoactivated rOps; TRPC channel, transient receptor potential C channel.

phototransduction cascade, which combines photoactivation of G-proteins by opsins with activation of TRPC channels (Fig. 4.13). Outgroup comparison suggests that TRPC dependent phototransduction probably evolved from CNG-dependent photo-transduction, since melatonin receptors – the G-protein coupled receptors most closely related to opsins – like "ciliary opsins" (c-opsins, cnidopsins/xenopsins, G_0 opsins) use CNG-dependent signaling mechanisms (Plachetzki et al. 2010). The first eumetazoans, thus probably used a "ciliary", CNG channel based mechanism for opsin-based photo-transduction (Fig. 4.13A). In contrast, TRP-channels including TRPC are intrinsically mechanosensitive (see discussion in Chapter 3), and thus have probably an ancient role in mechanotransduction (Kang et al. 2010; Quick et al. 2012; Venkatachalam, Luo, and Montell 2014; Sexton et al. 2016). However, they secondarily became responsive to a variety of stimuli that modulate PLC activity, which has been shown to activate nearby TRP channels via membrane stretching (Fig. 4.13A) (Hardie and Franze 2012; Liu and Montell 2015). As discussed in Chapter 3, to allow for stretch mediated activation of TRP-channels, all signal transduction components ($G\alpha_q$, PLC, TRP) needed to be co-localized in small cellular compartments such as cilia or microvilli, with one type of TRP channels, TRPC, showing microvillar localization.

When r-opsin originated by duplication of an ancestral *opsin* gene, its co-expression with $G\alpha_q$, PLC, and TRPC associated with the microvilli of a multimodal photo- and mechanoreceptive cell may have facilitated the evolution of new protein interactions with $G\alpha_q$ thereby establishing a new link between r-opsin and TRPC channels (Fig. 4.13B). This may explain the evolutionarily stable association between rhabdomeric phototransduction and the microvillar phenotype of rhabdomeric photoreceptors. If correct, this scenario for the evolution of new phototransduction suggests that rhabdomeric phototransduction mechanisms evolved in multimodal cells that initially combined "ciliary" CNG-based phototransduction and mechanotransduction. But even if this particular scenario turns out to be wrong, the reliance of "ciliary" and rhabdomeric phototransduction cascades on paralogous components, which arose by gene duplication ("ciliary" opsins vs. r-opsin; $G\alpha_{s/i/o}$ vs. $G\alpha_q$; GRK1/7/4/5/6 vs. GRK2/3), makes it unlikely that these evolved independently. This suggests that one of these (presumably the rhabdomeric) phototransduction cascades evolved in a photoreceptive cell type employing the other (presumably "ciliary") cascade. The evolution of new sensory transduction mechanisms, therefore, most likely preceded the evolution of distinct ciliary and rhabdomeric photoreceptive (and possibly mechanoreceptive) cell types that are specified by different CoRNs and specialized for sensory perception through a single signal transduction mechanism.

Before considering the question whether different photoreceptor cell types were present in the last common ancestors of Eumetazoans, we will have a brief look at the photoreceptors of ctenophores and sponges.

4.6.1.3 Ctenophores, Placozoans, and Sponges
Ctenophores have lamellate bodies formed by ingrown immotile cilia in their apical organ (Fig. 3.14C), which express one of the three opsins (opsin 2) identified in the genome of *Mnemiopsis,* suggesting that they serve as photoreceptors (Horridge 1964; Schnitzler et al. 2012). Another opsin (opsin 1) has been found to be expressed in the apical organ of the larva and both opsins are expressed in bioluminescent cells

(photocytes). The ctenophore opsins do not cluster well with bilaterian or cnidarian opsins and their phylogenetic relationship is currently unclear (Schnitzler et al. 2012; Feuda et al. 2014; Ramirez et al. 2016). In particular, the phylogenetic position of opsin 3 is contentious, while there is some support for opsin 1 and 2 to be related to the "ciliary" rather than rhabdomeric opsins. Although we do not know, which phototransduction mechanism is used by ctenophore opsins, the absence of r-opsins from their genome suggests that it may be of a CNG based "ciliary" rather than of a TRPC-based rhabdomeric kind.

Opsins have also been found in placozoans, but are absent from sponges (Feuda et al. 2012). The placozoan opsins are the phylogenetic outgroup of all other known opsins and they are unable to bind retinal, suggesting that they do not function as photopigments (Feuda et al. 2012). Assuming the traditional view of metazoan phylogeny with ctenophores as sister group of cnidarians, this suggests that opsin-dependent phototransduction may have evolved only in eumetazoans. If, alternatively, ctenophores are the most basal branching animals, opsin-dependent phototransduction must have been lost in sponges and placozoans. Although sponges lack sensory cells and neurons (see Chapter 3) and do not have opsins, sponge larvae are able to react to light (Leys et al. 2002). Their photosensitivity is based on cryptochromes, which are expressed in a ring of pigmented cells at the posterior end of the larva (Rivera et al. 2012).

4.6.2 TRANSCRIPTION FACTORS INVOLVED IN SPECIFYING PHOTOSENSORY CELL TYPES IN CNIDARIANS

The role of transcription factors in the specification of cnidarian mechanoreceptors has been reviewed in Chapter 3, but most of it applies to the specification of sensory cells in cnidarians in general. Except for cnidocytes, (which specifically require combined PaxA and MefA ativity; Babonis and Martindale 2017), there is currently little evidence for any transcription factors differentially required for the specification of mechano- or photoreceptors in cnidarians.

In bilaterians, transcription factors of the Pax4/6 family together with Otx appear to be core regulators of photoreceptors, while Pax2/5/8 family transcription factors, which arose as sister lineage of Pax4/6 by the duplication of a *PaxB*-like gene (Fig. 3.15), are required for specification of mechanoreceptors. However, as reviewed in Chapter 3, Pax4/6 appears to have been lost in cnidarians, while PaxB, which is related to the Pax2/5/8 family, probably resembles the ancestral protein in retaining all protein domains, some of which have been lost in the Pax2/5/8 or Pax4/6 lineages of bilaterians.

Cnidarian PaxB is expressed in both mechano- and photoreceptors in the rhopalia of cubozoans and in scattered ectodermal cells in many cnidarians, which probably include both photo- as well as mechanoreceptive cells (Gröger et al. 2000; Kozmik et al. 2003; Piatigorsky and Kozmik 2004; Matus et al. 2007). However, in the anthozoan *Nematostella*, PaxB is widely expressed in the general epidermis, but is not specifically enriched in any of its opsin expressing cells (Sebé-Pedrós et al. 2018). Photoreceptors in hydrozoan medusae do not express PaxB but instead depend on the only distantly related transcription factor PaxA (Suga et al. 2010). These observations indicate that the PaxB-like transcription factor of the last common eumetazoan ancestor, like PaxB in extant cnidarians, possibly played some general role in the specification of neuronal or sensory cells. After the duplication of this *PaxB*-line gene, the resulting Pax4/6 and Pax2/5/8 sister lineages then probably adopted specific functions in the specification of photo- and mechanoreceptors, respectively, only in bilaterians (Kozmik et al. 2003; Piatigorsky and Kozmik 2004).

Otx, which together with Pax4/6 has a central role for the specification of all photoreceptors in bilaterians, also is not restricted to photoreceptors in cnidarians. Widespread Otx expression in the rhopalium of the scyphozoan Aurelia probably encompasses both mechano- and photoreceptors (Nakanishi et al. 2010). Similarly, in *Nematostella* Otx is expressed in a large subset of neurons without any specific enrichment in opsin expressing cells, arguing against a specific role of Otx in photoreceptor specification (Sebé-Pedrós et al. 2018).

As shown in the last section, most cnidarian photoreceptors appear to rely on "ciliary" phototransduction mechanisms based on cnidopsin or other "ciliary" opsins in the wider sense, $G\alpha_s$ (or other unknown $G\alpha$ proteins) and CNG channels, with the exception of some anthozoan sensory cells that instead use r-opsin, $G\alpha_q$, and possibly other components of rhabdomeric phototransduction. However, there is little evidence that sensory cells using "ciliary" or rhabdomeric phototransduction or transducing mechanical or chemical stimuli, form distinct cell types, which are genetically individualized from each other, each regulated by its own CoRN of transcription factors.

Expression of the transcription factor Rx, which is specifically required for the ciliary type of photoreceptors in bilaterians, has only been analyzed in anthozoans but not in medusozoans. In *Nematostella,* it is expressed in a band of cells in the larva and in cells scattered throughout the tentacles in the adult (Mazza et al. 2010). However, the larval band of Rx expression does not correspond (although may be slightly overlapping) to the region on the aboral pole, where the cnidopsin acropsin2 is enriched (Mason et al. 2012). In the single cell RNA-seq study of adult *Nematostella*, high Rx expression is found in some secretory cells and in a very small number of neurons, but not in any neurons or sensory cells expressing "ciliary" or other opsins (Sebé-Pedrós et al. 2018). This argues against a role of Rx in specifying "ciliary" photoreceptors in cnidarians.

Similarly, there is so far no evidence that any of the transcription factors with roles in the specification of rhabdomeric photoreceptors in bilaterians (Atonal, Six1/2 or Six4/5, POU4 and possibly Gfi, BarH, and Prox) plays a corresponding role in cnidarians. Data are currently only available for Atonal, POU4 and Six1/2. As briefly summarized in the following paragraphs, these are all expressed in at least some sensory cells and/or neurons in cnidarians and appear to play general roles in the specification of subsets of sensory cells. However, there is no support for a specific role of any of these transcription factors in the specification of rhabdomeric photoreceptors.

As reviewed in Chapter 3, Atonal and Ascl appear to regulate neuronal/sensory differentiation in most neuronal and sensory cells, presumably including photo- as well as mechanoreceptive cells (e.g. Hayakawa, Fujisawa, and Fujisawa 2004; Seipel, Yanze, and Schmid 2004; Richards and Rentzsch 2015). This is also supported by the presence of Ascl in some opsin expressing sensory cells of *Nematostella*, while Atonal expression has not been reported in this study (Sebé-Pedrós et al. 2018).

Although data are limited, there is currently no evidence to suggest that Atonal would be specifically required for anthozoan photoreceptors depending on r-opsin, but not for photoreceptors employing a "ciliary" phototransduction mechanism.

POU4 is also present in a large subset of sensory and neuronal cells in *Nematostella* which include neurons expressing r-opsin and "ciliary" opsins among many other neuronal or sensory cell types including the mechanosensory cnidocytes (Sebé-Pedrós et al. 2018; Tournière et al. 2020). In the rhopalium of the scyphozoan *Aurelia*, POU4 appears to be localized predominantly to mechanoreceptors, but probably includes photoreceptors as well (Nakanishi et al. 2010). Moreover, terminal differentiation of cnidocytes and other sensory cells or neurons expressing POU4 is compromised in POU4 mutants (Tournière et al. 2020). This suggests that POU4 may play a central role in the specification of a large subset of neurons and sensory cells in cnidarians, including at least some mechanoreceptive cells (cnidocytes) and possibly photoreceptors, rather than being specifically required for rhabdomeric photoreceptors.

The expression of Six1/2 and Eya in medusozoans is found in the putative "ciliary" photoreceptors of the rhopalia as well as in mechanosensory cells of statocysts (Bebenek et al. 2004; Stierwald et al. 2004; Graziussi et al. 2012; Hroudova et al. 2012). In anthozoans, Six1/2 is restricted to a small subset of sensory cells or neurons, which do not express opsins (Sebé-Pedrós et al. 2018). While experimental evidence is lacking, this is compatible with a role of Six1/2 in the specification of a subset of neurons or sensory cells in cnidarians, but argues against any specific role for rhabdomeric photoreceptors.

In summary, this suggests that while cnidarians may be able to use both "ciliary" and rhabdomeric phototransduction mechanisms, they may have no distinct and genetically individualized "ciliary" and rhabdomeric photoreceptor cell types. Members of the CoRNs of ciliary (Rx) and rhabdomeric photoreceptors (e.g. Atonal, POU4, Six1/2) instead either play more general roles for the specification and differentiation of large subsets of neurons (Atonal, POU4) or may have adopted a narrow and possibly cnidarian-specific function in specific neuronal subtypes (Rx, Six1/2). Furthermore, there is no evidence for any consistent difference in the transcription factor profile between mechanoreceptors and photoreceptors in cnidarians. Expression of Atonal, POU4 and Six1/2 has been found in mechano- as well as photosensory cells and transcription factors that are differentially required for specification of mechano- (Pax2/5/8) or photoreceptors (Pax4/6, Otx) in bilaterians are either represented by a single gene (PaxB) in cnidarians or do not show such differential expression.

4.6.3   The Last Common Eumetazoan/Metazoan Ancestor

Taking these transcription factor profiles in cnidarians together with evidence for the existence of multimodal sensory cells, responding to both mechanical stimuli and light, this suggests a lack of genetic individualization of different sensory cell types in the last common ancestor of eumetazoans, which was largely maintained in extant cnidarians. This implies that mechanosensory cells (and the chemosensory cells repeatedly evolving from them), rhabdomeric photoreceptors, and ciliary photoreceptors as distinct cell types, each defined by its own CoRN, originated only in the bilaterian lineage. Importantly, this does not rule out that different types of neuronal or sensory cells with a well-defined CoRN may have evolved independently in cnidarians. Cnidocytes are one example of such a cell type without a homolog in bilaterians.

As discussed in Chapter 3, further inferences regarding the sensory equipment of the last common ancestor of all metazoans are complicated by the uncertain phylogenetic position of ctenophores and our lack of knowledge, which transcription factors specify their neurons or sensory cells. In placozoans and sponges, which have no neurons or sensory cells, transcription factors that are co-expressed and cross-regulated in the CoRNs of mechano- and photosensory cells of bilaterians, are expressed in different cell types (see Chapter 3). Moreover, Otx, which is crucial for specification of photoreceptors in bilaterians, appears to be absent from the sponge genome, while Rx expression has not been described (Larroux et al. 2008; Srivastava et al. 2010). If sponges are the most basally branching animals, this suggests that neuronal and specialized sensory cell types capable of photo- and mechanotransduction evolved only in eumetazoans and then further diversified into specialized sensory cell types in bilaterians. If ctenophores are the most basally branching animals, sponges, and placozoans must have either lost sensory cells/neurons or ctenophores must have evolved unique sensory cells and neurons independently of cnidarians. Further studies clarifying the phylogenetic position of ctenophores and characterizing the expression of transcription factors in their photoreceptors and other sensory cells will be required to shed light on these issues.

4.7   A SCENARIO FOR THE EVOLUTIONARY ORIGIN OF DIFFERENT SENSORY CELL TYPES IN BILATERIANS

Assuming that the last common ancestor of eumetazoans only had unspecialized and multimodal sensory cell types, how then and in which sequence did mechanoreceptors, rhabdomeric, and ciliary photoreceptors originate in stem bilaterians? If we assume with Arendt (Arendt 2008), that new cell types usually arise by duplication and divergence of an ancestral cell type, this would require two sequential branching events. For three cell types there are three possible trees, but only two of those receive support from the shared expression of CoRN members (Fig. 4.14A). In the first tree, a potentially multimodal ancestral cell type first gave rise to dedicated mechano- and photoreceptors, followed by the split of photoreceptors into a rhabdomeric and ciliary type (Fig. 4.14A$_1$). This sequence of branching events is supported by the shared dependence of both rhabdomeric and ciliary photoreceptors on Pax4/6 and Otx in contrast to the Pax2/5/8 dependence of mechanoreceptors. In the second tree, the ancestral cell type first split into ciliary photoreceptors and a second, multimodal lineage that subsequently gave rise to rhabdomeric photoreceptors and mechanoreceptors (Fig. 4.14A$_2$). This sequence of branching events is supported by the shared dependence of both rhabdomeric photoreceptors and mechanoreceptors on Atonal, POU4, and other transcription factors in contrast to the Rx dependence of ciliary photoreceptors. Thus, different transcription factors support different and conflicting scenarios of cell type evolution.

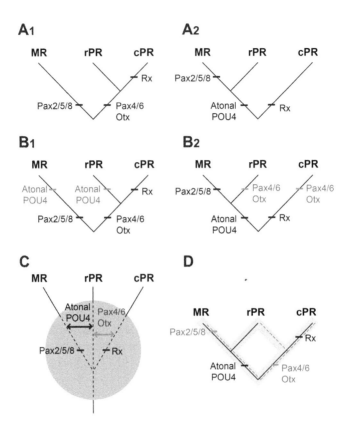

FIGURE 4.14 Scenarios for the evolution of mechanoreceptors and rhabdomeric and ciliary photoreceptors in bilaterians from an ancestral multimodal sensory cell type. (**A**) Different CoRN transcription factors support different sister group relationships of rhabdomeric photoreceptors (rPR), ciliary photoreceptors (cPR), and mechanoreceptors (MR). (**A₁**) Shared dependence of both rPRs and cPRs on Pax4/6 and Otx, suggests that these cell types are more closely related with each other than with MRs, which depend on Pax2/5/8 instead. (**A₂**) Shared dependence of both MRs and rPRs on Atonal and POU4, suggests that these cell types are more closely related with each other than with cPRs, which depend on Rx instead. (**B**) Dichotomous cell type trees as shown in (**A**) would imply convergent recruitment of either Atonal - POU4 (**B₁**) or Pax4/6 – Otx (**B₂**). (**C**) Alternatively, there may be no clear dichotomous branching pattern, because the three cell types may have been incompletely genetically individuated during their period of origin from a common ancestral sensory cell type (indicated by hatched lines; grey circle), with changes in some genes (e.g. *Atonal-POU4*) leading to concerted evolution of MRs and rPRs and changes in others (e.g. *Pax4/6-Otx*) leading to concerted evolution of rPRs and cPRs (double-headed arrows) (**D**) In a final scenario, rPRs may have evolved by the forging of a new CoRN by recombination of transcription factors from the CoRN of MRs (e.g. Atonal-POU4) and cPRs (e.g. Pax4/6-Otx). While expression and core regulatory function of transcription factors follows a dichotomous branching pattern (e.g. thin solid line for Atonal and POU4; thin grey hatched line for Pax4/6 and Otx), the evolution of cell types (broad grey lines encompassing the thin lines delineating the expression of core regulators) is not tree like.

There are several possible ways, in which this apparent conflict can be resolved (Fig. 4.14B–D). First, some apparently "shared" usages of the same transcription factor in the CoRNs of two cell types may reflect convergent recruitment of this transcription factor into each CoRN rather than shared origin from a common ancestor (Fig. 4.14B). For example, rhabdomeric photoreceptors and mechanoreceptors may have independently acquired POU4 dependency (Fig. 4.14B$_1$). Alternatively, rhabdomeric and ciliary photoreceptors may have independently recruited Pax4/6 as a core regulator (Fig. 4.14B$_2$). However, convergence becomes less likely as an explanation, when multiple CoRN members are shared, such as Pax4/6 and Otx between rhabdomeric and ciliary photoreceptors or POU4 and several other transcription factors (e.g. Atonal, Six1/2) between rhabdomeric photoreceptors and mechanoreceptors.

Second, multiple cell types may arise from a common precursor without a clear dichotomous pattern of branching. During a period of transition from a multimodal cell type to distinct mechanoreceptors and rhabdomeric and ciliary photoreceptors, there may have been incipient subtypes of cells that were only partly genetically individualized (see Chapter 2), but remain linked by shared dependence on different core regulators (Fig. 4.14C). For example, shared dependence on Pax4/6 may have led to concerted evolution of incipient rhabdomeric and ciliary photoreceptors, whereas shared dependence on POU4 and other transcription factors may have resulted in concerted evolution of rhabdomeric photoreceptors and mechanoreceptors. Only after stronger genetic individualization of these subtypes emerged, may the three cell types have become distinct and mostly independently evolving cell types. This may have happened, for example, when target genes increasingly required the presence of particular combinations of core regulatory transcription factors for activation. The combined expression of POU4 and Pax4/6 in rhabdomeric photoreceptors, will then, for example, allow the activation of a different set of target genes than the combination of POU4 with Pax2/5/8 in mechanoreceptors or the combination of Pax4/6 and Rx in ciliary photoreceptors.

Third, the assumption that cell types evolve predominantly by duplication and divergence may have to be reconsidered (Oakley 2017; Schlosser 2018). As explained in Chapter 2, novel cell types may also evolve by the recombination of pre-existing cell types, resulting in non-tree like patterns of cell type evolution (Fig. 4.14D). This raises the possibility that during the early evolution of sensory cells, ciliary photoreceptors and mechanoreceptors may have first originated by duplication and divergence from an ancestral cell type and rhabdomeric photoreceptors may have subsequently evolved by recombination of CoRN members from both ciliary photoreceptors (Pax4/6) and mechanoreceptors (Atonal, POU4 etc.).

The last two scenarios are not mutually exclusive and may both have contributed to the evolutionary origin of different sensory cell types in bilaterians. Unfortunately, all these events happened in extinct animals in the stem lineage of bilaterians without any living survivors that may have preserved any intermediary steps. So while we will be able to better judge the plausibility of these scenarios with increasing knowledge of the mechanisms of cell type specification and their evolutionary changes in other examples of cell type evolution, we will be unable to empirically verify any proposed sequence of events in the stem lineage of bilaterians in comparative studies.

5 Evolution of Neurosecretory Cell Types

The survey of sensory cell evolution in the last two chapters suggested that mechano-, chemo-, and photoreceptive cell types in all bilaterians may have evolved from a common sensorineural cell type present in the last common bilaterian ancestor. Thus, despite their vertebrate-specific specializations, the mechano- and chemosensory cells derived from cranial placodes appear to have a very long evolutionary history. In this chapter, I will ask the question, whether the neurosecretory cells derived from placodes can be traced back to remote ancestors in a similar fashion. These neurosecretory cells comprise the hormone-producing cells of the adenohypophysis and the neuropeptidergic cells of the olfactory placode (which produce, for example, gonadotropin releasing hormone (GnRH)). I hope to show that, while neurosecretory cells and neurons probably represent a single cell type family which originated in early eumetazoans, specialized neurosecretory cell types probably evolved many times independently. Apart from a small population of neurosecretory cells that are specified by Otp and other transcription factors in the vertebrate hypothalamus and in the anteromedial nervous systems of many other bilaterians, there is currently little evidence for neurosecretory cell types with a deep evolutionary history. The neurosecretory cells derived from the adenohypophyseal and olfactory placode most likely originated as novel cell types only in vertebrates or the last common tunicate-vertebrate ancestor.

5.1 NEURONS AND NEUROSECRETORY CELLS: AN ANCIENT CELL TYPE FAMILY

Before embarking on our brief journey through animal taxa, a few words are necessary to clarify some terminology. The terms "neurosecretory" and "neuroendocrine" are not used in a consistent manner in the literature so I want to briefly explain how I will employ them here (Fig. 5.1). Often used interchangeably, these terms may either be employed narrowly for neurons that also produce hormones or in a much wider sense to refer to neurons and other endocrine cell types that produce similar types of hormones (in particular peptide hormones) and/or share some other neuron-like properties. To disambiguate the terms, I will here use the term "neuroendocrine" in the former, narrow sense and "neurosecretory" in the latter, wider sense, thus making neuroendocrine cells a subset of neurosecretory cells. Let me explain this in a bit more detail.

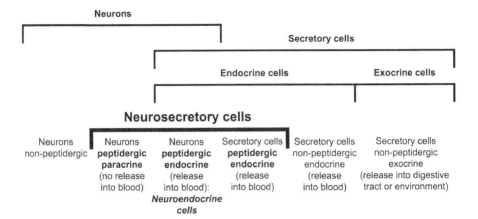

FIGURE 5.1 Terminology for neurosecretory cells. The term "peptidergic" is here used broadly to denote cells secreting protein-, peptide-, or amino acid-based hormones. See text for details.

Although most neurons interact with other cell types by releasing chemicals known as neurotransmitters, neurons and secretory cells are usually considered distinct cell types based on functional and morphological considerations. Neurons are equipped with cellular processes such as axons and dendrites enabling them to transmit electrical signals and communicate with other cells via specialized synapses, while secretory cells lack such processes and synapses. Secretory cells may have either exocrine or endocrine functions. Exocrine cells secrete enzymes or other products into ducts connected to the digestive tract or the outside world. In contrast, endocrine cells secrete their products directly into the blood (or hemolymph or other body fluids), where they will be distributed throughout the body and potentially act as hormones on distant cells.

The hormones secreted by endocrine cells are typically derived either from lipids or from proteins (reviewed in Hartenstein 2006; Norris and Carr 2020). Some molecules derived from single amino acids such as serotonin or the catecholamines (epinephrine, norepinephrine, dopamine), may also serve as hormones in addition to their function as classical neurotransmitters. Lipid-based hormones include, for example, the steroid hormones, which are derived from cholesterol. Protein-based hormones include the glycoprotein hormones (TSH, LH, FSH) and four-helix cytokine-like hormones (GH, PRL) of the anterior pituitary as well as neuropeptides. Apart from the neuropeptides produced by the anterior pituitary (ACTH, αMSH) there is a large diversity of neuropeptides produced by other cells in the brain, gut, and many other organs. Neuropeptides are made from larger precursor proteins (prohormones) that are cut into smaller pieces by special proteolytic enzymes (prohormone convertases). As explained in Chapter 8 of Volume 1 (fig. 8.4; Schlosser 2021), some prohormones such as proopiomelanocortin (POMC) can yield multiple peptides after processing; and the kinds of peptides produced can vary in a tissue-specific manner. After synthesis in the rough endoplasmic reticulum and processing in the Golgi apparatus, protein-based hormones are stored in relatively large dense-core vesicles (DCV) (Burgoyne and Morgan 2003). These differ from the small, translucent synaptic vesicles, in which classical neurotransmitters are stored by their large size and granular content; they are also usually distributed widely throughout the cell and not locally concentrated at synapses (Kim et al. 2006; Scalettar 2006).

It has long been known that cells that secrete neuropeptides (and possibly other protein- or amino acid-based hormones) are not restricted to the endocrine cells of the anterior pituitary and other endocrine glands. The same neuropeptides are also released by many neurons of the nervous system (CNS) and peripheral nervous system (PNS) as well as in specialized cells of the endodermally-derived digestive tract (enteroendocrine cells) (Norris and Carr 2020). The enteroendocrine cells of the gut and their cousins in pancreas and liver (which develop as outpocketings of the embryonic gut) release peptide hormones in response to changing nutrient levels in the digestive tract, which then regulate processes ranging from gut motility and glucose metabolism to appetite and other metabolic processes (Rindi et al. 2004; Gribble and Reimann 2016; Latorre et al. 2016).

Many neurons produce neuropeptides which then act on adjacent neurons or other target cells (i.e. in a paracrine fashion) and modulate their response to classical neurotransmitters released by the same or other neurons (reviewed in Bargmann 2012; Marder 2012). However, other neuropeptidergic neurons release neuropeptides into adjacent blood vessels which distribute them to more distant target cells. Neuropeptides in this case act as proper hormones on distant cells (i.e. in an endocrine fashion). Therefore, these neurons are classified as "neuroendocrine" cells (Fig. 5.1). In the vertebrate brain, many of these neuroendocrine cells are found in the hypothalamus (reviewed in Alvarez-Bolado 2019; Norris and Carr 2020). One population of hypothalamic neuroendocrine cells sends axons to the posterior pituitary where they release peptide hormones (adiuretin and oxytocin) into the blood stream. Another population sends its axons to the so-called median eminence, a stalk-like region where the anterior pituitary is attached to the hypothalamus (see fig. 1.11B in Schlosser 2021). There, the axons secrete a number of neuropeptides known as releasing hormones (RH) into the blood stream, which carries them via the hypophyseal portal vein to the anterior pituitary, where each RH stimulates hormone production of a specific endocrine cell type – GnRH, for example, stimulates gonadotropes, while TRH stimulates thyrotropes (fig. 1.11B in Schlosser 2021).

The endocrine cells of the anterior pituitary or other endocrine glands as well as the enteroendocrine cells of the digestive tract lack axons, dendrites and synapses. They are, therefore, not considered proper neurons and, thus, are not "neuroendocrine" cells in the sense defined here. However, in addition to the secretion of neuropeptides or other protein-based hormones, many endocrine cells also share other features with neurons. The endocrine cells of the anterior pituitary, for example, are excitable and form electrically and/or chemically synchronized networks of hormone release as explained in Chapter 8 of Volume 1 (Schlosser 2021). Some enteroendocrine cells of the gut, in turn, were recently shown to form synapses with vagal neurons via short cytoplasmic processes called neuropods and send information about intestinal glucose levels to the brain, suggesting that they are modified neurons (Bohorquez et al. 2015; Kaelberer et al. 2018). To emphasize the similarities between the various cells that produce protein-, peptide-, or amino acid-based hormones, I propose here to use the term "neurosecretory cells" for these cells. Neurosecretory cells, so defined, include (1) the subset of non-neuronal endocrine cells, for example, cells of the anterior pituitary and enteroendocrine cells, producing protein-, peptide-, or amino acid-based hormones (excluding those producing steroids or other lipid-based hormones); (2) neuroendocrine cells, i.e. neurons that produce such hormones and release them into the blood stream; (3) paracrine neurons that produce such hormones but do not release them into the blood stream (Fig. 5.1).

Although some open questions remain, there is evidence that all neurosecretory cells, including those that are not neurons, not only share neuron-like properties like excitability and neuropeptide-release, but also require proneural bHLH transcription factors (see Chapter 5 in Schlosser 2021) for specification and initiation of cytodifferentiation. These proneural factors are related to either *Drosophila* achaete and scute (Achaete/Scute-like proteins) or atonal (Atonal/Neurogenin/Neurod-like proteins). Members of both the Ascl and Atonal superfamilies of bHLH transcription factors have been identified in sponges (Simionato et al. 2007) indicating that they were present already in early metazoans.

Atoh1 in mammals and Ascl1 in zebrafish have been shown to be required for all secretory cell types in the digestive tract except for those producing digestive enzymes but including the mucus-producing goblet cells and enteroendocrine cells. In contrast, Neurogenin3 (Neurog3) is specifically required for the differentiation of most enteroendocrine cells (Yang et al. 2001; Jenny et al. 2002; López-Díaz et al. 2007; Flasse et al. 2013; Roach et al. 2013; Gribble and Reimann 2016). Downstream of Neurog3, other typically "neuronal" transcription factors such as NeuroD1 and Islet1 are required for differentiation of enteroendocrine cells (Gribble and Reimann 2016). Similarly, as discussed in Chapter 8 of Volume 1 (Schlosser 2021), Ascl1 and other, Atonal-related bHLH transcription factors are required for differentiation of endocrine cell types in the anterior pituitary, in a partly redundant fashion (Pogoda et al. 2006; Ando et al. 2018). Taken together, this suggests that in vertebrates proneural bHLH transcription factors play essential but possibly redundant roles for the differentiation not only of neurons but also of neurosecretory cells that are not neurons including the endodermally derived enteroendocrine cells.

In *Drosophila*, homologs of the vertebrate Ascl (*Drosophila* Scute, Asense) and Neurog (*Drosophila* Tap) transcription factors are likewise required for the specification of endodermal enteroendocrine cells (Amcheslavsky et al. 2014; Wang et al. 2015; Hartenstein et al. 2017). Although experimental data from other animals are missing, both neuropeptide expression and expression of proneural bHLH genes have been found in cells scattered throughout the endoderm in many bilaterians and cnidarians (Hartenstein 2006; Hartenstein et al. 2017). It has, therefore, been proposed that neurons and neurosecretory cells (including those that are not neurons) may comprise a single cell type family, which possibly originated from a common cell type in the last common ancestor of eumetazoans, followed by diversification into different neuronal and/or neurosecretory lineages (Fig. 5.2) (Hartenstein 2006; Hartenstein et al. 2017). Stepping back further in time, neuronal/neurosecretory cells

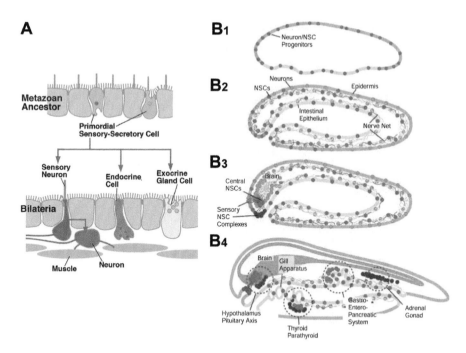

FIGURE 5.2 Evolutionary origin of neurosecretory cells. (**A**) Diagram illustrating hypothetical common origin of neurons, endocrine, and exocrine cells. (**B**) Scenario for the evolution of neurons and other neurosecretory cells (NSC). (**B₁**) Endocrine secretory cells are likely to predate the appearance of a nervous system, since they can be found in extant metazoa lacking nerve cells. (**B₂**) The first nervous system is thought to have been a basi-epithelial nerve net, similar to the one still found in present day cnidarians. At this stage, neurons and non-neuronal NSCs most likely had evolved into distinct lineages. (**B₃**) A central nervous system integrating multimodal sensory input evolved either once or several times independently in bilaterian animals. Specialized populations of neurosecretory cells (e.g. involved in the regulation of feeding and reproduction) evolved in the brain, pharynx, and gut. (**B₄**) Neurosecretory cells tend to form dedicated glands in the chordate lineage. ([A] Reprinted with permission from Hartenstein et al. 2017; [B] reprinted with permission from Hartenstein 2006.)

may even share a common cellular ancestor with exocrine cells given the many similarities in the mechanisms of vesicle formation and release (Burgoyne and Morgan 2003; Kim et al. 2006).

5.2 EVOLUTIONARY ORIGINS OF NEUROSECRETORY CELLS

While neurons and neurosecretory cells may well form an ancient cell type family, there is little evidence that within this family neurosecretory cells form a well-defined and evolutionarily stable subfamily, distinct from non-secretory neurons. Many neurosecretory cells express combinations of multiple neuropeptides and these combinations vary for different neurosecretory cells and are, most likely, evolutionarily

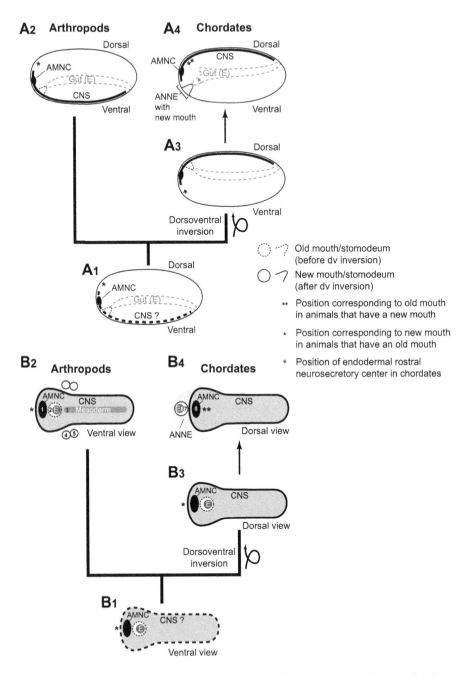

FIGURE 5.3 Comparison of neurosecretory centers in protostomes and extant chordates assuming dorsoventral inversion of the body in stem chordates. Lateral views (**A**) and dorsal or ventral views (**B**) of the hypothetical bilaterian ancestor (**A₁, B₁**) and of arthropods (**A₂, B₂**) and chordates (**A₃, A₄, B₃, B₄**) as representatives of protostomes and deuterostomes, respectively. The position of the anteromedial neurosecretory center (AMNC), the central nervous

FIGURE 5.3 (Continued)

system (CNS) and the mouth region– with ectodermal stomodeum surrounding the opening into the endodermal (E) gut – is shown. Dorsoventral (dv) inversion at the base of chordates is indicated. A_3 and B_3 show a hypothetical situation after dv inversion, but prior to the formation of a new mouth (whereas in reality the formation of a new mouth may have preceded dv-inversion or evolved in parallel to it). The outline of the CNS in the bilaterian ancestor is hatched because it is not clear whether a CNS was present in this ancestor that was inherited by protostomes and deuterostomes, or whether a CNS has evolved several times independently in chordates and some protostomian groups such as arthropods and annelids. Note that the old mouth inherited from the bilaterian ancestor penetrates the anterior CNS in many extant protostomes (e.g. arthropods, annelids). The new mouth forming after dorsoventral inversion is located in the anterodorsal non-neural ectoderm (ANNE), immediately anterior to the CNS. For orientation, the position of new and old mouths before and after dv inversion are indicated by single or double asterisks respectively. Numbers in B_2 indicate regions of origin for the following neurosecretory organs in arthropods: 1, pars intermedia and lateralis of brain (from AMNC); 2, stomatogastric nervous system (from stomodeal ectoderm surrounding old mouth); 3, corpora cardiaca (from anteroventral furrow of mesoderm); 4, corpora allata (from non-neural ectoderm); 5, prothoracic gland (from non-neural ectoderm). Numbers in B_4 indicate regions of origin for the following neurosecretory organs in chordates: 6, rostral neurosecretory center in amphioxus (Hatschek's pit); 7, adenohypophysis in vertebrates (from placode in stomodeal ectoderm surrounding new mouth); 8, hypothalamus in vertebrates (from AMNC). ([B] B_2 based on de Velasco et al. 2006 and Hartenstein 2006.)

flexible. Moreover, many neuropeptides and other protein hormones are produced by multiple different types of neurons and/or endocrine cells, each specified by different core regulatory networks (CoRNs) of transcription factors. This is best studied in the vertebrate hypothalamus, where the transcription factor profiles expressed by different neurosecretory cells have been mapped in detail (Morales-Delgado et al. 2011, 2014; Díaz, Morales-Delgado, and Puelles 2015) and their role has been assessed in functional studies (reviewed in Alvarez-Bolado 2019).

These studies have shown, to briefly sketch only one example, that the transcription factors Otp (the vertebrate homolog of *Drosophila* Orthopedia) and Sim1 as well as their downstream target POU3f2 (=Brn2) are required for specification of a large subset of neurosecretory cells in the hypothalamus including neuroendocrine cells (Michaud et al. 1998; Acampora et al. 1999; Wang and Lufkin 2000; Alvarez-Bolado 2019). While Otp is required together with Sim1 for the specification of one population of neurosecretory cells releasing the neuropeptide somatostatin (residing in the parvocellular nucleus of the hypothalamus), only Otp but not Sim1 is essential for another population (arcuate nucleus). However, cells expressing somatostatin or other neuropeptides in the gut develop in the absence of Otp or Sim1 (Simeone et al. 1994; Ema et al. 1996). This suggests that somatostatin producing cells in different locations are specified by different CoRNs. Space constraints do not allow me to elaborate this further here. For more details and additional examples, I refer the reader to previous studies (Morales-Delgado et al. 2011, 2014; Díaz et al. 2015; Alvarez-Bolado 2019).

Overall, this and many other case studies indicate that there is a complex and evolutionarily labile relationship between the transcription factors specifying neurosecretory and neuronal cell type identity and the types of hormones produced. This suggests that neurosecretory cells producing the same or related neuropeptides or other protein-based hormones may have evolved convergently multiple times from different neuronal or endocrine cell types.

Recent comparisons of neurosecretory cell types between different bilaterian taxa suggest, however, that there may be an important exception to this general rule after several core regulators of neurosecretory development in the vertebrate hypothalamus were found to be co-expressed in a group of anteromedial neurosecretory cells in other deuterostomes and protostomes. These core regulators comprise Nk2.1, Otp, and Sim1/2 which together with other transcription factors involved in anterior regional patterning (e.g. Pax4/6, Six3/6, Fezf1/2, Rx, Gsx) play central roles for the specification of neurosecretory cells in the hypothalamus (Kimura et al. 1996; Michaud et al. 1998; Acampora et al. 1999; Wang and Lufkin 2000; Manoli and Driever 2014; Orquera et al. 2019; Alvarez-Bolado 2019).

Several studies have now identified groups of neurosecretory cells expressing a similar profile of transcription factors in a topographically corresponding anteromedial position (see Fig. 5.3) in the brain of *Platynereis,* in the pars intermedia and lateralis of the *Drosophila* brain, and in the brain of centipedes (Tessmar-Raible et al. 2007; De Velasco et al. 2007; Hunnekuhl and Akam 2014). In the chordates *Ciona* and amphioxus, cells with a similar profile (NK2.1, Six3/6, Fezf1/2, Rx) are localized in the anterior neural tube although Sim and Otp in amphioxus and Nk2.1 in *Ciona* are expressed in more posterior domains (Venkatesh et al. 1999; Moret et al. 2005; D'Aniello et al. 2006; Joly et al. 2007; Kozmik et al. 2007; Del Giacco et al. 2008; Irimia et al. 2010; Vopalensky et al. 2012; Li, Lu, and Yu 2014; Albuixech-Crespo et al. 2017; Oonuma and Kusakabe 2019).

It was subsequently shown that neurosecretory (neuropeptidergic) cells in the apical organ of *Platynereis* and of other lophotrochozoan and cnidarian larvae also express Nk2.1, Otp, Six3/6, Fezf, and Rx as well as some additional transcription factors (Hbn, FoxQ2) that are absent from the mammalian genome (Fig. 5.4) (Chia and Koss 1979; Steinmetz et al. 2010; Nakanishi et al. 2012; Santagata et al. 2012; Sinigaglia et al. 2013; Conzelmann et al. 2013, Marlow et al. 2014; Williams et al. 2017). These cells later become incorporated into the anterior brain just rostral to the mouth opening, thereby contributing to the anteromedial neurosecretory brain center (Tessmar-Raible et al. 2007; Marlow et al. 2014; Tosches et al. 2014; Williams et al. 2017). The apical organ of larvae in deuterostomes (hemichordates

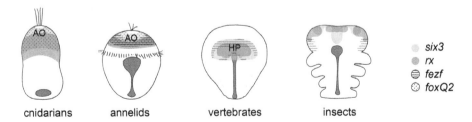

FIGURE 5.4 Expression of selected transcription factors in the apical organ and anteromedial neurosecretory brain centers of different eumetazoan embryos. The apical organ (AO) in cnidarian and annelid larvae and the hypothalamus primordium (HP) in the vertebrate embryo are indicated. (Reprinted with permission from Tosches and Arendt 2013.)

and echinoderms) also contains neurosecretory cells and expresses Nk2.1, Otp, Six3, Fezf, Rx, FoxQ2, and Hbn (Takacs et al. 2004; Dunn et al. 2007; Yankura et al. 2010; Marlow et al. 2014; Mayorova et al. 2016; Gonzalez, Uhlinger, and Lowe 2017; Valencia et al. 2019).

Taken together, this suggests that a group of anteromedial neurosecretory cells regulated by a conserved network of transcription factors and possibly derived from the apical organ may be an ancestral feature of eumetazoans (Fig. 5.3). In vertebrates, this anteromedial group of cells may have diversified and contributed to the neurosecretory cells of the vertebrate hypothalamus (Tessmar-Raible 2007; Tosches and Arendt 2013; Arendt, Tosches, and Marlow 2016).

In this book, however, I am mainly concerned with the neurosecretory cells of the vertebrate adenohypophysis, which are derived, like cranial sensory cells and neurons, from cranial placodes. While deep evolutionary roots have also been proposed for these cells (e.g. De Velasco et al. 2004; Hartenstein 2006; Wirmer, Bradler, and Heinrich 2012), I will use the remainder of this chapter to argue against this hypothesis. As I will show below, all hormones secreted by the vertebrate adenohypophysis are evolutionarily novel and have evolved only in the vertebrate lineage. Furthermore, while hormones distantly related to the vertebrate hormones have been found in other bilaterians, these are not expressed in cell types sharing a common profile of transcription factors with the vertebrate adenohypophysis outside of the chordates. Many of the CoRN transcription factors known to regulate differentiation of adenohypophyseal endocrine cells in a combinatorial fashion (including Pitx, Lhx3/4, Islet, Prop1, Pit1, Tbx19, FoxL2, GATA2; see Chapter 8 in Schlosser 2021), have only begun to be co-expressed in anterior endoderm in early chordates, where some of them may have adopted an essential role for specification of neurosecretory cells. They were then recruited to anterior non-neural ectoderm in the tunicate-vertebrate ancestor, and have adopted their unique roles in the specification of different types of neurosecretory cells only in vertebrates. Similarly, the neurons producing GnRH and other neuropeptides that are derived from the olfactory placode in vertebrates (see Chapter 6 in Schlosser 2021), probably were established as a specific individualized cell type only in the tunicate-vertebrate ancestor.

5.3 NEUROSECRETORY CELLS IN THE LAST COMMON TUNICATE-VERTEBRATE ANCESTOR

5.3.1 EVOLUTIONARY ORIGIN OF HORMONES PRODUCED IN ADENOHYPOPHYSEAL AND OLFACTORY NEUROSECRETORY CELLS

The neurosecretory cells of the vertebrate adenohypophysis produce hormones of three different families: peptide hormones, dimeric glycoprotein hormones and four-helix cytokine-like proteins (Fig. 5.5; for details see Chapter 8 in Schlosser 2021) (reviewed in Campbell, Satoh, and Degnan 2004; Kawauchi and Sower 2006; Norris and Carr 2020). The peptide hormones adrenocorticotropic hormone (ACTH) and alpha melanocyte stimulating hormone (α-MSH) are produced by the proteolytic processing from the same precursor protein POMC. They are released by cell types known as corticotropes (ACTH) and melanotropes (α-MSH). The four-helix

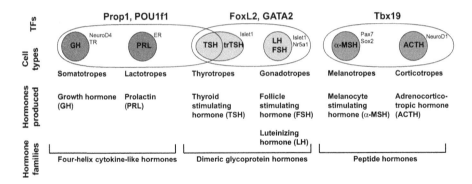

FIGURE 5.5 Neurosecretory cell types and hormone classes in the anterior pituitary of vertebrates. Hormones belonging to the same families are shown in the same shades of grey. Generally hormones belonging to the same family are specified by the same transcription factors (TFs): Prop1, POU1f11 (Pit1): four-helix cytokine like hormones GH and PRL; FoxL2, GATA2: dimeric glycoprotein hormones TSH, FSH, LH; Tbx19: peptide hormones MSH and ACTH. An exception to this rule is a subpopulation of TSH producing cells that are also dependent on Prop1 and POU1f1. trTSH, transitory thyrotropes. (Based on Zhu, Gleiberman, and Rosenfeld 2007; Kelberman et al. 2009; and Rizzoti 2015.)

cytokine-like hormones comprise growth hormone (GH) and prolactin (PRL). These are produced by somatotropes (GH) and lactotropes (PRL). Finally, the dimeric glycoprotein hormones (consisting of a common alpha subunit and specific beta subunits) include thyroid stimulating hormone (TSH), luteinizing hormone (LH), and follicle stimulating hormone (FSH). These are produced by thyrotropes (TSH) and gonadotropes (LH, FSH).

Genes encoding these adenohypophyseal hormones or their receptors could not be identified in genomes of tunicates, amphioxus, or other animals suggesting that they have evolved only in the vertebrate lineage (Dehal et al. 2002; Campbell et al. 2004; Holland et al. 2008; Putnam et al. 2008). One possible exception may be a GH-like protein and an associated receptor recently reported for amphioxus (Li et al. 2014, 2017). However, the phylogenetic relationships of this four-helix cytokine like protein is currently not strongly supported and further evidence is needed to confirm its position as the closest relative of vertebrate GH and PRL (Ocampo Daza and Larhammar 2018).

Although earlier studies reported the presence of cells immunoreactive for various adenohypophyseal hormones in tunicates and amphioxus, these findings may have reflected cross-reactivity of the antibodies used with other, possibly related hormones (e.g. Fritsch, Van Noorden, and Pearse 1982; Pestarino 1983, 1984; Thorndyke and Georges 1988; Nozaki and Gorbman 1992; Terakado et al. 1997; reviewed in Sherwood, Adams, and Tello 2005; Schlosser 2005). The isolation of POMC-related peptides from some protostomes may in turn be an artefact due to contamination with ingested vertebrate tissues (Duvaux-Miret et al. 1990; Salzet et al. 1997; Stefano, Salzet-Raveillon, and Salzet 1999).

Comparative studies indicate that the vertebrate-specific hormones of the adeno-hypophysis (Fig. 5.5) evolved from other members of the same hormone families, which can be traced back to the earliest metazoans (Campbell et al. 2004; Jékely 2013; Mirabeau and Joly 2013; Roch and Sherwood 2014). The α- and β-subunits of the *dimeric glycoproteins* (LH, FSH, TSH) of the anterior pituitary, GPH-α (also known as α-GSU) and GPH-β, evolved from the corresponding subunits, GPA2 and GPB5, of the glycoprotein thyrostimulin after gene duplications in the vertebrate lineage. GPH-β is still found in lampreys but gave rise to the β-subunits of LH, FSH, TSH in gnathostomes (see Chapter 8 in Schlosser 2021). Orthologs of GPA2 and GPB5 and of the glycoprotein-binding subfamily of G-protein coupled receptors (GPCR) can be dated back at least to the bilaterian ancestor (Park et al. 2005; Sudo et al. 2005; Dos Santos et al. 2009; Hausken et al. 2018).

Members of the *four-helix cytokine-like hormones* (GH, PRL) and their receptors have also been identified in protostomes although little is currently known about their evolutionary history (Huising, Kruiswijk, and Flik 2006). Finally, the *POMC-derived neuropeptides* (ACTH, aMSH, opioids) of the anterior pituitary bind to receptors of the γ-rhodopsin (opioid receptors) and α-rhodopsin (melanocortin receptors for ACTH and MSH) subfamilies of GPCRs; these subfamilies and corresponding neuropeptide ligands were present in the last common ancestor of bilaterians (Fredriksson and Schiöth 2005; Dores and Baron 2011; Mirabeau and Joly 2013; Sundstrom, Dreborg, and Larhammar 2010; Elphick, Mirabeau, and Larhammar 2018).

In contrast to the adenohypophyseal hormones, none of which are present outside of vertebrates, the neuropeptide GnRH, which is released by olfactory placode-derived neurosecretory neurons, is more ancient and shares a common evolutionary origin with adipokinetic hormone (AKH), corazonin (CRZ), and AKH/CRZ-related peptide (ACP) in protostomes (Roch, Busby, and Sherwood 2011; Sakai et al. 2017; Elphick et al. 2018). These neuropeptides signal via GnRH-type receptors and CRZ-type receptors, which probably originated in early bilaterians as subfamilies of β-rhodopsin type GPCRs (Roch et al. 2011; Roch, Tello, and Sherwood 2014; Sakai et al. 2017; Elphick et al. 2018).

5.3.2 NEUROSECRETORY CELL TYPES IN TUNICATES

In tunicates, the distribution of dimeric glycoproteins or four-helix cytokine-like proteins has not been described. However, many neuropeptides have been identified. These include GnRH and other peptides belonging to evolutionarily conserved families as well as species-specific peptides (Matsubara et al. 2016). In *Ciona*, there are six GnRH peptides encoded by two genes as well as four GnRH receptors (Adams et al. 2003; Roch et al. 2011; Sakai et al. 2017). These have multiple functions and are involved in increasing water flow, promoting gamete maturation and release, and initiating metamorphosis (Terakado 2001; Adams et al. 2003; Sakai et al. 2010; Sakai et al. 2012; Kamiya et al. 2014; Hozumi et al. 2020).

Neurosecretory cells releasing various neuropeptides and expressing prohormone convertases are found in a number of different tissues including gonads and gut, but are mostly concentrated in the CNS of the larva and the cerebral ganglion and neural gland of the adult (Fig. 5.6) (Thorndyke and Georges 1988; Bollner, Beesley, and Thorndyke 1992; Schlosser 2005; Hamada et al. 2011; Matsubara et al. 2016; Osugi, Sasakura, and Satake 2017, 2020). A small population of neurosecretory cells is also

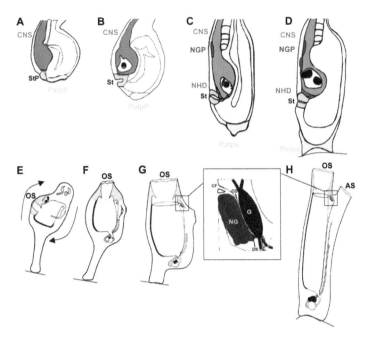

FIGURE 5.6 Development of ascidian adult cerebral ganglion and neural gland from larval central nervous system. Diagram of changes in the nervous system during larval development (**A–D**) and metamorphosis (**E–H**) of the ascidian *Ciona intestinalis*. Precursor cells arising from the larval central nervous system (CNS) give rise to the neural complex of the adult comprising cerebral ganglion (G) and neural gland (NG). The latter is connected via ciliary duct (CD) and ciliary funnel (CF) to the oral siphon (OS), which develops from the larval stomodeum (St). See text for details. AS, atrial siphon; CNS, larval central nervous system; DS, dorsal strand; NGP, neural gland primordium (NG may arise from entire larval CNS); NHD, neurohypophyseal duct; OS, oral siphon; PalpP, palp primordium; St, stomodeum; StP, stomodeum primordium. (Reprinted with permission from Joly et al. 2007; adapted from Bollner et al. 1992 and Chiba et al. 2004.)

found in the adult ciliary funnel and the ciliary duct connecting the ciliary funnel to the neural gland (Bollner et al. 1992). Fate mapping studies in *Ciona* indicate that cerebral ganglion and neural gland arise from the sensory vesicle and visceral ganglion of the larval CNS (Dufour et al. 2006; Joly et al. 2007; Horie et al. 2011). They also suggest that the anteriormost part of the larval CNS (the neurohypophyseal duct connecting the sensory vesicle to the oral siphon primordium) may contribute to the ciliated funnel, although previous studies suggested that ciliated funnel and duct arise from the larval oral siphon primordium (Fig. 3.2) (Manni et al. 2004, 2005).

GnRH secreting cells are predominantly localized to the larval brain, the adult cerebral ganglion and to a few other adult cell populations (dorsal strand, dorsal strand plexus) derived from the larval CNS (Fig. 5.6) (Tsutsui et al. 1998; Terakado 2001; Adams et al. 2003; Kavanaugh, Root, and Sower 2005; Kawada et al. 2009; Hamada et al. 2011). However, recent studies have found GnRH to be expressed in additional populations of neurons in the larva, the palp sensory neurons (PSNs) and aATENs of the peripheral nervous system (see Chapter 3; Fig. 3.3) (Kusakabe et al. 2012; Abitua et al. 2015; Horie, Hazbun et al. 2018). These neurons originate from cells in the non-neural ectoderm bordering the anterior neural plate (Horie, Horie et al. 2018).

Two additional cell types of the palp region (Figs. 3.2, 5.6), the so-called collocytes and axial columnar cells, probably derive from a common precursor with PSNs (Zeng et al. 2019). Collocytes are exocrine cells secreting mucus, which is thought to mediate substrate adhesion, like the cells of the vertebrate cement gland (Cloney 1977; Dolcemascolo et al. 2009; Zeng et al. 2019). The axial columnar cells harbor vesicles suggestive of a neurosecretory function (Zeng et al. 2019). Despite their primarily exocrine function, collocytes express many neuronal markers. Moreover, their profile of transcription factor expression clusters with PSNs, another PSN-related cell type (possibly corresponding to axial columnar cells) and other neurons in single-cell RNA-Seq analyses (Figs. 3.4, 3.5) (Cao et al. 2019; Zeng et al. 2019). This suggests that collocytes and possibly axial columnar cells are probably specialized neurosecretory cells. To evaluate how these cells may be related to rostral neurosecretory cells in amphioxus and vertebrates, we need to have a closer look at transcription factor expression profiles first.

5.3.3 TRANSCRIPTION FACTORS INVOLVED IN SPECIFYING TUNICATE NEUROSECRETORY CELL TYPES

Do any of the neurosecretory cells in *Ciona* belong to a cell type related to the neurosecretory cells in the adenohypophysis or the GnRH producing cells derived from the olfactory placode? The CoRN of transcription factors required for specifying adenohypophyseal neurosecretory cells in vertebrates has been well studied and was described in Chapter 8 of Volume 1 (fig. 8.5 in Schlosser 2021). To recapitulate briefly (Fig. 5.5), Pitx, Lhx3/4, and Islet1 are required for progenitor development of all neurosecretory lineages. Pitx and Lhx3/4 subsequently continue to play important roles for the differentiation of all neurosecretory lineages while Islet1 expression is only required for differentiation of gonadotropes and a subset of thyrotropes. Specification of all lineages releasing POMC-derived neuropeptides, i.e. melanotropes and somatotropes then additionally requires Tbx19. Specification of all lineages secreting dimeric glycoprotein hormones, i.e. gonadotropes and thyrotropes, in turn depends on FoxL2 in combination with GATA2. Finally, all lineages secreting four-helix cytokine-like hormones, i.e. somatotropes and lactotropes, as well as the majority of thyrotropes require Prop1 and POU1f1 (Pit1). Additional transcription factors (e.g. Nr5a1, ER, NeuroD4, NeuroD1, Pax7) help to specify individual cell types within these lineages, but I will ignore these here to avoid overcomplication of matters. Much less is known about the CoRN specifying the GnRH producing neurosecretory cells derived from the olfactory placode, although Islet1 may be involved here as well (see Chapter 8 in Schlosser 2021). Transcription factors that are specifically expressed in the olfactory, but not adenohypophyseal placode like FoxG, DMRT4/5, and Sp8 may help to distinguish olfactory placode-derived neurosecretory cells including the GnRH cells from those derived from the adenohypophyseal placodes (see Chapter 4 in Schlosser 2021).

Some of these transcription factors have clearly no comparable function in tunicates. Tbx19 is encoded by a gene that arose by duplication of the *Brachyury (T)* gene either in vertebrates or early chordates, but is absent from both *Ciona* and amphioxus genomes (Ruvinsky, Silver, and Gibson-Brown 2000; Papaioannou 2014; Inoue et al. 2017). *POU1f1* is also absent from the *Ciona* genome and Prop1 is expressed only in a rare CNS cell type (Cao et al. 2019). FoxL2/3 has an evolutionarily ancient role in the sexual differentiation of gonads, but its expression has not been described in tunicates, amphioxus or ambulacrarians (Bertho et al. 2016).

However, all other transcription factors involved in the specification of adenohypophyseal cells in vertebrates are expressed in tunicates in a region of ectoderm that I will term the anterior dorsal non-neural ectoderm (ANNE), located immediately adjacent to the anterior neural plate (Fig. 5.3). In tunicates, the ANNE comprises the palp forming region and the oral/preoral ectoderm, mostly arising from rows V and VI of ectodermal cells in the neural plate region (see Fig. 3.2). This is in striking contrast to amphioxus, where most of these transcription factors are confined to the endomesoderm (see below). The medial part of the ANNE domain (derived from a8.18 and a8.20; Fig. 3.2) gives rise to the palp region with the neurosecretory PSNs and collocytes, while the more lateral part (a8.26) gives rise to the GnRH producing aATEN sensory neurons (Horie, Hazbun et al. 2018). The expression of transcription factors in these cells is summarized and compared with vertebrates in Fig. 3.4.

Pitx is expressed in the palp region as well as in the oral siphon primordium (OSP), scattered cells in the CNS and broadly in the left endomesoderm (Boorman and Shimeld 2002; Christiaen et al. 2002; Yoshida et al. 2012; Cao et al. 2019). Lhx3/4 is expressed in the collocytes and several sensory neurons (but not PSNs), in addition to the posterior CNS and the early endoderm, where it is required for endoderm differentiation (Satou, Imai, and Satoh 2001; Ikuta and Saiga 2007; Kobayashi et al. 2010; Cao et al. 2019). Interestingly, Six1/2, Eya, Islet and GATA1/2/3, which are also expressed in the ANNE, likewise show endodermal expression (Giuliano et al. 1998; Mazet et al. 2005; Imai et al. 2004, 2006; Abitua et al. 2015; Ogura and Sasakura 2016; Cao et al. 2019). While ectodermal Six1/2 is restricted to sensory/neurosecretory cells derived from the ANNE (aATENs and PSNs), Islet and GATA1/2/3 (known as GATAb in *Ciona*) have a wider distribution. Islet is not only expressed in collocytes, PSNs and aATENs, but also in other sensory neurons and parts of the CNS (Fig. 3.4). GATAb, while being enriched in the ANNE, is widely expressed in the general non-neural ectoderm.

In vertebrates, Islet1, Six1/2 and Eya, apart from their role in the anterior pituitary (see Chapter 8 in Schlosser 2021), are similarly required for the differentiation of some enteroendocrine cells (Xu et al. 2002; Zou et al. 2006; Terry et al. 2014). In the pharyngeal endoderm, Six1 and Eya are expressed in the anlagen of the thymus and parathyroid glands, the ultimobranchial bodies (producing the calcitonin secreting C-cells of the thyroid), and in pulmonary neuroendocrine cells (PNEC), and appear to be required for their proper morphogenesis and differentiation (Xu et al. 2002; Zou et al. 2006; El Hashash et al. 2011; Travaglini et al. 2020). Furthermore, as we will see below, in amphioxus the expression of Pitx, Lhx3, Six1/2, Islet, and GATA1/2/3 is mostly endomesodermal and these transcription factors are co-expressed with another adenohypophyseal transcription factor, POU1f1, in a rostral outpocketing of the gut (Hatschek's left diverticulum). This raises the interesting question whether transcription factors originally involved in the specification of some endomesodermal cells in chordate ancestors were recruited to the anterior non-neural ectoderm in the tunicate-vertebrate ancestor (Fig. 5.7).

Recent findings concerning the *Ciona* ortholog of vertebrate FoxA1/2/3 (=HNF3α/β/γ) transcription factors support this hypothesis. Like in vertebrates, *Ciona* FoxA1/2/3 is strongly expressed in the developing endoderm and notochord and required for notochord development (Fig. 5.7) (Imai et al. 2004, 2006; Passamaneck et al. 2009; José-Edwards et al. 2015). In addition, however, FoxA1/2/3 is also transiently required (together with the ectodermal competence factor Dlx2/3/5) in early ectodermal precursors for the specification of the ANNE and the upregulation of GATA1/2/3 in this territory (Lamy et al. 2006; Imai et al. 2006).

FoxA transcription factors are known as key regulators of endoderm and notochord development in vertebrates and are thought to act as early pioneer factors endowing cells with the competence to differentiate into endodermal fates (Friedman and Kaestner 2006). At later stages of embryonic development, vertebrate FoxA1 and FoxA2 are specifically required for differentiation of enteroendocrine cells (Ye and Kaestner 2009). Together with transcription factor SPDEF, they are also essential for differentiation of mucus-secreting goblet cells in the endodermal digestive tract (Ye and Kaestner 2009; Chen et al. 2009, 2018; Noah et al. 2010; McCauley and Guasch 2015). However, FoxA factors are also expressed in stomodeum, olfactory epithelium, and adenohypophysis (Besnard et al. 2004; Giri et al. 2017; Chen et al. 2017). While their functions in these ectodermal territories are not well characterized, it has recently been shown that they play important roles for the development of endocrine cells in the adenohypophysis (Besnard et al. 2004; Giri et al. 2017); for differentiation of a subpopulation of small secretory cells in the amphibian epidermis (Dubaissi et al. 2014; Walentek et al. 2015); and for mouth development (Chen et al. 2017). They are also expressed together with SPDEF in putative goblet cells of ectodermally derived mucosae, suggesting that they are required for goblet cell differentiation in both endoderm and ectoderm (Gupta et al. 2011; Chen et al. 2018). Since FoxA and its downstream targets such as GATA1/2/3 are only endomesodermally expressed in amphioxus (see section 5.4 below), they probably were recruited to the ANNE domain only in the last common ancestor of tunicates and vertebrates (Fig. 5.7).

Interestingly, in vertebrates, another transcription factor, Hesx1 (=ANF1) is specifically expressed in the anterior endomesoderm in early embryos, where it is required for its own subsequent upregulation in the ANNE domain (Zaraisky et al. 1995; Thomas and Beddington 1996; Kazanskaya et al. 1997; see Chapter 4 in Schlosser 2021). Since ANF-family transcription factors have only evolved in vertebrates (Kazanskaya et al. 1997), they may have taken over some of the functions of other transcription factors like FoxA specifically in the anterior endomesoderm and ANNE domain.

In addition to mostly endomesodermal transcription factors like FoxA and GATA1/2/3, the ANNE domain in tunicates also retains the expression of FoxG, Sp6-9, and DMRT4/5 (DMRTa) (Fig. 3.4). These transcription factors are specifically expressed in the anterior-most neural plate and adjacent non-neural ectoderm in vertebrates and are required for the development of olfactory placodes and the

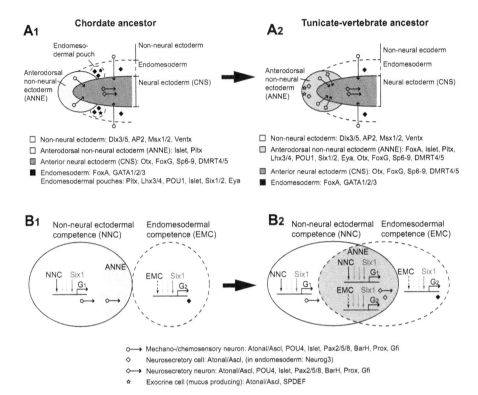

FIGURE 5.7 Model for evolution of new neurosecretory cell types in the anterior dorsal non-neural ectoderm (ANNE). Mechano-/chemosensory, neurosecretory (both neuronal and non-neuronal), and exocrine (mucus producing) cell types are depicted. Different shades of grey indicate derivation of cell types from endoderm (black), neural ectoderm (dark grey), anterodorsal non-neural ectoderm (ANNE; medium grey), or general non-neural ectoderm (white). Transcription factors involved in the CoRN of these cell types are indicated at the bottom of the fgure. (**A**) Schematic dorsal view of region surrounding the anterior central nervous system (CNS) of a hypothetical chordate ancestor (A_1) and tunicate vertebrate ancestor (A_2). The distribution of cell types in various regions of the ectoderm (solid outline) and underlying endo-mesoderm (hatched outline) is illustrated. Transcription factors expressed in these regions (and potentially involved in the specification of regional cell subtypes) are indicated next to the squares below. It is assumed here that there were bilateral endomesodermal pouches containing neurosecretory cells in the chordate ancestor, which were retained only unilaterally in amphioxus as Hatschek's pit (on the left side). During evolution of the tunicate-vertebrate ancestor, the ANNE recruited transcription factors from both anterior endo-mesoderm (FoxA, Pitx, Lhx3, GATA1/2/3, POU1, Six1/2, Eya, Islet) and anterior CNS (Otx, FoxG, SP6-9, DMRT) resulting in translocation of neurosecretory cells from the anterior endomesoderm to the ANNE and the evolution of new neurosecretory cell types. See text for details. (**B**) Model for evolution of novel and unique regional CoRNs in ANNE. (B_1) In the chordate ancestor, activation of differentiation genes G_1 of cell types developing from non-neural ectoderm (e.g. mechanosensory neurons) requires multiple core regulating transcription factors including transcription factors promoting non-neural competence (NNC) and Six1/2 (arrows). Conversely, activation of differentiation genes G_2 of endo-mesodermal types (e.g. enteroendocrine cells) requires multiple core regulating transcription factors including transcription factors promoting endomesodermal competence (EMC) and Six1/2 (arrows). The ANNE in

FIGURE 5.7 (Continued)

the chordate ancestor is a relatively unspecialized part of the general non-neural ectoderm (distinguished by expression of some anterior factors such as Six3/6 and Pax4/6). (B_2) After recruitment of endomesodermal competence factors and other transcription factors (including neural transcription factors; not shown) to the ANNE in the ancestors of tunicates and vertebrates, co-expression of EMC, NNC, and other transcription factors in this domain promotes co-expression of G_1 and G_2 in the same cells, thereby giving rise to new types of neurosecretory neurons and other neurosecretory cells.

telencephalon (e.g. Zembrzycki et al. 2007; Bellefroid et al. 2013; Kasberg, Brunskill, and Steven Potter 2013; Kumamoto and Hanashima 2017; see Chapter 4 in Schlosser 2021). In *Ciona*, DMRT4/5 is likewise expressed in the anterior CNS and the ANNE region and is required for the upregulation of FoxC and Six1/2 in the ANNE and for the development of palps, OSP and sensory vesicle (Imai et al. 2006; Tresser et al. 2010; Wagner and Levine 2012; Horie, Hazbun et al. 2018). Similarly, FoxG is transiently expressed in the anterior neural plate and OSP. In addition, it is persistently expressed in the palp region including PSNs and collocytes, where it is required for palp formation upstream of Islet and Sp6-9 (Horie, Hazbun et al. 2018; Cao et al. 2019; Liu and Satou 2019). In amphioxus, FoxG is expressed in the anterior neural tube but not in the non-neural ectoderm, while Sp6-9 and DMRT4/5 expression have not been described (see section 5.4 below). This suggests that FoxG, and possibly Sp6-9 and DMRT4/5 were recruited to the ANNE from the anterior neural plate only in the last common ancestor of tunicates and vertebrates (Fig. 5.7).

This scenario is further supported by Otx, which acts as a general anterior patterning factor in all germ layers throughout bilaterians (see Chapter 3; also see Chapter 4 in Schlosser 2021). Otx is expressed in both ANNE and the anterior neural tube (forebrain and midbrain) in vertebrates and tunicates and is required for their development, but is confined to the anterior neural tube in amphioxus (Williams and Holland 1996; Wada et al. 1996; Hudson and Lemaire 2001; Wada, Sudou, and Saiga 2004; Imai et al. 2004; Onai et al. 2009). It may, thus, have been restricted to the anterior CNS in ancestral chordates and have expanded into the ANNE domain only in the last common ancestor of tunicates and vertebrates.

5.3.4 THE LAST COMMON TUNICATE-VERTEBRATE ANCESTOR

In summary, these data suggest that in the lineage leading to tunicates and vertebrates, a new group of neurosecretory cells was established in the rostral ectoderm, which developed from the embryonic ANNE (Fig. 5.7A). The latter acquired a unique regional regulatory profile after recruiting transcriptional regulators from the anterior endomesoderm (FoxA, Pitx, Lhx3, GATA1/2/3, and possibly Six1/2, Eya and Islet) as well as from the anterior neural plate (FoxG, SP6-9, DMRT, Otx). At the same time, the ANNE continued to express transcription factors widely expressed in non-neural ectoderm (Dlx2/3/5, Msx1/2, TFAP2) and those promoting general, not germ-layer restricted anterior identity (Six3/6, Pax4/6), which already intersect in the ANNE in early chordates (reviewed in Schlosser, Patthey, and Shimeld 2014). Sensory cells that developed in the ANNE thereby may have acquired a novel and unique regional identity, while retaining the core regulatory network of transcription factors (e.g. POU4, Islet, Pax2/5/8, BarH, Prox, Gfi) also found in other sensory cells (e.g. pATEN, CESNs; see Chapter 3) developing from the non-neural ectoderm (Cao et al. 2019).

In addition, several of the transcription factors newly acquired by the ANNE have early functions as competence factors in progenitor cells and FoxA and other Fox family members are known to act as pioneer factors, which are able to bind to cis-regulatory regions even in compacted chromatin making these regions accessible for other transcription factors (Friedman and Kaestner 2006; Zaret and Carroll 2011). The expression of an endodermal competence factor like FoxA in the ANNE of the tunicate-vertebrate ancestor may have facilitated the redeployment of other transcriptional regulators from the endoderm, including those involved in the specification of neurosecretory cells in the rostral endomesodermal pouches (e.g. Pitx, Lhx3/4, POU1, Islet, Six1/2, and Eya) (Fig. 5.7A). The neurosecretory cells recruited from these endomesodermal pouches to the ANNE may subsequently have given rise to the neurosecretory axial columnar cells in tunicates and to the adenohypopyseal endocrine cells in vertebrates. The latter then further diversified in stem vertebrates by employing additional transcription factors (e.g. Tbx19, Nr5a1, ER, NeuroD4, NeuroD1, Pax7) in the specification of different subtypes. Another secretory cell type recruited from the endomesoderm in the last common tunicate-vertebrate ancestor may have retained a function in mucus secretion and may have given rise to the collocytes in tunicates and the cement gland in vertebrates.

The recruitment of competence factors from the endomesoderm and possibly the anterior neural plate may also have endowed the ANNE with the competence to activate certain differentiation gene batteries, which ancestrally were activated only in endomesodermal or neural plate-derived cell types (Fig. 5.7B). These may have included gene batteries of neurosecretory cells (e.g. encoding GnRH, other neuropeptides, prohormone convertases etc.) and of the mucus producing goblet cells. The co-activation of endomesodermal and anterior neural competence factors with competence factors for non-neural ectoderm in the ANNE may then have promoted the co-expression of gene batteries for neurosecretory and sensory differentiation in the same cell, regulated by competence factors for endomesoderm/anterior neural ectoderm and non-neural ectoderm, respectively (Fig. 5.7B).

Consequently, some novel neurosecretory cell types of the ANNE probably originated in the stem lineage of tunicates and vertebrates not by simple duplication and divergence from pre-existing sensory neurons or neurosecretory cells, but rather by recombination (see Chapter 2) of CoRN members from sensory cells (derived from non-neural ectoderm) and neurosecretory cells (derived from endoderm or anterior neural ectoderm), which were present in ancestral chordates (Fig. 5.7B). One of these new cell types, a neurosecretory sensory neuron, may subsequently have given rise to the GnRH secreting sensory neurons in tunicates and to the GnRH secreting olfactory neurons in vertebrates. Some recombination between the CoRNs of neurosecretory cells and sensory cells may also have played a role in the origin of other cell types such as the neurosecretory axial columnar cells and collocytes in tunicates (which cluster with sensory neurons in single-cell RNA-Seq studies; Cao et al. 2019) and the adenohypophyseal endocrine cells of vertebrates.

5.4   NEUROSECRETORY CELLS IN THE LAST COMMON CHORDATE ANCESTOR

5.4.1   NEUROSECRETORY CELL TYPES IN AMPHIOXUS

Like in vertebrate and tunicates, neurosecretory cells in amphioxus have been identified in both the neural tube and the digestive tract (reviewed in Sherwood et al. 2005). In addition, neurosecretory cells seem to be concentrated in Hatschek's pit, an endomesodermally derived mucus-producing structure located anterior to the mouth. Hatschek's pit develops from an outpocketing of the gut on the left side anterior to the first somites known as Hatschek's left diverticulum (Fig. 5.8A). In amphioxus, many mesodermal structures (e.g. the anterior somites) arise from similar outpocketings of the gut. Opposite of Hatschek's left diverticulum, a corresponding outpocketing on the right side (Hatschek's right diverticulum) will expand to form the head coelom. In contrast, Hatschek's left diverticulum will bud off the pharynx and fuse with the surface ectoderm to form the preoral pit, which gives rise to Hatschek's pit in the adult (Hatschek 1884; Glardon et al. 1998; Kaji et al. 2016). Hatschek's pit, in the narrow sense, forms an epithelium of secretory cells, which is surrounded by the secretory cells of the wheel organ (Fig. 5.8). Whether the latter are also derived from endoderm or from ectoderm surrounding the preoral pit has not been determined. For simplicity, I will here use the term "Hatschek's pit" to refer to the larger structure comprising both wheel organ and Hatschek's pit in the narrow sense in the adult or to the preoral pit in the larva.

One of the functions of Hatschek's pit appears to be secretion of mucus for trapping food particles, together with other mucus secreting organs like the endodermal endostyle and club-shaped gland (Tjoa and Welsch 1974; Lacalli 2008). Mucus is probably released by the large vesicles observed in the apical part of the secretory cells in Hatschek's pit (Fig. 5.8B). In addition to this exocrine function, some of the secretory cells appear to also have an endocrine function. This is suggested by the prevalence of small vesicles in the basal parts of these secretory cells, which are located next to blood spaces (Fig. 5.8B) (Tjoa and Welsch 1974; Sahlin and Olsson 1986) and by their immunoreactivity for many neuropeptides or other protein hormones (see below).

Hatschek's pit resembles the vertebrate pituitary because of its presumptive neurosecretory functions, its rostral position and its attachment to the ventral side of the brain (Gorbman, Nozaki, and Kubokawa 1999). Moreover, the opening of Hatschek's diverticulum into the ectoderm is reminiscent of pores that connect the anterior coelom (protocoel; Fig.1.15) in ambulacrarians to the outside and form the hydropore in echinoderms and the proboscis pores in hemichordates (Goodrich 1917; Benito and Pardos 1997). Similar connections between the adenohypophysis and adjacent endomesodermal head cavities have also been found in some vertebrates (Goodrich 1917). Hatschek's pit has, therefore, been often suggested to be homologous to the adenohypophyseal placode, which gives rise to the anterior pituitary of vertebrates (reviewed in Patthey, Schlosser, and Shimeld 2014; Schlosser 2015). This will be discussed further below and in the next chapter.

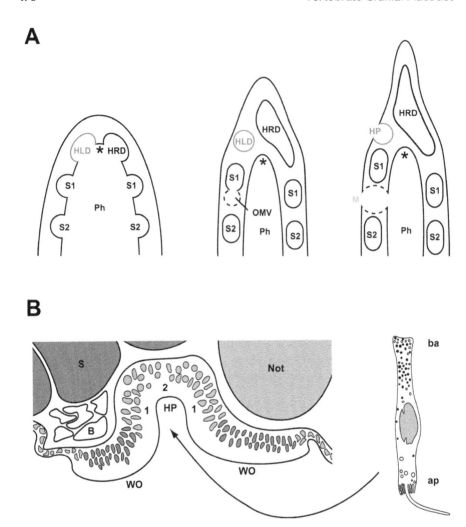

FIGURE 5.8 Hatschek's pit in amphioxus: development and distribution of neurosecretory cells. (**A**) Schematic horizontal section through the pharyngeal region (Ph) of amphioxus showing the formation of endomesodermal pouches giving rise to Hatschek's left and right diverticula (HLD, HRD) and the somites (S). Hatschek's pit (HP) develops from Hatschek's left diverticulum. The mouth (M) is here shown to develop from the oral mesovesicle (OMV) budding off the first somite (S1) as proposed by Kaji et al. 2016. Asterisks indicate the approximate position, in which the mouth forms in vertebrates and tunicates. (**B**) Cross-section through Hatschek's pit (HP) shown in left panel, single cell from Hatschek's pit in the right panel. All cells in Hatschek's pit are ciliated and have large vesicles on the apical side with putative exocrine function. In addition, some cells (particularly in region 1) contain numerous small vesicles with putative endocrine (neurosecretory) function and are located next to blood vessels (B). ap, apical; ba, basal; Not, notochord; S, somite; WO, wheel organ. ([A] Adapted from Soukup and Kozmik 2016; [B] reprinted and modified from Schlosser 2017. Left panel adapted from Sahlin and Olsson 1986, right panel adapted from Welsch and Welsch 1978.)

However, Hatschek's pit in amphioxus differs from the adenohypophysis of verte-brates, not only because of its endomesodermal origin, but also because of its asymmetric development on the left side. It shares this asymmetric localization with other structures including the mouth, which also develops on the left side, and has been suggested to be derived either from a gill slit or from a pore connecting an anterior coelom budding off the first somite to the outside (Fig. 5.8A) (Stach 2002; Yasui and Kaji 2008; Kaji et al. 2016; Schlosser 2017; Holland 2018). Although its developmental and evolutionary origin is still unresolved, it is widely agreed that the amphioxus mouth is not homologous with the mouth of other chordates and probably evolved in the context of increasingly asymmetric development after the lineage of amphioxus diverged from the last common ancestor of chordates (Yasui and Kaji 2008; Schlosser 2017; Holland 2018). If this interpretation is correct, ancestral chordates probably were largely symmetrical and were equipped with a midline mouth subsequently lost in amphioxus and possibly with paired endomesodermal pouches containing neurosecretory cells similar to Hatschek's left diverticulum in amphioxus (Fig. 5.8A).

Because the neurosecretory cells of Hatschek's pit are exposed to the external environment, they may respond to environmental stimuli. While apically, the cells of Hatschek's pit resemble sensory cells in bearing a cilium surrounded by microvilli (Fig. 5.8B), they do not contain synaptic vesicles or form axonal processes (Tjoa and Welsch 1974; Sahlin and Olsson 1986). Furthermore, while some nerve fibers have been found in the vicinity of Hatschek's pit, there is currently no support that the pit itself is innervated (Kaji et al. 2001). It has, therefore, been proposed that the cells of Hatschek's pit may release hormones directly into the blood stream in response to environmental stimuli, thereby regulating reproduction, feeding and other body functions in an endocrine way (Nozaki and Gorbman 1992; Gorbman 1999), similar to enteroendocrine cells.

As reviewed above, only GnRH but none of the hormones released by the vertebrate adenohypophysis are present in amphioxus based on its genome sequence (Holland et al. 2008; Putnam et al. 2008). However, other representatives of the three adenohypophyseal hormone families – neuropeptides, dimeric glycoprotein hormones and four-helix cytokines – appear to be present and are expressed in Hatschek's pit and elsewhere in amphioxus.

First, genes encoding many *neuropeptides* and their GPCRs could be identified in the amphioxus genome (Holland et al. 2008; Putnam et al. 2008; Mirabeau and Joly 2013; Osugi et al. 2016). Cells immunoreactive for various neuropeptides such as FMRF amide, neuropeptide Y and somatostatin were identified in the neural tube, gut and Hatschek's pit of amphioxus (Reinecke 1981; Chang et al. 1984; Nozaki and Gorbman 1992; Massari, Candiani, and Pestarino 1999; Castro, Manso, and Anadón 2003; reviewed in Sherwood et al. 2005). Immunoreactivity for GnRH was found mostly in the neural tube, while an earlier report of GnRH immunoreactivity in Hatschek's pit (Fang, Huang, and Chen 1999), could not be confirmed in further studies (Castro et al. 2006; Roch et al. 2014). One GnRH peptide and four GnRH receptors were previously isolated from amphioxus (Tello and Sherwood 2009; Roch et al. 2014). Two of these receptors are vertebrate-like GnRH-type receptors, while the other two belong to the CRZ-type family (Roch et al. 2011, 2014; Sakai et al. 2017; Elphick et al. 2018). Since amphioxus GnRH activates only one of its two CRZ-type but none of its GnRH-type receptors, additional GnRH peptides are probably present in amphioxus, but remain to be identified (Roch et al. 2014). One additional GnRH-like peptide was described but could not be confirmed in genomic analyses (Chambery et al. 2009).

Second, while amphioxus lacks genes encoding the subunits of the *heterodimeric glycoproteins* FSH, LH or TSH, orthologs for the two subunits (GPA2, GBP5) of the related heterodimeric glycoprotein thyrostimulin and for their GPCRs are present in the genome (Dos Santos et al. 2009; Roch and Sherwood 2014). Although cells immunoreactive for the gonadotropins FSH and LH were originally described in Hatschek's pit (Chang et al. 1984; Nozaki and Gorbman 1992), these findings are probably due to cross-reactivity of the antibodies used with other glycoprotein hormones. Recently, the genes encoding GPA2 and GPB5 were indeed found to be expressed in Hatschek's pit as well as in the endoderm/ gut, neural tube and gonads of amphioxus (Dos Santos et al. 2009; Tando and Kubokawa 2009; Wang et al. 2018).

Third, and finally, a member of the *four-helix cytokines* has recently been identified in the amphioxus genome, although it's phylogenetic relationship to vertebrate GH and PRL remains to be clarified (Li et al. 2014; Ocampo Daza and Larhammar 2018). This protein has been suggested to play a dual role for growth and osmoregulation in amphioxus (Li et al. 2014, 2017). It is expressed in Hatschek's pit and several other tissues of amphioxus including gut and gonads (Li et al. 2014), while previous studies found no evidence for immunoreactivity with GH and PRL antibodies in Hatschek's pit (Nozaki and Gorbman 1992).

Taken together, this indicates that Hatschek's pit, apart from its exocrine, mucus-secreting function, serves as a rostral neurosecretory structure. Its neurosecretory cells co-express hormones from all three classes – neuropeptides, heterodimeric glycoproteins, and four-helix cytokine-like proteins. Other neurosecretory cells releasing these hormones are found in the gut (all three classes) and the CNS (neuropeptides, dimeric glycoproteins). However, the one known GnRH protein of amphioxus appears to be confined to neurosecretory cells in the neural tube.

5.4.2 TRANSCRIPTION FACTORS INVOLVED IN SPECIFYING AMPHIOXUS NEUROSECRETORY CELL TYPES

Whereas in tunicates many of the transcription factors specifying adenohypophyseal and olfactory neurosecretory cells in vertebrates are expressed in the ANNE, this is not the case in amphioxus. Instead most of these transcriptional regulators including Pitx, Lhx3/4, Islet, Six1/2, Eya are expressed widely in the anterior endomesoderm (Pitx being confined to the left side), and subsequently are maintained strongly in the endomesodermally derived Hatschek's left diverticulum and later in Hatschek's pit (Jackman, Langeland, and Kimmel 2000; Yasui et al. 2000; Wang et al. 2002; Kozmik et al. 2007). In addition, POU1f1 is expressed specifically in Hatschek's left diverticulum and pit (Candiani et al. 2008). While the exact position of the boundary between the endomesodermally derived Hatschek's pit and the surrounding ectoderm has not been mapped, most transcription factors expressed in Hatschek's left diverticulum and pit, later appear to be confined to its central, probably endomesodermal part. A notable exception is Pitx, which is also present in the left ectoderm surrounding Hatschek's pit, while both Pitx and Islet are additionally expressed in a separate domain in non-neural ectoderm immediately anterior to the neural plate corresponding to the ANNE (Jackman et al. 2000; Yasui et al. 2000). Other transcription factors expressed in Hatschek's pit such as Otx, Pax4/6, and Six3/6 are also found in other anterior tissues derived from all germ layers (Williams and Holland 1996; Glardon et al. 1998; Kozmik et al. 2007).

Apart from these general anterior transcription factors, Pitx and Islet, the anterior non-neural ectoderm in amphioxus expresses other transcription factors, which are widely expressed throughout non-neural ectoderm (Dlx2/3/5, TFAP2, Msx1/2, Ventx: Holland et al. 1996; Sharman, Shimeld, and Holland 1999; Kozmik et al. 2001; Meulemans and Bronner-Fraser 2002; Gostling and Shimeld 2003; Yu et al. 2007, 2008). However, none of the endodermal (FoxA, GATA1/2/3) and neuroectodermal (FoxG, SP6-9, DMRT4/5) transcription factors, which are co-expressed with these transcription factors in the ANNE of tunicates and the anterior placodal region of vertebrates, were found in either Hatschek's pit or the anterior non-neural ectoderm in amphioxus. Amphioxus FoxA and GATA1/2/3 are widely expressed in endoderm at early embryonic stages (Shimeld 1997; Terazawa and Satoh 1997; Zhang and Mao 2009). However, at later embryonic stages, FoxA only persists in post-pharyngeal endoderm, floorplate, notochord, and GATA1/2/3 in post-pharyngeal endoderm and cerebral vesicle, while neither of them is maintained in Hatschek's pit. Amphioxus FoxG expression is restricted to its anterior neural tube (Toresson et al. 1998), while the expression of SP6-9 and DMRT4/5 have not been described (Wang et al. 2012; Dailey, Kozmikova, and Somorjai 2017).

5.4.3 THE LAST COMMON CHORDATE ANCESTOR

Taken together with the evidence from tunicates and vertebrates, these findings in amphioxus suggest that the last common ancestor of all chordates may already have evolved a rostral concentration of neurosecretory cells in the rostral endoderm under the control of general endodermal competence factors (FoxA and GATA1/2/3) and transcriptional regulators such as Pitx, Lhx3/4, Islet, POU1f1, Six1/2, Eya specifically localized in anterior endomesodermal pockets (Fig. 5.7A). Of these, only Pitx and Islet had an additional expression domain in the ectoderm immediately rostral to the neural plate (ANNE). Since Pitx is required for proper mouth formation in vertebrates (Chen et al. 2017), this expression domain may have defined the region where the new chordate mouth developed. The latter was retained in tunicates and vertebrates, but lost with the formation of another mouth opening from some pharyngeal pore or gill slit on the left side in the amphioxus lineage. Moreover, FoxG (and possibly SP6-9 and DMRT4/5, given their distribution in protostomes) was probably confined to the anterior brain and Otx to anterior brain and endomesoderm in the chordate ancestor.

Since strong left-right asymmetries probably originated only in the amphioxus lineage (see Chapter 1), in the last common ancestor of chordates these neurosecretory cells may have resided in a pair of outpocketings from the rostral endoderm that fused with the overlying ectoderm (Fig. 5.7A). Alternatively, neurosecretory cells may have already been concentrated in the left of these endomesodermal pouches like in extant amphioxus. To decide between these two possibilities, we need to learn more about when Pitx acquired its central role in neurosecretory specification. Pitx is known to be involved in generating left-right asymmetries downstream of nodal signaling not only in chordates but also in echinoderms and mollusks indicating that it has a pan-bilaterian role in left-right patterning (Boorman and Shimeld, 2002, Duboc et al. 2005; Yoshida and Saiga 2008; Grande and Patel 2009a, 2009b, Molina et al. 2013). Accordingly, Pitx expression is confined to the right side of ambulacrarians and protostomes but to the left side in chordates due to dorsoventral inversion. Therefore, if Pitx already had adopted a central role in the specification of neurosecretory cell in the rostral endoderm in the last common ancestor of chordates, these neurosecretory cells may well have been confined to the left side. Unfortunately, we currently lack crucial information about the function of Pitx in amphioxus and whether it is required for the specification of neurosecretory cells in Hatschek's pit that would help us to decide between these two alternative scenarios.

The neurosecretory cells in the rostral endomesodermal pouch/pouches of the chordate ancestor probably secreted a cocktail of protein hormones including neuropeptides, dimeric glycoprotein hormones, and four-helix cytokine-like hormones allowing the regulation of metabolic and reproductive processes in response to environmental cues. They probably resembled enteroendocrine cells and did not have a neuronal phenotype, lacking synapses, dendrites, and axonal processes. A new regulatory environment was then established in the ANNE in the lineage leading to tunicates and vertebrates by the recruitment of transcription factors from the endoderm (FoxA, GATA2/3) and anterior neuroectoderm (FoxG, SP6-9, DMRT4/5) of ancestral chordates and the redeployment of the anterior endomesodermal group of transcription factors (Pitx, Lhx3/4, Islet, POU1f1, Six1/2, Eya) in this region of the non-neural ectoderm. This may have led to the recruitment of neurosecretory cells to the rostral ectoderm and the origin of a new type of neurosecretory neurons as proposed above (Fig. 5.7A, B).

5.5   NEUROSECRETORY CELLS IN THE LAST COMMON DEUTEROSTOME ANCESTOR

Relatively little is known about neurosecretory cell types in the ambulacrarians. Apart from the apical organs, neurosecretory cells secreting various neuropeptides have been identified in the nervous system and gut of hemichordate and echinoderm larvae (Nezlin and Yushin 2004; Mayorova et al. 2016; Wood et al. 2018). GnRH was found in the nervous system of echinoderms and putatively neurosecretory cells scattered throughout the rostral epidermis in hemichordates (Cameron et al. 1999; Rowe and Elphick 2012).

We currently do not know whether any of these neurosecretory cell populations is enriched in the rostral endoderm comparable to Hatschek's pit in amphioxus. However, Six1/2, Eya, and Islet are enriched in the anterior endomesoderm of echinoderms (Yankura et al. 2010; Materna et al. 2013; Slota, Miranda, and McClay 2019; Valencia et al. 2019) and Six1/2 and Eya are specifically expressed in the developing gill slits of the pharyngeal endoderm in hemichordates (Gillis, Fritzenwanker, and Lowe 2012). Pitx is expressed on the right side in all germ layers in sea urchins including the anterior endoderm (Duboc et al. 2005; Hibino, Nishino, and Amemiya 2006) – due to the dorsoventral inversion of chordates, this corresponds to the left side in chordates. While POU1f1 expression has not yet been described in ambulacrarians, no endodermal expression was reported for Lhx3/4, which is instead found in ventral ectoderm in hemichordates (Lowe et al. 2006; Yasuoka et al. 2009). This raises the possibility that Six1/2, Eya1, Islet, and possibly Pitx may already have had a specific role in anterior endoderm in the last common ancestor of deuterostomes, possibly in relation to gill slit formation and/or the differentiation of cell types specific for the anterior endoderm. The latter may have included gill-associated and/or other pharyngeal neurosecretory cells as they are known from vertebrates (Jonz et al. 2016; Hockman et al. 2017). This will be discussed further in Chapter 6.

Like in chordates, FoxA and GATA1/2/3 as well as GATA4/5/6, are predominantly expressed in the endomesoderm in ambulacrarians (Davidson et al. 2002; Oliveri et al. 2006; Darras et al. 2011; Solek et al. 2013; Materna et al. 2013). Extensive functional experiments in sea urchins have shown that GATA4/5/6 plays a key role in the specification of endomesoderm upstream of GATA1/2/3, while FoxA is a key regulator for endoderm development (Davidson et al. 2002; Oliveri et al. 2006). FoxA also is expressed in the ectodermal "stomodeum" and required for mouth development in sea urchins (Oliveri et al. 2006). However, their "stomodeum" forms at a different position and is not equivalent to the stomodeum of chordates, assuming that chordates have formed a new mouth and stomodeum after dorsoventral inversion (Fig. 5.3: the position corresponding to the ambulacrarian "stomodeum" in chordates is marked by ** in A_4 and B_4). This suggests that rostral ectodermal expression of FoxA in echinoderms and chordates may either have evolved independently or that ectodermal FoxA expression shifted to a new position anterior to the neural plate with the evolution of a new mouth in chordates (presumably secondarily lost in amphioxus when the latter evolved yet another mouth from some pharyngeal pore or gill slit on the left side). The expression of FoxG and Sp6-9 in the anterior ectoderm of ambulacrarians resembles the one in the anterior neuroectoderm of chordates (Lemons et al. 2010; Saudemont et al. 2010; Yankura et al. 2010). However, echinoderm DMRT4/5 is expressed in anterior endoderm of echinoderms, different from chordates (Slota et al. 2019).

In summary, we know very little about the core regulators of neurosecretory cells in the last common ancestor of deuterostomes. The presence of Six1/2, Eya, Islet, and Pitx in the anterior endoderm of ambulacrarians and chordates suggests that these transcription factors may already have played a specific role in this domain, possibly promoting the formation of pharyngeal gill slits. It is well possible that they may also be involved in the specification of neurosecretory cells that arise specifically from the endoderm associated with pharyngeal gill slits, but this needs to be confirmed in further studies.

5.6 NEUROSECRETORY CELLS IN THE LAST COMMON BILATERIAN ANCESTOR

5.6.1 NEUROSECRETORY CELL TYPES IN PROTOSTOMES

Neurosecretory cells producing neuropeptides are found throughout the central and peripheral nervous systems and in the digestive tract of protostomes (Fig. 5.9). These cannot be described here in any detail and the reader is referred to previous reviews for further details (e.g. Nijhout 1998; Nassel 2002; Hartenstein 2006; Tessmar-Raible 2007; Jékely 2013; Hartenstein et al. 2017; Williams and Jékely 2019). In addition, the two subunits of glycoprotein hormones related to those of vertebrate thyrostimulin have been identified in neurosecretory cells in the brain of insects and mollusks (Sellami, Agricola, and Veenstra 2011; Heyland et al. 2012), while nothing is known about the expression and function of four-helix cytokine-like hormones.

As discussed above, some of these cells form a neurosecretory center in the anteromedial forebrain that may be derived from the apical organ and be evolutionarily related to the vertebrate hypothalamus. However, I will argue in this section that there is no compelling evidence for neurosecretory cells of protostomes being closely related to the particular neurosecretory cells of vertebrates derived from cranial placodes.

Based on their innervation from the anteromedial neurosecretory center and some similarities in position and transcription factor expression, two peripheral endocrine glands in arthropods, the corpora cardiaca and the corpora allata, have been proposed to be homologs of the adenohypophysis (De Velasco et al. 2004; Wirmer et al. 2012). The corpora allata are derived (like the adjacent prothoracic glands) from non-neural ectoderm of head appendages (Figs. 5.3, 5.9) (Hartenstein 2006). Their neurosecretory cells produce a lipid-derived hormone, juvenile hormone, which regulates metamorphosis, and release it into the hemolymph. Apart from general anterior transcription factors like Six3/6, they have not yet been shown to express any other transcription factors that would support their homology with the neurosecretory cells of the adenohypophysis suggesting that they have evolved independently.

The corpora cardiaca have a different embryonic origin and arise from the anterior ventral furrow, the anteriormost part of the endomesoderm in insects (Figs. 5.3, 5.9) (De Velasco et al. 2004, 2006). The neurosecretory cells of the corpora cardiaca release a GnRH-related peptide hormone, AKH into the hemolymph, which has important metabolic functions (e.g. mobilizing lipids to provide energy for flight). The anterior ventral furrow in *Drosophila* expresses Six1/2 which is required for the development of the corpora cardiaca (Cheyette et al. 1994; De Velasco et al. 2004). This raises the possibility that neuropeptidergic cells specified by Six1/2 may have been present in the anterior endomesoderm of the last common ancestor of protostomes and deuterostomes. However, neurosecretory cells regulated by Six1/2 in the anterior endomesoderm have so far

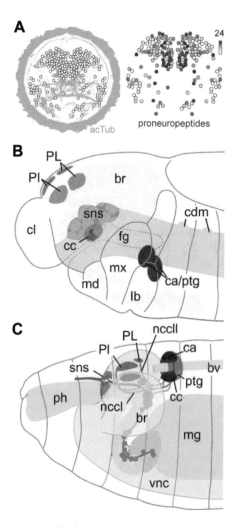

FIGURE 5.9 Neurosecretory cells in protostomes. (**A**) Neuropeptidergic cells in the apical nervous system of *Platynereis*. Left panel: positions of cells (nodes) from single cell RNA-Seq from 2-day-old larvae projected on axon scaffold (acetylated tubulin immunostaining, grey, anterior view). Right panel: Map of combined expression of 80 proneuropeptides. (**B, C**) Neurosecretory cells in *Drosophila*. Diagrams of two stages of embryonic development of *Drosophila* heads in dorsolateral view (anterior to the left). (**B**) Primordia of the neuroendocrine centers of the brain, the pars intercerebralis (PI) and pars lateralis (PL), form ectodermal thickenings in the anteromedial neuroectoderm of the head. Primordia of the corpora allata (ca) and prothoracic gland (ptg) originate from the lateral ectoderm of the segments carrying the mouth parts (lb, labium; mx, maxilla; and md, mandible). Primordia of corpora cardiaca (cc) and the stomatogastric nervous system (sns) invaginate from the anteroventral furrow of mesoderm and the stomodeum, respectively in the area of the foregut (fg). (**C**) At a later stage, the corpora allata, prothoracic gland, and corpora cardiaca have coalesced into the ring gland that surrounds the anterior end of the dorsal blood vessel (bv). Neurosecretory axons from the PI and PL reach the ring gland via the nccI and nccII nerves respectively. ([A] Reprinted with permission from Williams et al. 2017; [B-C] reprinted with permission from Hartenstein 2006.)

not been identified in other protostomes (see below). Therefore, neurosecretory cells regulated by Six1/2 in the anterior endomesoderm may have, alternatively, evolved convergently in insects and chordates. Current evidence is insufficient to resolve this question.

Another potential candidate for a homolog to the adenohypophysis is the stomatogastric nervous system containing many neuropeptidergic neurons innervating muscles of the gut, which is found in arthropods, annelids and possibly other protostomes (Hartenstein 1997; Purschke 2015). In insects, it arises from the stomodeum and expresses Lhx3/4 like the vertebrate adenohypophysis (see section 5.6.2 below); subsequently it becomes closely connected to the corpora cardiaca (Fig. 5.9B,C) (Hartenstein 1997; De Velasco et al. 2004). Its embryological origin in annelids or other groups has not been described. Importantly, the stomodeum of insects and other protostomes forms at the center of the anterior CNS and marks the original mouth opening of bilaterians. Thus, assuming that there was a dorsoventral inversion of the body plan in the chordate lineage (Chapter 1), the stomodeum of protostomes corresponds positionally to the stomodeum of ambulacrarians (see above) but not to the stomodeum of chordates. The latter was established at a new position in ANNE rostral to the CNS after dorsoventral inversion (Fig. 5.3). Consequently, homology of the stomatogastric nervous system with the adenohypophysis is unlikely, if one accepts dorso-ventral inversion.

A study in the beetle *Tribolium* has proposed an anteromedial "head placode" as yet another head structure in insects that may be homologous to the vertebrate adenohypophysis (Posnien, Koniszewski, and Bucher 2011). While there is no direct evidence for neurosecretory cells originating from this domain in *Tribolium*, based on its position and shared expression of some transcription factors it likely corresponds to the pars intercerebralis and/or lateralis of other insects, putative homologs of the vertebrate hypothalamus (Fig. 5.3; see above). Apart from expressing general anterior transcription factors such as Otx, Six3/6, and Pax4/6, this "head placode" co-expresses Six4/5 with Pitx, FoxG1, and Lhx3 similar to the vertebrate adenohypophysis. However, the co-expression of these factors in anterior neuroectoderm is at variance with their co-expression in rostral endomesoderm in amphioxus and ANNE in tunicates and vertebrates. Moreover, a similar ectodermal domain co-expressing these transcription factors has not been identified in other protostomes (see below). Taken together, this provides little support for the proposed homology and suggests instead that the similarities in transcription factor expression to the vertebrate adenohypophysis may result from convergent evolution.

5.6.2 Transcription Factors Involved in Specifying Placode-Derived Neurosecretory Cell Types in Vertebrates Play Different Roles in Protostomes

Although several head structures in insects have been proposed to be homologs of the vertebrate adenohypophysis, overall the expression patterns of transcription factors provide little evidence that protostomes possess any neurosecretory cells homologous to the placode-derived neurosecretory cells of vertebrates. In particular, transcription factors that are co-expressed in Hatschek's pit in amphioxus and have adopted a central role for the specification of adenohypophyseal and possibly olfactory neurosecretory cells in vertebrates (Pitx, Lhx3/4, Islet, POU1f1, Six1/2, Six4/5/, Eya), do not show comparable distributions in protostomes.

POU1f1 has been lost in both insects, nematodes and many lophotrochozoans, indicating its dispensability for specification of neurosecretory cell in these taxa, while its expression in annelids has not been described (Jacobs and Gates 2003; Gold, Gates, and Jacobs 2014). Few data exist for FoxL2/3 where only gonadal expression is well documented (Bertho et al. 2016). Pitx controls chirality in snails and possibly plays an ancient role in left-right patterning (Grande and Patel 2009a, 2009b). It is also expressed in several subpopulation of the CNS in both lophotrochozoans and ecdysozoans (Vorbruggen et al. 1997; Posnien et al. 2011; Vergara et al. 2017) and has been implicated in the specification of subsets of neurons including some neurosecretory cells (Jin, Hoskins, and Horvitz 1994; Shiomi et al. 2007; Ohtsuka et al. 2011; März, Seebeck, and Bartscherer 2013; Hobert 2016).

Lhx3/4 is likewise expressed in parts of the CNS including motorneurons and some neurosecretory cells in lophotrochozoans and ecdysozoans (Thor et al. 1999; Nomaksteinsky et al. 2013; Focareta, Sesso, and Cole 2014; Achim et al. 2015; Vergara et al. 2017; Rybina et al. 2019). It is also found in the stomodeum of insects which gives rise to the stomatogastric nervous system (Fig. 5.9B, C) (De Velasco et al. 2004). Islet is widely expressed not only in the PNS (reviewed in Chapter 3) but also in the CNS of protostomes where it is known to be involved in subtype specification of motor neurons and probably other neurons (reviewed in Hobert and Westphal 2000; Thor and Thomas 2002; Allan and Thor 2015).

Six1/2, Six4/5, and Eya are predominantly expressed in rhabdomeric photoreceptor and other putative sensory cells in protostomes (see Chapters 3 and 4), but have additional domains in the CNS and endomesoderm (Bonini, Leiserson, and Benzer 1993; Cheyette et al. 1994; Serikaku and O'Tousa 1994; Seo et al. 1999; Kumar and Moses 2001; Arendt et al. 2002; Mannini et al. 2004; Passamaneck, Hejnol, and Martindale 2015). As discussed above, in insects, Six1/2 is also present in the stomodeum and in the anterior-most part of the endomesoderm, where it is required for the development of the neurosecretory corpora cardiaca (De Velasco et al. 2004). Since endomesodermal expression of Six1/2 is not anteriorly confined in brachiopods (lophotrochozoans) and acoel flatworms (Chiodin et al. 2013; Passamaneck et al. 2015), a role of Six1/2 in the specification of neurosecretory cells from the anterior endomesoderm may have been independently acquired in insects and chordates rather than reflecting an ancestral, pan-bilaterian condition. However, this remains to be confirmed in further studies. Apart from Six1/2 in the anterior endomesoderm of insects, none of the transcription factors discussed is expressed in anterior endomesoderm or in the anterior non-neural ectoderm of protostomes. This suggests that the unique regulatory environment required for the specification of rostral neurosecretory cells (as found in Hatschek's pit of amphioxus and in the ANNE of tunicates and vertebrates) did not exist in the last common ancestor of bilaterians and only evolved in the deuterostomian or chordate lineage.

This conclusion is further supported by the distribution of transcription factors FoxA and GATA1/2/3, FoxG, SP6-9, and DMRT4/5. Like in deuterostomes, FoxA,

GATA4/5/6, and GATA1/2/3 play important roles for endomesoderm development in protostomes (reviewed in Stainier 2002; Martindale and Hejnol 2009; de-Leon 2011; Nielsen, Brunet, and Arendt 2018). Together with Brachyury, FoxA1/2/3 is initially expressed around the blastopore and subsequently is maintained in endoderm and stomodeum, while GATA4/5/6 and GATA1/2/3 are widely expressed in endomesoderm. This suggests that FoxA and GATA1/2/3 probably had an ancestral role in endomesoderm development in the last common ancestor of bilaterians, while their specific role in the ANNE only evolved in the lineage leading to tunicates and vertebrates.

Apparently at variance with this interpretation, a previous study in *Platynereis* found only GATA4/5/6 in endomesoderm, while GATA1/2/3 was expressed mostly in neuroectoderm (Gillis, Bowerman, and Schneider 2007). Because expression of GATA1/2/3 in the CNS or in neurons was also reported in *Drosophila* and *Caenorhabditis elegans* (Brown and Castelli-Gair Hombria 2000; Smith, McGarr, and Gilleard 2005), this study suggested that GATA1/2/3 may have played an ancestral bilaterian role in ectoderm rather than endomesoderm specification. However, GATA1/2/3 is also widely expressed in endomesoderm in *Drosophila*, other polychaetes and mollusks (Brown and Castelli-Gair Hombria 2000; Yue et al. 2014; Wong and Arenas-Mena 2016). Moreover, ectodermal GATA1/2/3 appears mostly restricted to the neuroectoderm in protostomes (e.g. *Drosophila*, *Platynereis*) but to non-neural ectoderm in tunicates and vertebrates. Taken together with the exclusively endomesodermal expression of GATA1/2/3 in ambulacrarians and amphioxus, this suggests that this transcription factor played an ancestral bilaterian role in endomesoderm development, but adopted additional functions in the ectoderm independently in protostomes and the tunicate/vertebrate clade.

Finally, FoxG, SP6-9 and DMRT4/5, which were suggested to be expressed in the anterior CNS of ancestral chordates, are also known to be present in the CNS of protostomes (Grossniklaus, Pearson, and Gehring 1992; Tomer et al. 2010; Schaeper, Prpic, and Wimmer 2010; Bellefroid et al. 2013; Picard et al. 2015), suggesting that they played some roles in CNS development of ancestral bilaterians, while adopting a new role in the ANNE only in tunicates and vertebrates.

5.6.3   THE LAST COMMON BILATERIAN ANCESTOR

In summary, a unique regulatory environment required for specification of rostral neurosecretory cells in either anterior endomesoderm or anterior non-neural ectoderm did most likely not exist in the last common ancestor of bilaterians. Pitx, Lhx3/4, Islet, Six1/2, Six/4/5, and Eya, among other transcription factors, may have played some role in the developing nervous system of the bilaterian ancestor and were possibly involved separately in the specification of various neuronal and/or neurosecretory subtypes in the ectoderm and/or endoderm, however without cooperating in specification of a particular cell type.

FoxG, SP6-9, and DMRT4/5 may also have each contributed to specification or patterning of different neuronal cell populations, while FoxA and GATA1/2/3 probably had an early role in blastopore development and promoting endodermal competence. Pitx, Lhx3/4, Islet, Six1/2, Six/4/5, and Eya were then probably recruited to the anterior endomesoderm in the deuterostomian or chordate lineage, possibly establishing a new CoRN driving the specification a special rostral population of neurosecretory cells (as found in Hatschek's pit of amphioxus). Only in the ancestor of tunicates and vertebrates were FoxA, GATA1/2/3, FoxG, SP6-9, DMRT4/5, and the rostral neurosecretory cells regulated by these transcription factors recruited to the ANNE followed by the diversification of neurosecretory cells in this domain in vertebrates.

5.7   NEUROSECRETORY CELLS IN THE LAST COMMON EUMETAZOAN AND METAZOAN ANCESTORS

5.7.1   Neurosecretory Cell Types in Cnidarians, Ctenophores, Placozoans, and Sponges

Neurosecretory cells in cnidarians are found scattered throughout ecto- and endoderm with either neuronal or epithelial phenotypes (Fig. 5.10A) (Lesh-Laurie 1988; Thomas and Edwards 1991; Grimmelikhuijzen and Westfall 1995; Koizumi, Sato, and Goto 2004; Hartenstein 2006; Galliot et al. 2009; Rentzsch, Layden, and Manuel 2017). Cells that secrete protein hormones have also been described along the tentacles, around the mouth and in the apical organ of ctenophores, but whether any of them has a neuronal phenotype has not been determined (Moroz et al. 2014). While placozoans and sponges lack neurons, secretory cells that release neuropeptides have been described in their body wall (Fig. 5.10B) (Renard et al. 2009; Smith et al. 2014; Nakanishi, Stoupin et al. 2015). The flask cells of sponge larvae contain many secretory vesicles and mediate larval settling and metamorphosis in response to sensory cues (Nakanishi, Stoupin et al. 2015).

However, the secretory products released by neurosecretory cells of these basal metazoans and their receptors are overall very different from bilaterians. With exception of small amidated peptides (RFamide, RYamide, and Wamide), glycoprotein-type, insulin-type, and bursicon-type hormones and their receptors, no homologs of the bilaterian neuropeptides or their receptors have been found in these basal metazoans, suggesting that most families of neuropeptides and their cognate receptors originated only in bilaterians (Jékely 2013; Roch and Sherwood 2014; Elphick et al. 2018). However, a variety of cnidarian-specific and ctenophore-specific neuropeptides have been identified in these two clades (Grimmelikhuijzen, Williamson, and Hansen 2004; Anctil 2009; Moroz et al. 2014). Simple amidated peptides and putative GPCR receptors have also been found in placozoans, but not in sponges (Srivastava, Simakov et al. 2010; Jékely 2013; Krishnan and Schiöth 2015; Nikitin 2015). However, sponges possess proneuropeptide processing enzymes as well as cystine knot hormones related to the glycoprotein hormones and their putative receptors (Srivastava, Simakov et al. 2010; Roch and Sherwood 2014). Thus, the origin of neurosecretory cells (but not neurons) can be dated back to the origin of metazoans with significant diversification in eumetazoans and especially bilaterians.

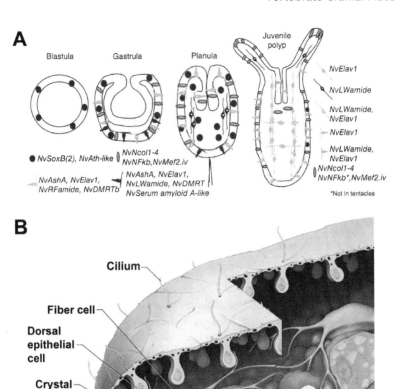

FIGURE 5.10 Neurosecretory cells in cnidarians and placozoans. (**A**) Neurons and neurosecretory cells in different developmental stages of the cnidarian *Nematostella vectensis* (oral pole facing up). Expression of several transcription factors and neuropeptides (e.g. RFamide, LWamide) are indicated. Note the widespread expression of neuropeptides in many types of neurons. (**B**) Diagram of a placozoan, showing ciliated epithelial cells, secretory cells (gland cells) and other cell types. ([A] Reprinted with permission from Rentzsch, Layden, and Manuel 2017; [B] reprinted with permission from Smith et al. 2014.)

5.7.2 TRANSCRIPTION FACTORS INVOLVED IN SPECIFYING PLACODE-DERIVED NEUROSECRETORY CELL TYPES IN VERTEBRATES PLAY DIFFERENT ROLES IN CNIDARIANS, CTENOPHORES, AND SPONGES

I have argued in the previous section that those transcription factors (including Pitx, Lhx3/4, Islet, POU1f1, Six1/2, Six4/5, Eya) that coordinate the specification of a specific rostral subpopulation of neurosecretory cells in chordates (derived from Hatschek's pit in amphioxus and from the adenohypophyseal and olfactory placodes in vertebrates), probably served different and separate functions in the last common bilaterian ancestor. Even though the information on the expression of these and other relevant transcription factors in basal metazoans is limited, a similar pattern emerges for these taxa.

Pitx is expressed in lateral buds and in the upper body column of *Hydra* presumably including neurons or neurosecretory cells (Watanabe et al. 2014). Lhx3/4 is expressed in the apical organ of ctenophores known to harbor sensory neurons and neurosecretory cells, while its expression in cnidarians has not been described (Srivastava, Larroux et al. 2010; Simmons, Pang, and Martindale 2012). Islet (see Chapter 3 for more detail) is found in many neuronal cell types of cnidarians at low levels and is strongly expressed in a subset of them including neuropeptidergic cells (Sebé-Pedrós, Saudemont et al. 2018). In ctenophores it is expressed in both sensory cells of the apical organ and the non-sensory comb cells carrying motile cilia (Simmons et al. 2012; Sebé-Pedrós, Chomsky et al. 2018), whereas in sponges it is enriched in pinacocytes and amoeboid-neuroid cells (Musser et al. 2019). The latter cells also express genes involved in dense-core vesicle secretion suggestive of a neurosecretory function (Musser et al. 2019). Similarly, Six1/2 and Eya (see Chapter 3) are expressed in subsets of sensory cells and neurons including some neurosecretory neurons in ectoderm and possibly endoderm of cnidarians (Stierwald et al. 2004; Graziussi et al. 2012; Hroudova et al. 2012; Sebé-Pedrós, Saudemont et al. 2018; Nakanishi, Camara et al. 2015). In sponges, they are strongly expressed in some pinacocytes and amoeboid-neuroid cells (Musser et al. 2019). FoxL2/3 is again expressed in subsets of neurons and cnidocytes in cnidarians and in some pinacocytes and amoeboid-neuroid cells in sponges (Sebé-Pedrós, Chomsky et al. 2018; Musser et al. 2019). Finally, POU1f1 is found in a subset of sensory cells in the rhopalia of the cnidarian *Aurelia* (Nakanishi et al. 2010), while it is widely expressed in all cell types of sponges (Musser et al. 2019).

Our survey of bilaterian taxa suggests that additional transcription factors such as FoxG, SP6-9, and DMRT4/5 that were possibly confined to the CNS in in early chordates, may have also contributed to the specification of different neuronal cell populations in ancestral bilaterians. In cnidarians, the expression of FoxG and SP6-9 has not been described (Schaeper et al. 2010; Shimeld et al. 2010), whereas a DMRT4/5 ortholog (DMRTb) was recently localized in scattered neuronal cells in both ectodermal and endodermal layers of *Nematostella* (Parlier et al. 2013). In sponges, both FoxG and SP6-9 are enriched in amoeboid-neuroid cells while DMRT4/5 expression has not been described. (Musser et al. 2019).

Let us finally turn to FoxA and GATA1/2/3, which were proposed to have played an early role in blastopore development and in promoting endodermal competence in early bilaterians. In cnidarians, FoxA expression also initiates around the blastopore and later continues in the pharynx and the gastral filaments, where it is expressed in both exocrine cells secreting digestive enzymes and in neurosecretory (insulin secreting) cells (Martindale, Pang, and Finnerty 2004; Steinmetz et al. 2017; Sebé-Pedrós, Saudemont et al. 2018). A recent study has revealed that the pharynx and gastral filaments are derived from the outer, "ectodermal" layer in *Nematostella*; however, based on the expression profile of multiple transcription factors, it has suggested these tissues to be homologous to the endoderm of bilaterians, while the remaining gastrodermis may be homologous to the bilaterian mesoderm (Steinmetz et al. 2017; Steinmetz 2019). If supported by further evidence, this interpretation would require us to abandon the traditional view that the cnidarian gastrodermis is homologous to bilaterian endoderm and mesoderm (e.g. Martindale et al. 2004). There is only a single GATA transcription factor in cnidarians, which duplicated into GATA1/2/3 and GATA4/5/6 in bilaterians. Cnidarian GATA is widely expressed in the gastrodermis (in regions proposed to be homologous to bilaterian endo- and mesoderm) and was found to be enriched in neurons including neuropeptidergic neurons (Martindale et al. 2004; Steinmetz et al. 2017; Sebé-Pedrós, Saudemont et al. 2018). While FoxA could not be identified in sponges or ctenophores and expression of GATA in ctenophores remains to be described, GATA is enriched in choanocyte and/or amoeboid-neuroid cells of sponges (Leininger et al. 2014; Musser et al. 2019). Since choanocytes form the inner layer of sponges, they may possibly be forerunners of the endomesoderm of eumetazoans (Leininger et al. 2014).

5.7.3 THE LAST COMMON EUMETAZOAN/METAZOAN ANCESTOR

Taken together this suggests that a role of Pitx, Lhx3/4, Islet, Six1/2, Six/4/5, Eya, DMRT4/5, and possibly FoxL2 and POU1f1 in the specification of different neuronal subtypes, including some neurosecretory cells, can be dated back to the eumetazoan ancestor (while there is insufficient evidence for FoxG and SP6-9). Possibly some of these transcription factors were already promoting neurosecretory cell types in metazoan ancestors prior to the origin of proper neurons. Moreover, while details of germ layer homology between sponges, cnidarians and bilaterians may still be contentious, FoxA and GATA appear to have played an early role in endomesodermal development in the last eumetazoan or possibly metazoan ancestor and possibly were also required for the differentiation of exocrine and neurosecretory cells within this tissue. However, while each of these transcription factors may have played some role in the specification of neuronal or neurosecretory subtypes, there is no evidence to suggest that these transcription factors co-regulated any particular neurosecretory cell type in the eumetazoan or metazoan ancestors.

6 Evolutionary Origin of Vertebrate Cranial Placodes

So far, in this book I have attempted to trace the evolutionary history and origin of the sensory and neurosecretory cell types that arise from cranial placodes in vertebrates. I have argued that mechano- and chemosensory cells are closely related cell types and are possibly evolutionarily derived from a sensorineural cell type already present in the last common ancestor of Eumetazoans (Chapter 3). This ancient sensorineural cell type probably also gave rise to photoreceptors (Chapter 4). In contrast, specialized types of neurosecretory cells probably evolved many times independently within the large family of cell types comprising neurons and endocrine cells releasing protein-based hormones (Chapter 5). In particular, the neurosecretory cell types that arise from the adenohypophyseal and olfactory placodes in vertebrates only evolved as new cell types within the chordate lineage (Chapter 5).

In this last chapter, I now address the question of how cranial placodes evolved as evolutionarily novel structures in vertebrates by redeploying pre-existing and sometimes evolutionarily ancient cell types. I will try to answer this question by integrating insights from comparative studies (Chapters 3–5), which are summarized in Fig. 6.1, with our knowledge on vertebrate placode development (Schlosser 2021). In the first section, I will briefly summarize the evolutionary history of placodally derived sensory and neurosecretory cells. In the second section, I will discuss, how these cell types may have become concentrated in non-neural ectoderm adjacent to the neural plate and how they may have become segregated into different cranial placodes in the vertebrate lineage due to changing responsiveness to factors patterning the ectoderm along the dorsoventral and anteroposterior axes. In the third section, I will then sketch how an increase in progenitor expansion may have endowed placodes with the capacity to form larger and more complex sense organs. In the fourth section, I will present a scenario of placode evolution. And in the fifth and final section, I will draw some general conclusions on the evolution of novelties as illustrated by placodes and placodal cell types.

6.1 EVOLUTION OF PLACODAL CELL TYPES – CONSERVATION AND NOVELTY

Neurons and/or neurosecretory cell types that required Ascl- and/or Atonal-related transcription factors for their specification probably originated in the stem lineage of eumetazoans and soon diversified into a few different cell types including a specialized sensory cell (Fig. 6.1). A core regulatory network (CoRN) of POU4, Islet, and PaxB-like transcription factors may have been specifically required to establish

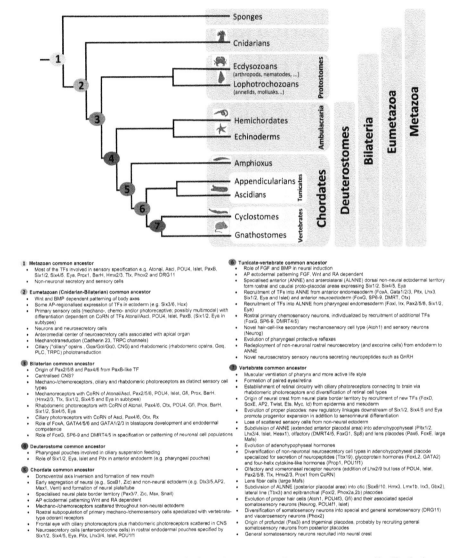

FIGURE 6.1 Key steps in the evolution of sensory and neurosecretory cells. Evolutionary changes of ectodermal patterning mechanisms and origin of new cell types are mapped onto the phylogenetic tree of metazoans. Important nodes are numbered and the key depicts, which characters can be traced to each node. Character origination events whose placement on the tree is controversial are indicated by a question mark. AP, anteroposterior; DV, dorsoventral; TF, transcription factor. (Modified from Schlosser et al. 2014.)

the identity of this primitive sensory cell type and distinguish it from other types of neurons. Other transcription factors including Six1/2, Six4/5, and their Eya cofactor probably played some roles in defining subtypes of sensory cells. Equipped with an apical cilium surrounded by microvilli and an axon, this cell type initially may have responded to a variety of stimuli including mechanical and chemical stimuli and light. This multimodal receptor diversified into multiple more specialized receptor cells in each eumetazoan lineage, giving rise to a large and multibranched cell type family. In ancestral eumetazoans, it may have transduced mechanical stimuli using Cadherin-23 and transient receptor potential (TRP) channels and used "ciliary" opsins (of the c-opsin, cnidopsin/xenopsin, or group 4 opsin families) and cyclic nucleotide gated (CNG) channels for phototransduction. Later, a different phototo-transduction mechanism using r-opsins and TRP channels evolved in some cells of this type.

In the last common ancestor of bilaterians, this ancestral sensory cell type gave rise to a dedicated mechanoreceptor cell type and two types of photoreceptors, ciliary, and rhabdomeric (Figs. 4.14, 6.1). The mechanoreceptor was specified by a CoRN of Pax2/5/8, POU4, Islet, Gfi, and the NK-related transcription factors Prox, BarH, with other NK factors (Hmx2/3, Tlx) as well as Six1/2, Six4/5, and Eya playing a role for specification of some of its subtypes. This mechanoreceptor, while employing some conserved mechanotransduction components (Protocadherin-15, Cadherin-23, MyosinVII, TRPC channels) gave rise to highly diverse mechanosensory cell types in the various bilaterian lineages. Moreover, chemoreceptors relying on different chemotransduction mechanisms repeatedly evolved convergently from this cell type. Rhabdomeric photoreceptors were specified by a CoRN resembling the one of mechanoreceptors (POU4, Gfi, Prox, BarH, Six1/2, Six4/5, Eya, but not Islet), but requiring Pax4/6 instead of Pax2/5/8, while ciliary photoreceptors were specified by Rx in combination with Pax4/6. These two photoreceptor cell types were maintained in most bilaterian lineages, although rhabdomeric photoreceptors came to predominate in the eyes of protostomes and ciliary photoreceptors in the eyes of deuterostomes. Other cell types like retinal neurons, pigment cells and lens cells may have evolved from these photoreceptors (e.g. vertebrate lens cells, which require large Maf proteins for specification like vertebrate ciliary photoreceptors).

In early chordates (Fig. 6.1), mechano- and chemoreceptors specified by the same pan-bilaterian mechanosensory CoRN and, thus, members of the same cell type family, probably were distributed as scattered cells throughout the non-neural ectoderm. Based on differences in the expression profile of transcription factors, several subtypes of sensory cells were probably present. The subtypes included primary sensory cells, located in the rostral ectoderm that expressed vertebrate-type odorant receptors and, thus, probably served as chemoreceptors. Other primary sensory cells presumably served mostly mechanoreceptive functions. There were also secondary mechano- or chemosensory cells (without axon), innervated by a sensory neuron, which transmitted stimuli to the central nervous system (CNS). Six1/2, Six4/5, and Eya probably contributed to the specification of subtypes of sensory receptors in the non-neural ectoderm. In the endoderm, however, Six1/2, Six4/5, and Eya cooperated with other transcription factors (Pitx, Lhx3/4, Islet, POU1f1) in the specification of a rostral population of endodermal neurosecretory cells (enteroendocrine cells) some of which also may have had exocrine functions in mucus secretion.

In the last common ancestor of tunicates and vertebrates (Fig. 6.1), different types of primary and secondary sensory cell types specified by the pan-bilaterian mecha-nosensory CoRN could still be found distributed widely throughout the non-neural ectoderm. Secondary sensory cells and their associated sensory neurons had evolved from primary sensory cells, by becoming differentially dependent on Atonal versus Neurogenin, respectively. Because we lack data from amphioxus, it is currently unclear, whether this happened already in the chordate ancestors or only in the lineage leading to tunicates and vertebrates. In the tunicate-vertebrate ancestor, multiple transcriptional regulators were then recruited into the anterior dorsal non-neural ectoderm (ANNE) from the anterior endomesoderm (FoxA, GATA1/2/3, Pitx, Lhx3, Six1/2, Eya, and Islet) and neuroectoderm (FoxG, SP6-9, DMRT, Otx) as well as into the anterolateral dorsal non-neural ectoderm (ALNNE) from the anterolateral (pharyngeal) endomesoderm (FoxI, Irx, Pax2/5/8, Six1/2, Eya) as will be discussed in more detail in section 6.2 below.

Six1/2 and Eya were now widely expressed in a new domain in the dorsal non-neural ectoderm and, thus, conferred Six1/2 and Eya-dependent identity on all cells that develop within this region (with Six4/5 probably playing a partly redundant role to Six1/2). Other transcription factors that are regionally confined to the ANNE or ALNNE, respectively, may have helped to specify region-specific subtypes of sensory and/or neurosecretory cells within these domains.

The co-expression of Six1/2 and Eya in the dorsal non-neural ectoderm with transcription factors (including competence factors) previously confined to anterior endoderm and neuroectoderm favored the establishment of new cell types defined by CoRNs that combine the identities of cell types previously restricted to non-neural ectoderm (sensory cells), endoderm (enteroendocrine cells), and neuroectoderm (some neurosecretory cells). This may have facilitated (1) the evolution of a specialized subpopulation of rostral primary chemosensory neurons individualized by their dependence on transcription factors initially confined to the anterior CNS (such as FoxG and possibly SP6-9, DMRT4/5) in early chordates, (2) the evolution of novel types of neurosecretory sensory neurons secreting neuropeptides such as gonadotropin releasing hormone (GnRH) (confined to the CNS in early chordates), and (3) the redeployment of non-neuronal rostral neurosecretory cells which release a cocktail of protein-based hormones and were confined to the endoderm in early chordates (retained in Hatschek's pit in amphioxus), to the ANNE. While some of these neurosecretory cells may have preserved an additional exocrine function in mucus secretion, cells may subsequently have specialized for either exocrine or neurosecretory functions.

While Six1/2- and Eya-independent sensory cell types continued to develop outside the dorsal non-neural ectoderm in the ancestor of tunicates and vertebrates, such cells were lost in the vertebrate lineage, where Six1/2, Six4/5, and Eya adopted more central roles as CoRN members of sensory cells. In vertebrates, secondary sensory cells and their associated sensory neurons probably persisted as hair cells and the special somatosensory neurons (SSNs) that innervate them. General somatosensory neurons (GSNs) and viscerosensory neurons (VSNs) may subsequently have evolved as subtypes of SSNs that lost their associated secondary sensory cells. After the evolution of the neural crest in stem vertebrates, the CoRN for GSNs probably was redeployed in a subpopulation of neural crest cells, leading to the formation of dorsal root ganglia. Also, in the vertebrate lineage, the CoRN of the rostral population of primary chemosensory cells may have diverged further from the CoRN of other sensory cells and sensory neurons (becoming dependent on Lhx2/9 but independent of POU4, Islet, Pax2/5/8, Tlx, Hmx2/3, Prox) and diversified into several subtypes of olfactory and vomeronasal sensory cells. The neurosecretory neurons secreting GnRH and other neuropeptides persisted but largely lost their sensory function after migrating into the CNS (except for some neurons of the terminal nerve). Finally, the CoRNs of the rostral neurosecretory cells diversified into the many cell types of the vertebrate adenohypophysis.

This brief recapitulation of Chapters 3–5 highlights both the role of conservation and of novelty (due to duplication/divergence as well as recombination) in the evolution of the sensory and neurosecretory cells that develop from cranial placodes in vertebrates. It also reminds us that the central role of Six1/2 and Eya as upstream regulators of all cranial sensory cells and sensory neurons in the peripheral nervous system (PNS), evolved only after vertebrates branched off from their last common ancestor with tunicates. However, the evolutionary origin of vertebrate cranial placodes as novel structures, which give rise to new sensory and endocrine organs of the vertebrate head, cannot be fully understood by only considering the evolution of its constituent cell types. There are at least three other aspects that need to be considered. First, as alluded to already in the paragraphs above and in Chapter 5, the evolution of placodes also involved changes in the distribution of sensory and neurosecretory cells resulting from altered regulatory linkages to dorsoventral and anteroposterior patterning mechanisms. Second, in contrast to the scattered sensory or neurosecretory cells developing in the non-neural ectoderm of other chordates, vertebrate cranial placodes give rise to high density arrays of sensory receptors, neurosecretory cells, and supporting cells, presumably resulting from changes in the regulation of proliferation of sensory or neurosecretory progenitors and their subsequent differentiation. And third, in contrast to the simple arrangement of sensory and neurosecretory cells in the surface ectoderm of other chordates, many vertebrate cranial placodes undergo elaborate morphogenetic movements to form complex sensory organs and the anterior pituitary. Whereas space constraints do not allow me here to address the third point concerning the evolution of morphogenetic mechanisms, I will briefly discuss changes in ectodermal patterning and in the regulation of progenitor expansion in the following two sections.

6.2 CHANGES IN ECTODERMAL PATTERNING DURING EVOLUTION OF CRANIAL PLACODES

Evolutionary changes in ectodermal patterning mechanisms have been extensively reviewed in previous publications (Holland 2005; Arendt et al. 2008; Petersen and Reddien 2009; Holstein, Watanabe, and Ozbek 2011; Niehrs 2012; Holland et al. 2013; Schlosser, Patthey, and Shimeld 2014; Arendt, Tosches, and Marlow 2016; Kozmikova and Yu 2017; Zhao, Chen, and Liu 2019; Liu and Satou 2020). I will, therefore, not attempt an exhaustive survey here but rather highlight some of the important changes during the evolution of vertebrate cranial placodes (summarized in Figs. 6.1, 6.2). Because another novel vertebrate ectodermal structure, the neural crest, occupies an ectodermal territory adjacent to placodes, I will also briefly consider changes in ectodermal patterning that may have facilitated the evolution of the neural crest.

Amphioxus   **Tunicates**   **Vertebrates**

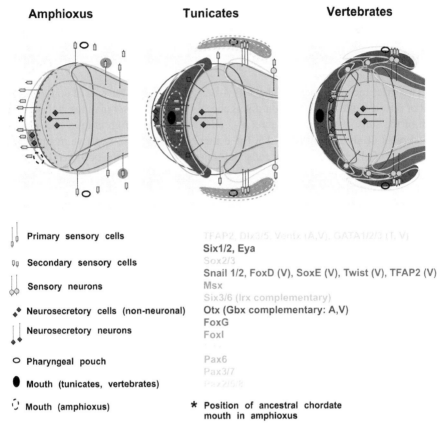

Legend (left)	Legend (right)
Primary sensory cells	TFAP2, Dlx3/5, Vent, (A,V), GATA1/2/3 (T, V)
	Six1/2, Eya
Secondary sensory cells	Sox2/3
	Snail 1/2, FoxD (V), SoxE (V), Twist (V), TFAP2 (V)
Sensory neurons	Msx
	Six3/6 (Irx complementary)
Neurosecretory cells (non-neuronal)	**Otx (Gbx complementary: A,V)**
	FoxG
Neurosecretory neurons	FoxI
Pharyngeal pouch	Pax6
	Pax3/7
Mouth (tunicates, vertebrates)	Pax2/5/8
Mouth (amphioxus)	* **Position of ancestral chordate mouth in amphioxus**

FIGURE 6.2   Schematic overview of transcription factor expression domains and origin of differentiated cell types around the neural plate in chordate embryos. The neural plate is indicated in green and epidermis in yellow. Domains of Six1/2 and Eya expression are shown in red. Pale red domains in amphioxus (Hatschek's pit and sensory cells) and tunicates (atrial siphon primordia) indicate expression domains only established at later developmental stages. Colored outlines enclose transcription factor expression domains except for Msx1 in all taxa and FoxI in vertebrates, which are expressed peripheral to the lines shown. Hatched lines indicate transcription factor expression domains that are only established at later developmental stages. The Snail1/2 expression domains in amphioxus and tunicates are shown by a blue outline, whereas the neural crest domain in vertebrates, which co-expresses Snail1/2 with an array of other transcription factors, is shown in solid blue. Expression domains of Irx and Gbx are not shown, but are posteriorly abutting the domains of Six3/6 and Otx expression, respectively. Some expression domains are found only in a subset of taxa as indicated (A, amphioxus; T, tunicates; V, vertebrates). See text for details. (Modified from Schlosser, Patthey, and Shimeld 2014.)

6.2.1 DORSOVENTRAL PATTERNING

6.2.1.1 Bilaterian Origins of Dorsoventral Patterning Mechanisms

A gradient of bone morphogenetic protein (BMP) signaling plays important roles in patterning the dorsoventral axis in many protostomes and deuterostomes, indicating an ancient role of BMP in dorsoventral patterning (e.g. Sasai et al. 1995; Lowe et al. 2006; Denes et al. 2007; Yu et al. 2007). This gradient is established by expression of high levels of BMP antagonists on one side of the embryo. Whereas in protostomes and hemichordates, BMP antagonists are enriched on the ventral side (where the mouth forms), they are concentrated on the dorsal side of chordates. Together with other evidence, this supports the idea that the dorsoventral axis was probably inverted and a new mouth formed in the stem lineage of chordates (Arendt and Nübler-Jung 1994). Moreover, in both protostomes (e.g. arthropods, annelids) and chordates, the CNS was shown to originate on the side with low BMP levels (i.e. on the ventral side in protostomes and on the dorsal side in chordates) (reviewed in Arendt et al. 2008; Schlosser et al. 2014). It has, therefore, been initially suggested that BMP has an anti-neurogenic effect (De Robertis and Sasai 1996; Arendt and Nübler-Jung 1997), until several studies established that high BMP levels are compatible with neuronal differentiation in the PNS in various taxa (Lowe et al. 2006; Denes et al. 2007; Lu, Luo, and Yu 2012).

It is more likely that the gradient of BMP, instead of directly repressing differentiation of specific cell types, rather specifies different territories of transcription factors, which then serve as molecular addresses along the dorsoventral axis (e.g. Niehrs 2010). As was discussed in Chapter 3 of Volume 1 (Schlosser 2021), in vertebrates this involves the upregulation of transcription factors promoting non-neural ectodermal competence (e.g. Dlx3/5, GATA2/3, TFAP2, FoxI1-4, Ventx1/2) ventrally and those promoting neural competence (e.g. Zic, SoxB1) dorsally (Fig. 6.2 and fig. 3.5 in Schlosser 2021). Initially overlapping in a "neural plate border region" during early embryonic development, some of these transcription factors subsequently resolve into mutually exclusive territories (figs. 3.4, 3.5 in Schlosser 2021). Additional signals impinging on this region from adjacent tissues then induce the Six1 expressing pre-placodal ectoderm in the dorsal part of the non-neural ectoderm and the neural crest in the ventral (lateral) part of the neural ectoderm.

As discussed in Chapter 1, it is still controversial whether the regulatory network partitioning the ectoderm into neural (CNS) and non-neural (epidermal) territories is evolutionarily ancient and can be traced back to ancestral bilaterians or whether centralization of the nervous system has evolved independently in the chordates, arthropods, annelids, and mollusks. On the one hand, there are similarities in the distribution of some transcription factors, which argue for a common bilaterian origin of the CNS. For example, SoxB1 is enriched on the neural side and Dlx on the non-neural side in insects, annelids, and chordates and other transcription factors have similar relative positions along the dorsoventral (medial to lateral, NK2.2, Gsx-NK6-Pax4/6, and Msx-Pax3/7) and anteroposterior axis (rostral to caudal, Otx, Pax2/5/8, and Hox) (Arendt and Nübler-Jung 1999; Lichtneckert and Reichert 2005; Hirth and Reichert 2007; Arendt et al. 2008, 2016; Urbach and Technau 2008; Schlosser et al. 2014). On the other hand, extensive phylogenetic surveys have revealed some substantial differences in the distribution of transcription factors between different phyla and have highlighted that most bilaterian phyla have a diffuse nervous system with simple ganglia, arguing for an independent origin of the centralized nervous systems in arthropods, annelids, and chordates (Roth and Wullimann 1996; Holland 2003, 2016; Moroz 2009; Northcutt 2012; Schlosser et al. 2014; Martin-Durán et al. 2018). Fortunately, for our discussion of placodal origins, we can leave this question open here and concentrate on the chordates.

6.2.1.2 Establishment of Neural Versus Non-neural Ectoderm in Chordates

The numerous similarities in development, structure, and function leave no doubt that the CNSs of amphioxus, tunicates, and vertebrates are homologous and are derived from a dorsal neural tube in the last common ancestor of chordates. Comparisons between the developmental mechanisms segregating the ectoderm into neural and non-neural territories (and then further subdividing these along the dorsoventral axis) in amphioxus, tunicates, and vertebrates, will give us some insights into how the ectoderm was patterned in ancestral chordates and how ectodermal patterning changed in the course of chordate evolution. We must proceed with caution here, since modifications of early development probably have evolved in both the amphioxus and tunicate lineage (see Chapter 1). However, overall, amphioxi belong to a slowly evolving lineage and apart from some left-right asymmetries they have largely maintained, like vertebrates, the ancestral regulative development of chordates (reviewed in Garcia-Fernàndez and Benito-Gutiérrez 2009; Bertrand and Escriva 2011; Schubert, Escriva et al. 2006). In contrast, tunicates comprise a fast evolving lineage and appear to have drastically modified their early development by increasingly substituting cell lineage-dependent patterning mechanisms for long-range inductive interactions (reviewed in Lemaire, Smith, and Nishida 2008; Lemaire 2009).

In amphioxus, like in vertebrates, the dorsal blastopore lip (dorsal mesoderm) of the embryo acts as an organizer, which is required for proper dorsoventral patterning of both endomesoderm and ectoderm and secretes various signaling molecules including BMP inhibitors and Nodal (Yu et al. 2007; Onai et al. 2010; Kozmikova et al. 2011, 2013; Morov et al. 2016; Le Petillon et al. 2017). Like in vertebrates, inhibition of BMP in amphioxus leads to a ventral expansion of the neural competence factor SoxB1. However, different from vertebrates, neurons do not differentiate in this expanded SoxB1 domain and Nodal is required for full neural induction, while fibroblast growth factor (FGF) is dispensable (Bertrand et al. 2011; Le Petillon et al. 2017).

In tunicates, the organizer and BMP-dependent patterning of the ectoderm has been completely lost (Darras and Nishida 2001a, 2001b, Lemaire et al. 2008). Instead, segregation of neural and non-neural ectoderm depends on cytoplasmic determinants in the posterior neural plate (derived from A-line blastomeres) and on FGF signaling in the anterior neural plate (derived from a-line blastomeres) (Figs. 1.11, 3.2B) (Hudson and Lemaire 2001; Bertrand et al. 2003; Hudson et al. 2003; Rothbächer et al. 2007). Consequently, the early neural marker Otx (Wada et al. 1996) is activated only in those a-line cells (left and right a6.5) that border the FGF expressing A-line cells at the 32-cell stage. These cells give rise to two rows of cells at the 110-cell stage (subsequently forming rows V/VI and III/IV, respectively, of the so-called "neural plate"; Fig. 3.2B). FGF signals from the adjacent A-line cells will then consolidate the neural fate in the posterior of these rows (III/IV) by promoting expression of Zic (ZicL in *Ciona*) and repressing FoxC, while the anterior row will form the ANNE (Wagner and Levine 2012). However, the central cells of row III/IV (a8.19) will ultimately be excluded from the neural tube and will join the ANNE to form the non-neural oral siphon primordium (Nishida 1987; Veeman et al. 2010). Possibly Dlx and Pitx transcription factors help to protect these cells from neuralizing transcription factors although this remains to be confirmed experimentally (see Schlosser et al. 2014).

These findings in amphioxus and tunicates suggest that dorsoventral patterning by a BMP gradient set up by the release of BMP inhibitors from the dorsal mesoderm (organizer) is an ancient chordate trait. However, it has been embellished and modified in different ways in the different chordate lineages, with Nodal acquiring a more central role in the amphioxus lineage and FGF in the tunicate-vertebrate lineage. Despite these differences in early signaling requirements, the resulting expression domains of many early ectodermal transcription factors are remarkably conserved between the different chordates (Fig. 6.2). Although data are incomplete, in both amphioxus and tunicates at least some of the putative neural (SoxB1) and non-neural transcription factors (Dlx3/5 – known as DllB in tunicates; TFAP2) begin to be expressed before gastrulation in widely overlapping domains throughout the ectoderm (Holland et al. 1996; Cattell et al. 2012; Oda-Ishii et al. 2016; Imai et al. 2017). During gastrulation, neural (SoxB1, Zic) and non-neural (Dlx3/5, TFAP2, Msx, Ventx in amphioxus) transcription factors become increasingly restricted to the dorsal and ventral ectoderm, respectively, until they form sharp boundaries at neural plate stages (Fig. 6.2) (Holland et al. 1996, 2000; Sharman, Shimeld, and Holland 1999; Kozmik et al. 2001; Meulemans and Bronner-Fraser 2002; Wada and Saiga 2002; Gostling and Shimeld 2003; Miya and Nishida 2003; Imai et al. 2004, 2006, 2017; Irvine et al. 2007; Meulemans and Bronner-Fraser 2007; Yu et al. 2007, 2008; Wagner and Levine 2012).

In tunicates, GATA factors also play an important role in promoting ectodermal competence, with GATA4/5/6 promoting general ectodermal competence and GATA1/2/3 promoting non-neural ectodermal competence in a positive feedback loop with Dlx3/5 (Bertrand et al. 2003; Imai et al. 2006; Irvine et al. 2011). Whereas the role of GATA1/2/3 in non-neural ectoderm is shared with vertebrates (see Chapter 3 in Schlosser 2021; Fig. 6.2), the early ectodermal function of GATA4/5/6 appears to have evolved specifically in tunicates. FoxI transcription factors, which likewise play a role in mediating non-neural ectodermal competence in vertebrates, are also expressed in parts of the non-neural ectoderm (including the prospective atrial primordia) in ascidians (Mazet et al. 2005). While FoxI has not been found outside the deuterostomes and its expression in amphioxus has not been described, it is expressed predominantly in the endoderm, including gill slits in other deuterostomes (Mazet, Yu et al. 2003; Tu et al. 2006; Fritzenwanker et al. 2014). This suggest that FoxI, like GATA factors, had an ancestral function in the endoderm and acquired new functions in the non-neural ectoderm, possibly including a role for non-neural competence, only in the tunicate-vertebrate lineage.

In summary, the many similarities in transcription factor expression suggest that neural and non-neural competence territories in the embryonic ectoderm may have been largely specified in a vertebrate-like manner in the chordate ancestor, although functional data are missing. However, GATA1/2/3 and FoxI may have a acquired a new role in non-neural competence only in the stem lineage of tunicates and vertebrates. In the next sections, I will explore how specialized territories in the neural and non-neural ectoderm originated during chordate evolution, which then gave rise to the neural crest and the pre-placodal ectoderm (PPE) in vertebrates.

6.2.1.3   Patterning of Neural Ectoderm in Chordates

While the lateral neural plate in amphioxus is enriched for some of the transcription factors (e.g. Pax3/7, Zic, Msx, Dlx3/5, Snail) defining the lateral neural plate and/or neural crest in vertebrates (Fig. 6.2), most neural crest specifiers (e.g. FoxD, SoxE, TFAP2, Twist, Ets, Myc, Id) are instead expressed in mesoderm, epidermis, or other parts of the neural plate and no neural crest-like migratory cells emerge from this region (Holland et al. 1996, 1999; Langeland et al. 1998; Yasui et al. 1998; Sharman et al. 1999; Meulemans and Bronner-Fraser 2002, 2004; Yu, Holland, and Holland 2002; Gostling and Shimeld 2003; Yu et al. 2008). The special regulatory environment in the lateral neural plate probably plays an important role in the specification of dorsal neural subtypes such as the intramedullary sensory neurons or photoreceptive Joseph cells (Lacalli 2004; Wicht and Lacalli 2005; Pergner and Kozmik 2017).

In ascidians, the lateral neural plate is enriched for Pax3/7, Zic, Msx, Ets, and Snail, while most neural crest specifiers (e.g. FoxD, TFAP2, Twist, Myc, Id) are expressed in other tissues or are not employed in early development at all (SoxE) (Fig. 6.2) (Ma et al. 1996; Wada, Holland, and Satoh 1996; Corbo et al. 1997; Wada and Saiga 2002; Imai, Satoh, and Satou 2003; Mazet, Hutt et al. 2003; Imai et al. 2004; Squarzoni et al. 2011; Wagner and Levine 2012). Thus, apart from specific Ets expression, the transcription factor expression profile in the lateral neural plate of tunicates closely resembles the one in amphioxus.

Nevertheless, several recent studies suggest that tunicates, but not amphioxus, have neural crest-like cells that originate from the lateral neural plate. First, there is the precursor of the ocellus in the ascidian sensory vesicle (Chapter 4), which is derived from the lateral neural plate, expresses FoxD and pigmentation genes and becomes migratory after overexpression of Twist (Abitua et al. 2012). Based on these features, these cells may represent a cell population in the lateral neural plate, from which the neural crest evolved in the vertebrate lineage (Abitua et al. 2012; Ivashkin and Adameyko 2013). Second, there are the bipolar tail neurons (BTNs) introduced in Chapter 3. These are Pax3/7, Snail, and Neurog expressing migratory sensory neurons which originate near the posterior neural plate border and send their dendrites to peripheral sensory cells (CESN) and their axons to motor neurons in the sensory vesicle (Stolfi et al. 2015). Due to this combination of properties, BTNs have been suggested to be neural-crest like cells resembling the neurons of the dorsal root ganglia, while the Atoh1 expressing CESNs have been proposed to be Merkel-like cells (Stolfi et al. 2015). However, as discussed in Chapter 3, based on their transcription factor profile, BTNs and CESNs resemble other sensory neurons originating from non-neural ectoderm in ascidians (e.g. PSNs, aATENs, pATENs; see Figs. 3.4,

3.5) (Cao et al. 2019). Moreover, both BTNs and CESNs originate from the same precursor cell (b8.18; see Fig. 3.2B) (Pasini et al. 2006; Stolfi et al. 2015; Horie et al. 2018), which abuts the Zic expressing neural plate and does not itself express Zic1 (Wada and Saiga 2002; Imai et al. 2004, 2006). This indicates that these cells arise from non-neural ectoderm rather than from the lateral neural plate. Together with their origin from a common precursor and their differential expression of Neurog (in BTNs) and Atoh (in CESNs), this suggests that CESNs and BTN may correspond to the hair cells and their associated sensory neuron in vertebrates rather than to any neural crest derived cell type and Merkel cells, respectively (see Chapter 3 for further discussion). Third and finally, a population of migratory cells that give rise to pigmented cells were initially proposed to be neural crest-like cells, but have subsequently been shown to derive from the mesodermal lineage (A7.6 cells) rather than from the neural plate border (Jeffery, Strickler, and Yamamoto 2004; Jeffery et al. 2008).

Taken together, the similarities in neural plate patterning between amphioxus and tunicates indicate that the lateral neural plate in ancestral chordates was already distinguished by a unique profile of transcription factors from the medial neural plate. This allowed the formation of specialized cell populations in this domain such as the pigmented ocellus in tunicates. While there may have been minor changes to the patterning of this domain in the stem lineage of tunicates and vertebrates, the neural crest only evolved in the vertebrate lineage by recruitment of many additional transcription factors into this domain, which probably endowed these cell with migratory properties and the ability to differentiate into many different cell types (reviewed in Meulemans and Bronner-Fraser 2004; Sauka-Spengler and Bronner-Fraser 2008; Medeiros 2013; Green, Simoes-Costa, and Bronner 2015; Cheung et al. 2019; Eames, Medeiros, and Adameyko 2020; York and McCauley 2020).

6.2.1.4 Patterning of Non-neural Ectoderm in Chordates

In contrast to the neural ectoderm, there is no indication that differential expression of transcription factors subdivides the non-neural ectoderm of amphioxus along the dorsoventral axis. In particular, Six1/2, Six4/5, and Eya are not expressed in the dorsal non-neural ectoderm like in tunicates and vertebrates, but rather are found in a few scattered sensory cells at later embryonic stages (Kozmik et al. 2007). However, early endomesodermal expression of these transcriptional regulators in amphioxus is clearly enriched on the dorsal side, suggesting that Six1/2, Six4/5, and Eya may already have responded to dorsally localized inductive signals in ancestral chordates (Kozmik et al. 2007). Thus, whereas regionalization of the neural ectoderm along the dorsoventral axis can be traced back at least to chordate ancestors, the non-neural ectoderm in ancestral chordates was probably more uniform.

A more elaborate subdivision of non-neural ectoderm appears to have evolved in the last common ancestor of tunicates and vertebrates. In tunicates, the dorsal non-neural ectoderm (NNE) has been shown to form placode-like thickenings surrounding the neural plate (Manni et al. 2004, 2005; Bassham and Postlethwait 2005; Mazet et al. 2005; Kourakis, Newman-Smith, and Smith 2010; Veeman et al. 2010). The palp region and the invaginating oral siphon primordium arise in early embryos from ectoderm at the anterior border of the neural plate (ANNE), while the invaginating

atrial siphon primordia arise at much later larval stages from ectoderm at the anterolateral border of the neural plate (ALNNE) (Fig. 3.2B). Several transcription factors including Pax3/7 are enriched in the dorsal non-neural ectoderm of ascidians
although their function in this domain is not well understood (Wada et al. 1996,
1997; Mazet, Hutt et al. 2003; Imai et al. 2004; Pasini et al. 2006). Other transcription factors are enriched in the anterior (Six1/2, DMRT4/5, FoxC, FoxG, SP6-9) or
posterior (Msx) part of the dorsal non-neural ectoderm, as will be discussed further
in section 6.2.2 on anteroposterior patterning below (Aniello et al. 1999; Imai et al.
2004, 2006; Tresser et al. 2010; Wagner and Levine 2012; Abitua et al. 2015; Horie
et al. 2018; Cao et al. 2019; Liu and Satou 2019). These anteroposterior restrictions
are probably maintained by cross-regulatory interactions between Msx and Six1/2 or
DMRT4/5 (Horie et al. 2018).

It is particularly noteworthy, that Six1/2 and Eya are specifically expressed in
a crescent-shaped domain in the anterior part of the dorsal non-neural ectoderm
(ANNE) in both ascidians and appendicularians resembling the PPE, the region that
gives rise to all cranial placodes in vertebrates (see Chapter 3 in Schlosser 2021).
Later, Six1/2 and Eya continue to be expressed in the palp region (Eya) and oral
siphon primordium in tunicates (Bassham and Postlethwait 2005; Mazet et al. 2005;
Abitua et al. 2015; Horie et al. 2018). In *Ciona*, Six1/2 is activated downstream of
DMRT4/5 in this region and requires both Dlx3/5 and Zic like in vertebrates (Imai
et al. 2006; Wagner and Levine 2012; Horie et al. 2018). Whereas a cocktail of
BMP-inhibitors, Wnt-inhibitors and FGF is required to induce Six1 in the PPE of
vertebrates (see Chapter 3 in Schlosser 2021), so far only BMP-inhibitors have been
shown to be required in ascidians (Abitua et al. 2015). In contrast to vertebrates, Wnt
does not repress Six1/2 in *Ciona* (Feinberg et al. 2019), while FGF is required for
specification of anterior neural tube (via repression of FoxC) rather than the ANNE
(Wagner and Levine 2012).

It is tempting to speculate that the responsiveness of Six and Eya to dorsally
localized inductive signals already in ancestral chordates (where they controlled dorsal endodermal expression), may initially have helped to confine their expression to
the dorsal part of anterior non-neural ectoderm (i.e. the ANNE). Dorsally restricted
ectodermal transcription factors like Zic may then have acquired new functions as
upstream regulators of Six and Eya at a later stage of evolution.

One notable difference between the ANNE in tunicates and the corresponding
anterior part of the PPE in vertebrates is that the mouth of tunicates forms in the dorsal part of the ANNE very close to the neural plate, while it is located more ventrally
in vertebrates. This suggests that there may have been lineage specific shifts in the
position of the mouth towards the dorsal or ventral side of this domain in tunicates
and vertebrates, respectively.

However, the scenario presented so far is incomplete, since it does not account for
ectodermal expression of Six1/2 and Eya outside of the ANNE in tunicates. At larval
stages, Six1/2 and Eya as well as Six4/5 are also upregulated in the anterolateral part
of the dorsal non-neural ectoderm (ALLNE) which invaginates to form the atrial
siphon primordium (Mazet et al. 2005). Since the atrial siphon primordium gives rise
to sensory cells of the cupular and capsular organs and co-expresses a combination
of transcription factors (e.g. Pax2/5/8 and FoxI) resembling the posterior placodal

area in vertebrates (see Chapter 3), it has been proposed to correspond to the region of ectoderm giving rise to the posterior (otic, lateral line and epibranchial) placodes of vertebrates (Jefferies 1986; Wada et al. 1998; Mazet, Hutt et al. 2003; Mackie and Singla 2004; Manni et al. 2004; Mazet et al. 2005; Kourakis and Smith 2007; Kourakis et al. 2010; Graham and Shimeld 2013). This hypothesis is supported by the expression patterns of additional transcription factors, as I will show in section 6.2.2 on anteroposterior patterning below.

In summary, the expression of Six1/2 and Eya in the ANNE and ALLNE together with the expression of other transcription factors in these domains suggests that the anterior and posterior part of the PPE of vertebrates evolved from corresponding domains of the dorsal non-neural ectoderm in the last common ancestor of tunicates and vertebrates. In contrast, the lateral neural plate with its co-expression of Zic, Pax3/7, Msx, Ets, and Snail is the most likely precursor region for the vertebrate neural crest. Taken together with the many developmental differences between neural crest and placodes in vertebrates in particular relating to their developmental fates and the underlying ectodermal competence (see Schlosser 2021), this supports previous proposals that placodes and neural crest have an independent evolutionary origin from dorsal non-neural and lateral neural ectoderm, respectively (Schlosser 2008).

6.2.2 ANTEROPOSTERIOR PATTERNING

6.2.2.1 Bilaterian Origins of Anteroposterior Patterning Mechanisms

As discussed in Chapter 4 in Volume 1 (Schlosser 2021), Wnt, FGF, and retinoic acid (RA) signaling play prominent roles in patterning the vertebrate ectoderm along the anteroposterior axis, with Wnt inhibitors secreted from anterior sources setting up a Wnt gradient of increasing concentrations toward the posterior pole. Comparison with other metazoans indicates that the role of Wnt in anteroposterior patterning is evolutionarily ancient and probably dates back to eumetazoans, whereas RA has adopted this role only in chordates and FGF only in the vertebrate-tunicate lineage (Campo-Paysaa et al. 2008; Itoh and Ornitz 2011). A role of canonical Wnt signaling in promoting posterior fates (or oral fates in cnidarians) has been shown for cnidarians, protostomes (e.g. flatworms, polychaetes, some arthropods) and in the anterior region of hemichordates (Miyawaki et al. 2004; McGregor et al. 2008; Petersen and Reddien 2008; Duffy et al. 2010; Niehrs 2010; Holstein et al. 2011; Pani et al. 2012; Marlow, Matus, and Martindale 2013; Marlow et al. 2014). However, Wnt-dependent patterning has been replaced by alternative mechanisms for anteroposterior patterning several times independently in different taxa (one famous example is the bicoid gradient in *Drosophila*).

Similar to BMP along the dorsoventral axis, the Wnt gradient specifies different territories of transcription factors along the anteroposterior axis and some of these transcription factors appear to have a conserved distribution in both neural and non-neural ectoderm (as well as in other germ layers) along the anteroposterior axis in bilaterians. In particular, posterior expression of the direct Wnt targets Gbx, Irx, and Cdx as well as of Hox transcription factors and anterior expression of Otx, Emx, Six3/6, FoxQ2, Rx and Fezf, has been observed in ecdysozoans, lophotrochozoans and deuterostomes (e.g. Hirth et al. 2003; Lowe et al. 2003; Aronowicz and Lowe

2006; Irimia et al. 2010; Tomer et al. 2010; Steinmetz et al. 2010, 2011; Posnien, Koniszewski, and Bucher 2011; Pani et al. 2012; Santagata et al. 2012; Sen, Reichert, and VijayRaghavan 2013; Marlow et al. 2014; Gonzalez, Uhlinger, and Lowe 2017; Valencia et al. 2019). In addition, FoxG is associated with the anterior neural ectoderm in many taxa (see Chapter 5). Pax4/6 and Pax2/5/8 also tend to have regionalized expression in neural and non-neural ectoderm along the anteroposterior axis in many bilaterians. However, apart from an anterior Pax4/6 domain and a Pax2/5/8 domain located between anterior Otx and posterior Gbx expression, there is no clear evidence that any of these domains is evolutionarily conserved (Grossniklaus, Pearson, and Gehring 1992; Hirth et al. 2003; Lowe et al. 2003; Urbach and Technau 2003; Denes et al. 2007; Pani et al. 2012).

6.2.2.2 Anteroposterior Ectodermal Patterning in Chordates

Similar to vertebrates, amphioxus employs a Wnt gradient established by anteriorly secreted Wnt inhibitors as well as posteriorizing RA signaling to pattern neural and non-neural ectoderm along the anteroposterior axis (Escriva et al. 2002; Holland 2002; Schubert et al. 2004; Holland et al. 2005; Schubert, Holland et al. 2006; Yu et al. 2007; Onai et al. 2009, 2012; Koop et al. 2010). However, in contrast to vertebrates, FGF in amphioxus is not required as a posteriorizing signal in the ectoderm, but rather contributes to the maintenance of anterior neural fates (Beaster-Jones et al. 2008; Bertrand et al. 2011; Le Petillon et al. 2017). In tunicates, the importance of these global signals for anteroposterior ectodermal patterning has declined and they have been partially superseded by cell-autonomous patterning mechanisms. With exception of the posterior tail epidermis, graded Wnt signaling plays no role for anteroposterior patterning in tunicates (Pasini et al. 2012). However, both RA and FGF signaling help to pattern the ectoderm along the anteroposterior axis in ascidians, while RA signaling pathway genes have been lost in appendicularians (Katsuyama et al. 1995; Katsuyama and Saiga 1998; Hudson et al. 2003; Nagatomo and Fujiwara 2003; Nagatomo et al. 2003; Canestro and Postlethwait 2007; Hudson, Lotito, and Yasuo 2007; Kanda, Wada, and Fujiwara 2009; Pasini et al. 2012; Racioppi et al. 2014).

Most of the transcription factors that show regionalized expression in ectoderm along the anteroposterior axis in vertebrates in response to Wnt or RA signaling show a similar distribution in amphioxus. This includes anterior expression of Otx, Fezf, Six3/6, and Pax4/6 and posterior expression of Gbx, Cdx, Irx, and Hox (Williams and Holland 1996; Brooke, Garcia-Fernàndez, and Holland 1998; Glardon et al. 1998; Castro et al. 2006; Kozmik et al. 2007; Beaster-Jones et al. 2008; Kaltenbach et al. 2009; Irimia et al. 2010; Koop et al. 2010; Pascual-Anaya et al. 2012; Vopalensky et al. 2012). Among those, the expression boundaries of Otx, Cdx, and Hox have been shown to shift in response to Wnt (Otx, Cdx) or RA (Hox) signaling (Onai et al. 2009; Koop et al. 2010). Apart from Fezf, all these transcription factors are expressed in multiple germ layers and apart from Otx and Fezf, all of them show regionalized expression in the non-neural as well as neural ectoderm.

In tunicates, some of these transcription factors have been lost from the genome (Gbx) or have a diminished role in anteroposterior patterning (e.g. Cdx, Hox), possibly in relation to the modified mode of early tunicate development (Wada et al.

2003; Ikuta, Satoh, and Saiga 2010). In particular, the clustering of *Hox* genes in the genome has been lost and the spatial and temporal coordination of their expression has been compromised (Ikuta et al. 2004, 2010; Seo et al. 2004). However, together with Irx, Hox transcription factors are still confined to posterior regions of the ecto-derm, while Otx, Six3/6, and Emx (Fezf has not been described) are enriched in the anterior ectoderm including the palp region, oral siphon primordium, and the anterior neural plate (Wada et al. 1996; Oda and Saiga 2001; Ikuta et al. 2004; Imai et al. 2004; Mazet et al. 2005; Moret et al. 2005; Pasini et al. 2012). The conserved distribution of these transcription factors suggests that they already played a role in patterning the ectoderm along the anteroposterior axis in the chordate ancestor, a role that was largely maintained in extant chordates even though some of the upstream signaling mechanisms may have been altered in the tunicate lineage.

6.2.2.3 Patterning of the Anterior and Anterolateral Non-neural Ectoderm

Besides transcription factors with a general anteroposterior patterning function in multiple germ layers, we have to consider whether additional transcription factors that are specifically involved in distinguishing anterior from posterior placodes in vertebrates have any comparable patterning functions in amphioxus or tunicates. As discussed in Chapter 5, in amphioxus several of the transcription factors (Pitx, POU1f1, Lhx3, Islet as well as Six1/2 and Eya) specifying an anterior group of plac-odes are co-expressed in an anterior outpocketing of the endoderm on the left side, Hatschek's diverticulum, which subsequently gives rise to Hatschek's pit, a struc-ture with exocrine and neurosecretory function (Jackman, Langeland, and Kimmel 2000; Yasui et al. 2000; Boorman and Shimeld 2002; Wang et al. 2002; Kozmik et al. 2007; Candiani et al. 2008). Of these transcription factors, only Pitx and Islet also show some expression in the anterior non-neural ectoderm. Because Pitx has been implicated in mouth formation in vertebrates, the transient Pitx expression in amphioxus anterior to the neural plate may be a remnant of the region where the midline mouth formed in ancestral chordates (Yasui et al. 2000; Chen et al. 2017). Additional transcription factors defining anterior placodes in vertebrates are restricted to the endomesoderm (GATA1/2/3, FoxE), the anterior neural tube (FoxG) or have not been analyzed in amphioxus (DMRT4/5, SP6-9) (Toresson et al. 1998; Yu et al. 2002; Zhang and Mao 2009; Wang et al. 2012; Dailey, Kozmikova, and Somorjai 2017).

In tunicates, many of the transcription factors specifying the adenohypophyseal and olfactory placodes in vertebrates, including GATA1/2/3, Pitx, Lhx3, Islet, Otx, FoxG, SP8, and DMRT4/5 are co-expressed with Six1/2 and Eya in the ANNE (see Chapter 5) (Boorman and Shimeld 2002; Christiaen et al. 2002; Imai et al. 2004; Mazet et al. 2005; Tiozzo et al. 2005). One exception is FoxE, which defines the lens placode in vertebrates and which is only expressed in the endomesoderm of tunicates (Ogasawara and Satou 2003).

Together with the data from amphioxus, this suggests that several of the transcrip-tion factors defining the identity of anterior placodes in vertebrates, may initially have functioned in the anterior endoderm in ancestral chordates, where some of them (Pitx, POU1f1, Lhx3, Islet, Six1/2, Eya) may have contributed to the specifica-tion of rostral neurosecretory cells. Others may have been expressed in other parts of the endoderm (FoxA, GATA1/2/3/, FoxE) or the anterior neural plate (e.g. FoxG). Pitx may also have had an additional expression domain in the non-neural ectoderm ANNE, where it possibly helped to regulate formation of the new chordate mouth. In the last common ancestor of tunicates and vertebrates, most of these endodermal and neuroectodermal transcription factors (except FoxE) then adopted new roles in defining the ANNE, from which anterior placodes then evolved in vertebrates, with adenohypophyseal, olfactory, and lens placodes only becoming individualized in the vertebrate lineage.

In contrast to some of the markers of anterior-most placodes, neither Six1/2, Six4/5, and Eya nor any of the transcription factors defining the identity of more posterior (otic, lateral line, and epibranchial) placodes (e.g. Pax2/5/8, SoxB1, Tbx1/10, and Irx) are expressed in the anterolateral neural ectoderm of amphioxus (Kozmik et al. 1999; Mahadevan, Horton, and Gibson-Brown 2004; Kozmik et al. 2007; Kaltenbach et al. 2009). However, in tunicates Six1/2, Six4/5, Eya, FoxI, Pax2/5/8, and Irx but not Tbx1/10 are expressed in the ALLNE, when this invaginates to form the atrial siphons at late larval stages (Wada et al. 1998; Imai et al. 2004; Takatori et al. 2004; Mazet et al. 2005).

Interestingly, many of the transcriptional regulators defining the posterior placodal area in vertebrates are also expressed in the pharyngeal endomesoderm and the developing gill slits throughout chordates (Irx, Tbx1/10, SoxB1) or even deuterostomes (Six1/2, Six4/5, Eya, FoxI, Pax2/5/8) (Holland 2005; Schlosser 2005; Kozmik et al. 2007; Meulemans and Bronner-Fraser 2007; Gillis, Fritzenwanker, and Lowe 2012; Fritzenwanker et al. 2014; Nakayama and Ogasawara 2017). This suggests that they may have played an ancestral role in the development of gill slits and/or in the differentiation of neurosecretory cells developing from the pharyngeal endoderm, but adopted new roles in the adjacent ALLNE in the last common ancestor of tunicates and vertebrates before becoming individualized into otic, lateral line and epibranchial in the vertebrate lineage.

The last common ancestor of tunicates and vertebrates may, therefore, have sported two ectodermal domains expressing Six1/2 and Eya, an anterior one in the ANNE and a more posterior one in the ALLNE, which were distinguished from each other by the co-expression of different transcription factors (e.g. Pitx, Lhx3, FoxG, SP8, DMRT4/5, Otx anteriorly; Pax2/5/8, FoxI and Irx posteriorly). Many of these transcriptional regulators including Six1/2 and Eya may have originally functioned in the anterior endoderm in ancestral chordates before becoming recruited to the ectoderm in the tunicate/vertebrate stem lineage. The ANNE may subsequently have evolved into the palp region and oral siphon primordium in tunicates and into the anterior part of the PPE in vertebrates, from which adenohypophyseal, olfactory and lens placodes originated. The ALLNE, in turn, may have given rise to the atrial siphon primordium in tunicates and the posterior part of the PPE in vertebrates, from which otic, lateral line and epibranchial placodes originated. It is possible that ALLNE formation was subsequently delayed relative to ANNE development in tunicates due to their evolution of a non-feeding larva and the postponement of pharyngeal development into late (premetamorphic) larval stages (Baker and Schlosser 2005). Since ANNE and ALLNE are discrete ectodermal regions that still lack many defining features of placodes (e.g. the massive expansion of neuronal progenitor cells discussed below), I refer to them as rostral and caudal proto-placodal domains (Schlosser et al. 2014). Whether the profundal and trigeminal placodes in vertebrates originated from the anterior or posterior proto-placodal region is currently unclear, since they share expression of Irx and FoxI1-3 with posterior placodes but expression of Otx with anterior placodes (see section 6.4 below for further discussion).

6.3 CHANGES IN THE REGULATION OF PROGENITOR DEVELOPMENT AND SENSORY DIFFERENTIATION DURING EVOLUTION OF CRANIAL PLACODES

As discussed in Chapter 5 in volume 1 (Schlosser 2021), Six1 and Eya1 in vertebrates promote differentiation of sensory cells and neurons by directly activating the bHLH genes *Atoh1* and *Neurog1* and multiple other neuronal differentiation genes, while Six4 acts in partly redundant fashion to Six1 (Ahmed et al. 2012; Ahmed, Xu, and Xu 2012; Riddiford and Schlosser 2016). Six1/2 also directly activates Atonal in *Drosophila* (Zhang et al. 2006). Together with their expression in subpopulations of sensory or neurosecretory cells in many cnidarians, protostomes and deuterostomes (see Chapters 3–5), this suggests that these transcriptional regulators have an evolutionarily ancient role in initiating differentiation in subsets of sensory cells.

When Six1/2, Six4/5, and Eya were recruited to the dorsal non-neural ectoderm (ANNE and ALNNE) in tunicate/vertebrate ancestors, where they are co-expressed with some of the transcription factors defining anterior and posterior placodes, they probably maintained their ancient evolutionary role in promoting sensory or neurosecretory differentiation (Fig. 6.3). However, these anterior and posterior proto-placodal domains in tunicates only give rise to sparse populations of scattered sensory and neurosecretory cells. Whereas the neurosecretory cells of the palp region, the GnRH expressing aATENs of the larva and the cells of the coronal organ in the oral siphon of the adult, develop from the anterior proto-placodal territory, other larval sensory neurons (possibly pATENs) and the cupular organs in the atrial siphon of the adult develop from the posterior proto-placodal territory (see Chapters 3 and 5).

Because only a subset of the cells expressing Six1/2, Six4/5, and Eya in these domains differentiate into sensory neurons and neurosecretory cells in tunicates, these proteins probably need to cooperate with additional, unknown transcription factors that are only present in some cells to initiate differentiation of these sensory and neurosecretory cells in tunicates (Fig. 6.3). In contrast, most cells with sustained expression of Six1/2 and Eya in the PPE and its derivative placodes in vertebrates appear to differentiate into sensory cells, sensory neurons or neurosecretory cells (see Chapter 5 in Schlosser 2021), whereas those cells that express Six1/2 and Eya only transiently differentiate into lens or epidermis (Pieper et al. 2011). This suggests that Six1/2 and Eya (and probably Six4/5) may have become sufficient to drive sensory and/or neurosecretory differentiation in the non-neural ectoderm of vertebrates, possibly by gaining control of any additional transcription factors originally required for this process in the common ancestor of tunicates and vertebrates (X in Fig. 6.3). As a consequence, almost the entire set of cells in the Six1/2/and Eya expressing

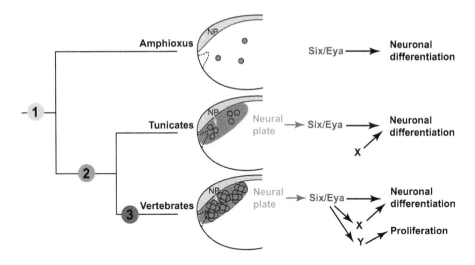

FIGURE 6.3   Scenario for regulatory evolution upstream and downstream of *Six1/2* and *Eya* genes in chordates. It is proposed that the role of Six1/2 and Eya (dark grey) in vertebrate placodes evolved in 3 steps. Step 1: In the chordate ancestor, Six1/2 and Eya were expressed in sensory cells scattered throughout the non-neural ectoderm regulating their neuronal/sensory differentiation and had additional functions in the endoderm (not shown). This condition is maintained in amphioxus. Step 2: In the common ancestor of tunicates and vertebrates, Six1/2 and Eya acquired new upstream regulators and were recruited to the border of the non-neural ectoderm with the neural plate. However, only scattered sensory cells differentiate from that region, indicating that Six1/2 and Eya probably cooperate with other factors (X) that may only be available in a few cells to promote neuronal/sensory differentiation. This condition is maintained in tunicates. Step3: In the vertebrate lineage, Six1/2 and Eya acquired new target genes (e.g. X and Y) allowing them to adopt additional roles in patterning and proliferation control of neuronal/sensory progenitors. Neurula stage embryos are shown with the neural plate (NP) indicated in light grey. Cells (circles) and regions expressing Six1/2 and Eya are highlighted in dark grey. The position of the mouth is indicated by a hatched line. (Reprinted from Schlosser, Patthey, and Shimeld 2014, modified from Schlosser 2011.)

PPE will differentiate into neurons and sensory cells in vertebrates, in contrast to the small subset of Six1/2 and Eya expressing cells in tunicates. However, more evidence is needed to provide support for this hypothesis.

Furthermore, in contrast to amphioxus, tunicates and many other bilaterians, the Six1, Six4, and Eya expressing progenitors of sensory neurons and neurosecretory cells in the vertebrate PPE divide extensively to give rise to the large, high density arrays of sensory, neuronal, and neurosecretory cells comprising the cranial sensory organs, ganglia, and anterior pituitary. As reviewed in Chapter 5 in Volume 1 (Schlosser 2021), Six1 in conjunction with Eya promotes this expansion of progenitors possibly via the direct transcriptional activation of *SoxB1*, *Hes*, and other genes promoting progenitor maintenance and proliferation although the precise mechanisms are still unclear (e.g. Li et al. 2003; Schlosser et al. 2008; Zou et al. 2008; Chen, Kim, and Xu 2009; Riddiford and Schlosser 2016). Because Six1/2 and Eya are only expressed in individual sensory cells in amphioxus and do not appear to drive progenitor expansion in amphioxus or tunicates, this new role in progenitor proliferation probably evolved only in the vertebrate lineage (Fig. 6.3).

In conclusion, the evolution of proper placodes in vertebrates from Six1/2, Six4/5, and Eya expressing proto-placodal territories probably involved rewiring of the gene regulatory network not only upstream but also downstream of these transcriptional regulators as previously proposed (Schlosser 2007; Schlosser et al. 2014). While maintaining their ancient role in sensory and neuronal differentiation, Six1/2, Six4/5, and Eya probably acquired additional roles in promoting progenitor expansion along with new target genes involved in the regulation of sensory, neuronal, or neurosecretory differentiation. In addition, they may have also acquired new regulatory linkages to regulators of morphogenetic processes permitting invagination of placodes and cell migration, a topic beyond the scope of this book. Overall, these transcriptional regulators, which initially functioned as initiators of differentiation in a subset of sensory cells, thereby adopted a much more comprehensive role for sensory and neurosecretory development in vertebrates, resembling the "intercalary evolution" proposed for Pax4/6 in eye evolution and for evolution of several other transcription factors (Gehring and Ikeo 1999; Peter and Davidson 2015).

6.4   A SCENARIO FOR PLACODE EVOLUTION

Since the mid-19th century, several different structures in amphioxus and tunicates have been proposed as potential homologs of various vertebrate placodes based on similarities in position, overall morphology, properties of cell types, gene expression, or mechanisms of development. These previous proposals have been extensively reviewed elsewhere (Holland and Holland 2001; Holland 2005; Schlosser 2005; Patthey, Schlosser, and Shimeld 2014) and will be summarized here only briefly in Table 6.1. Instead, I will here propose a scenario of placode evolution that integrates and summarizes the main points discussed in this and previous chapters (indicating the expression of some core transcriptional regulators in the tissues or cell types mentioned in parentheses) (Fig. 6.4). I will attempt to embed them into a narrative of plausible functional changes in the spirit of and building on the "New Head hypothesis" (Northcutt and Gans 1983). This will necessarily involve some speculation but will be informed by what we currently know about the biology of amphioxus and tunicates (see Chapter 1). I will divide the course of placode evolution within the chordates into six main steps, the first three leading up to the evolution of proper placodes in early vertebrates and the second three leading to the diversification of placodes in the vertebrate lineage. The scenario assumes that step 1 clearly preceded step 2 and this preceded steps 3–6. However, steps 3–6 all occurred in the stem lineage of vertebrates and it remains currently unresolved, whether these steps occurred in parallel or consecutively and if so, in which temporal order.

TABLE 6.1

A Summary of Proposals for Homologues of Placodes or Placode-Derivatives in Tunicates and Amphioxus

Placode	Authors	Species	Hypothesis/key findings	Comments
Olfactory	de Quatrefages (1845)	Amphioxus	Identification of anterior primary sensory neurons and rostral never clusters in adult	Subsequently suggested as possible olfactory or mechanosensory organ (see for example Baatrup (1982), Baker and Bronner-Fraser (1997), Sharman et al. (1999)
	Sharman et al. (1999)	Amphioxus	Expression of *Msx* in larval rostrum and tentative suggestion of olfactory homologue as per de Quatrefages (1845)	Still lack data connecting larval gene expression and adult neural structures
Adenohypophyseal	Julin (1881)	Ascidians	Comparison of ascidian neural complex with vertebrate pituitary	Subsequently supported by peptide antibody staining Chang et al. (1982) and neuroanatomy Gorbman et al. (1999).
	Hatschek (1881)	Amphioxus	Hatschek's pit as a possible adenohypophysis homologue	
	Nozaki and Gorbman (1992)	Amphioxus	Immunoreactivity for adenohypophyseal hormones in Hatschek's pit	Proposed homology of Hatschek's pit with adenohypophysis
	Burighel et al. (1998), Manni et al. (1999)	*Botryllus schlosseri* (ascidian)	Delamination of neurons in the ascidian neural complex similar to placode neuron delamination	Proposed homology of ascidian neurohypophyseal duct with adenohypophyseal placode
	Boorman and Shimeld (2002), Yasui et al. (2000),	Amphioxus	Expression of *Pitx* in the pre-oral pit of amphioxus, which gives rise to part of Hatschek's pit	Proposed homology of pre-oral pit with adenohypophysis
	Boorman and Shimeld (2002), Christiaen et al. (2002)	*Ciona*	Expression of *Pitx* in ascidian oral siphon rudiment and ciliary funnel	Proposed to support homology of ciliary funnel with adenohypophysis

Otic/Lateral line	Bone and Ryan (1978)	*Ciona*	Identification of sensory cell clusters in ascidian atrium with some similarity to vertebrate otic/lateral line hair cells	Homology to hair cells questionable as these ascidian cells are primary sensory cells
	Jefferies (1986)	Ascidians	Proposed homology of the ascidian atrial openings with vertebrate otic placode	Based on comparative anatomy integrated with palaeontological data
	Wada et al. (1998)	*Halocynthia roretzi* (ascidian)	Expression of ascidian *pax2/5/8* in atrial primordia compared to orthologue expression in vertebrate otic placode	First use of molecular data to support an hypothesis of placode homology
	Burighel et al. (2003)	Various ascidians	Further characterization of neural complex and identification of secondary sensory neurons in the coronal organ of the oral siphon of diverse ascidian species	Proposal of homology between coronal organ and otic and lateral line placodes.
	Kourakis et al. (2010), Kourakis and Smith, (2007)	*Ciona*	Development of ascidian atrial primordium regulated by FGF signaling and details of underlying cell biology	These papers do not propose novel homologies but data are discussed in the context of *Ciona* placode homologues
	Lacalli and Hou (1999)	Amphioxus	Detailed characterization of sensory neurons in the surface ectoderm	Only scattered cells identified, not placode-like territories
Others	Mazet et al. (2005)	*Ciona*	Expression of several placode marker genes in ascidian atrial and oral siphon rudiments	Proposal of two ancestral placode territories.
	Bassham and Postlethwait (2005)	*Oikopleura* (non-ascidian tunicate)	Expression of placode marker genes in ectodermal sensory structures and proposition of olfactory and adenohypophyseal homologues	
	Benito-Gutierrez et al. (2005), Kozmik et al. (2007), Mazet et al. (2004)	Amphioxus	Expression of various sensory neuron markers (Coe, Six, Eya, Trk) in scattered amphioxus ectoderm cells	Likely to be sensory neurons as identified by Lacalli and Hou (1999)

Source: Reprinted from Patthey, Schlosser, and Shimeld 2014

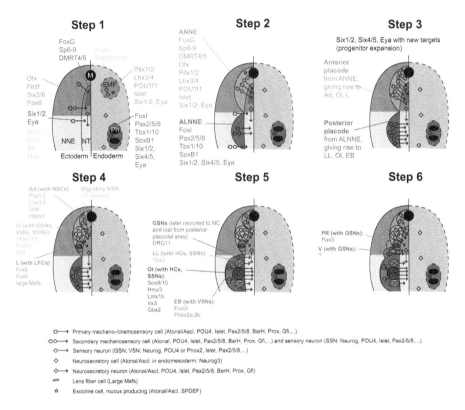

FIGURE 6.4 A scenario for placode evolution in six steps. Anterior parts of embryos are shown schematically in dorsal views with ectoderm shown to the left and endoderm to the right (anterior to the top). Expression domains of transcriptional regulators and distribution of different types of sensory and neurosecretory cells are indicated. Neural crest evolution (during step 3) is not illustrated. See main text for details. M, mouth; EMP, endomesodermal pouch; NNE, non-neural ectoderm; NT, neural tube; Ph, Pharynx with pharyngeal slits; ANNE, anterior dorsal non-neural ectoderm; ALNNE, anterolateral dorsal non-neural ectoderm; Ad, adenohypophyseal placode; Ol, olfactory placode; L, lens placode; LFC, lens fiber cell; OSN, olfactory sensory neuron; VSN, vomeronasal sensory neuron; NSN, neurosecretory neuron (GnRH); NSC, neurosecretory cells (non-neuronal); LL, lateral line placodes; Ot, otic placode; EB, Epibranchial placodes; HC, hair cell; SSN, special somatosensory neuron; GSN, general somatosensory neuron; VSN, viscerosensory neuron; PR, profundal placode; V, trigeminal placode.

6.4.1 First Step: Concentration of Neurosecretory Cells in Rostral Endoderm

The first step, resulting in the concentration of neurosecretory cells in the rostral endoderm, was taken in stem chordates (Fig. 6.4). After dorsoventral inversion and formation of a new mouth immediately rostral to the neural plate, these evolved into amphioxus-like animals, but with a largely symmetrical body plan. They were able to move freely, if still quite sluggishly, powered by alternating contractions of trunk and tail muscles. Like their deuterostome ancestors, they were filter feeders using cilia to create a flow of water through the pharyngeal region and capturing food particles in mucus. Ciliary movement in the pharynx was probably modulated by efferent neurons from the CNS innervating the pharynx (as in extant amphioxus and tunicates, Bone 1961; Mackie et al. 1974)

In these animals, primary mechano- and chemosensory cells (POU4, Islet, Pax2/5/8, Gfi, Prox, BarH; Six1/2, Six4/5, and Eya in subset) were scattered throughout the non-neural ectoderm (with some chemosensory cells expressing odorant receptors clustered rostrally) similar to extant amphioxus. These sensory cells projected to a motor center in the prospective hindbrain, which controlled trunk muscles, thereby modulating swimming and/or eliciting simple escape responses (Lacalli 1996, 2001; Wicht and Lacalli 2005). This motor center subsequently evolved into the primary motor center in amphioxus and the visceral ganglion in tunicates. Secondary sensory cells and sensory neurons of unknown function were probably also present (Fig. 6.5A), possibly homologous to those found in vertebrates (with an Atoh-dependent sensory cell and a Neurog-dependent sensory neuron).

Neurosecretory cells in the endoderm (as in extant amphioxus and tunicates: Sherwood, Adams, and Tello 2005; Osugi, Sasakura, and Satake 2020) modulated metabolic body functions in response to nutrient levels in an endocrine and/ or paracrine fashion (Fig. 6.5A). In areas, where the digestive tract had openings in the oral and pharyngeal regions, some of these neurosecretory cells were in contact with the environment and may have also responded to changes in environmental conditions. A specialized population of neurosecretory cells was probably concentrated rostrally in a pair of endomesodermal pouches (either bilaterally or only on the left side as discussed in Chapter 5). These pouches (retained in amphioxus as Hatschek's left diverticulum/pit; Pitx, Lhx3/4, POU1f1, Islet, Six1/2, Eya) were connected to the exterior serving both exocrine (mucus secretion) and endocrine functions, releasing protein hormones in response to environmental stimuli. Their rostral position near the mouth facilitated quick responses to chemical changes in the stream of water flowing towards the head. This probably promoted the specialization of these neurosecretory cells for the modulation of feeding (via control of ciliary movements) and of metabolic and reproductive processes (Fig. 6.5A).

Regionally appropriate development of sensory and neurosecretory cells was controlled by networks of transcription factors (Fig. 6.4). Already during gastrulation, there was a clear segregation of neural plate (Zic, SoxB1) and non-neural ectoderm (Dlx3/5, TFAP2, Msx, Ventx) in response to a dorsoventral BMP gradient. The non-neural ectoderm which was competent to give rise to sensory cells was relatively uniform along the dorsoventral axis. However, the neural plate had a molecularly distinct lateral part (Pax3/7, Zic, Msx, Dlx3/5, Snail) without giving rise to a migratory and multipotent neural crest. Along the anteroposterior axis, the ectoderm was subdivided (like other germ layers) by the anteriorly (Otx, Fezf, Six3/6, Pax4/6) or posteriorly (Gbx, Cdx, Irx, and Hox) restricted expression of several transcription factors in response to Wnt and RA signaling. However, most transcription factors showing a placodal expression in vertebrates were not expressed in the non-neural ectoderm but instead in the general endomesoderm (FoxA, GATA1/2/3) or were

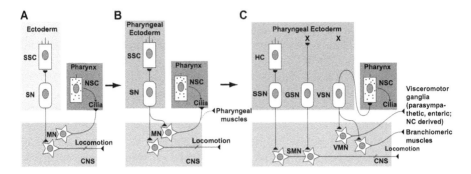

FIGURE 6.5 Model for evolution of somatosensory and viscerosensory neurons. (**A**) In early chordates, primary sensory cells (not shown) as well as secondary sensory cells (SSC) and their associated sensory neurons (SN) contacted somatic motor neurons (SMN) in the CNS that (indirectly) controlled locomotion by trunk muscles. Other motor neurons controlled ciliary beating in the pharynx. Neurosecretory cells (NSC) in the pharynx (endoderm) modulated ciliary beating and other metabolic functions in response to nutrient levels. (**B**) In the ancestors of tunicates and vertebrates, primary sensory cells (not shown) and SSCs/ SNs located in pharyngeal ectoderm (ALNNE) near the pharyngeal slits developed new connections to motor neurons innervating the pharynx, modulating beating of cilia and pharynx-associated muscles, allowing the formation of protective pharyngeal reflexes. (**C**) In the pharyngeal ectoderm (posterior placodal area derived from the ALNNE) of vertebrates, SSCs and SNs evolved into hair cells (HC) and special somatosensory neurons (SSN), respectively. In addition, some SNs lost their associated SSCs (X). One subpopulation of these sensory neurons evolved into general somatosensory neurons (GSNs): Dendrites maintained contact with ectodermal cells, while axons projected to somatomotor neurons (SMN) innervating locomotory muscles of the trunk. Another subpopulation of these sensory neurons evolved into viscerosensory neurons (VSNs): Dendrites formed new contacts with endodermal cells (in particular neurosecretory cells sensing nutrient and gas levels), while the axons project to motor neurons innervating the branchiomeric muscles ventilating the pharynx. Some visceromotor neurons (VMN) subsequently gained additional functions by connecting to neural crest (NC) derived parasympathetic and enteric ganglia. See text for details.

enriched in the anterior-most endomesoderm (Six1/2, Six4/5, Eya, Pitx, POU1F1, Lhx3, Islet, FoxE), pharyngeal endoderm (Six1/2, Six4/5, Eya, FoxI, Pax2/5/8, Tbx1/10, SoxB1) or anterior neuroectoderm (FoxG, Sp6-9, DMRT4/5). In these domains, they probably helped to promote regional identity for neurosecretory and other cells.

6.4.2   SECOND STEP: FORMATION OF ANTERIOR AND POSTERIOR PROTO-PLACODAL TERRITORIES

The second step resulting in an anterior (ANNE) and posterior (ALNNE) proto-placodal area in the non-neural ectoderm was taken in the stem lineage leading up to the last common ancestor of tunicates and vertebrates (Fig. 6.4). This was most likely a much more amphioxus - than tunicate-like animal but was possibly distinguished by a larger and more efficient pharyngeal apparatus, still powered by ciliary movement. Together with the increasing rostral concentration of specialized sensory and neurosecretory cells, this may have allowed it to become a more actively exploring filter feeder. The increased efficiency of pharyngeal food trapping may, on the one hand, have been a first step towards the increased mobility in vertebrates but, on the other hand, may have facilitated the evolution of sessility (with an elaborate pharyngeal basket) and of a specialized non-feeding larva in tunicates.

The increasing specialization of the rostral pole involved the redeployment of primary and secondary sensory cells from non-neural ectoderm and of neurosecretory and mucus-producing cells from the endoderm to the ANNE located anterior to the neural plate in the very anterior part of the dorsal non-neural ectoderm (Fig. 6.4). The recruitment of transcription factors from the general endomesoderm (FoxA, GATA1/2/3), anterior endomesoderm (Six1/2, Six4/5, Eya, Pitx, POU1F1, Lhx3/4, Islet, FoxE), and anterior neuroectoderm (FoxG, Sp6-9, DMRT4/5) resulted in the co-activation of multiple transcription factors regulating competence and/or cell-type identity in the same cells which facilitated the evolution of new types of cells and their clustering in this domain (Fig. 6.4; see Chapter 5). These new cell types included non-neural neurosecretory cells (as found in tunicate palps and vertebrate adenohypophysis) and neurosecretory sensory cells releasing GnRH and other neuropeptides (as in the tunicate ATENs and in olfactory placode derived GnRH neurons in vertebrates). The sensory cells in the ANNE conveyed information about mechano- and chemosensory stimuli to the brain, where they continued to affect the locomotor response of trunk muscles. Some of the rostral sensory cells may also have been involved in mediating protective reflexes allowing the modulation of ciliary beating in the pharynx and the contraction of muscles associated with the mouth to prevent the ingestion of large particles (as shown for the coronal organs in the oral siphon of ascidians; Mackie et al. 1974, 2006). Finally, sensory cells also sent side branches to adjacent sensory and neurosecretory cells (Fig. 6.6A) (as in ascidians; Ryan, Lu, and Meinertzhagen 2018). In addition, the neurosecretory subpopulation of sensory cells modulated the activity of their target cells – including sensory and neurosecretory cells in the rostral epithelium derived from the ANNE and in the brain – by the paracrine release of GnRH in response to environmental cues (as shown for vertebrates: Park and Eisthen 2003; Kawai, Oka, and Eisthen 2009).

The close association of the ALNNE with the pharynx suggests that the evolution of this second special domain of the non-neural ectoderm, may have been linked to the fusion of epithelia during gill slit formation; the formation of invaginating primordia of an atrium surrounding the pharynx (Katz 1983; Jefferies 1986; Kourakis et al. 2010); and/or to the evolution of a more sophisticated neural control of pharyngeal function in the last common ancestor of tunicates and vertebrates (Fig. 6.5B). The latter may have required the evolution of new central connections between sensory cells – either primary sensory cells or secondary sensory cells (Atoh1, POU4, Islet) and their associated neurons (Neurog, POU4, Islet) – located near the pharyngeal slits and the motor neurons innervating the pharynx and modulating the beating of pharyngeal cilia and the contraction of pharynx-associated muscles (as in extant tunicates: Mackie et al. 1974, 2006). This may have allowed the evolution of protective reflexes of the pharyngeal apparatus in response to mechanical or chemical stimuli in the water current passing through the pharyngeal slits, in line

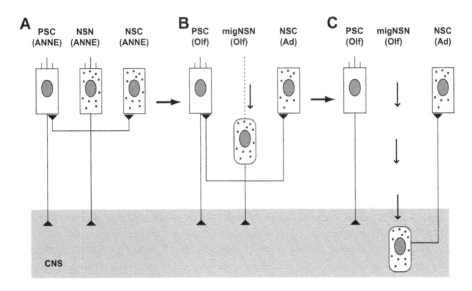

FIGURE 6.6 Evolution of migratory GnRH neurons in the olfactory placode. (**A**) Neurosecretory neurons (NSN) (expressing GnRH and other neuropeptides) in the anterior dorsal non-neural ectoderm (ANNE) initially were primary sensory cells that modulated activity of adjacent primary sensory cells (PSCs) and non-neuronal neurosecretory cells (NSC). (**B**) Subsequently NSNs arising from the olfactory placode (Olf) became migratory (mig NSN) but still maintain sensory connection to the olfactory placode via their dendrite (hatched). These migNSNs continue to modulate PSCs (also derived from Olf) and NSCs (derived from the adenohypophyseal placode: Ad) via their axons. This situation persists in the vertebrate terminal nerve. (**C**) Some NSNs invade the CNS and lose their dendritic connection to the olfactory placode giving rise to the NSNs of the vertebrate hypothalamus that modulate the release of glycoprotein hormones from the NSCs in the adenohypophysis.

with a previous proposal (Baker, O'Neill, and McCole 2008). Although, the neural organization of the brain was radically reorganized when ascidians evolved a radical metamorphosis, in which the larval nervous system is completely remodeled into a relatively unstructured cerebral ganglion, the circuitry underlying these protective reflexes appears to have been maintained (Mackie et al. 2006). However, only primary sensory cells were retained in the tunicate atrial siphon primordium (ALNNE), while secondary sensory cells and their associated sensory neurons were lost.

Recruitment of transcription factors from the adjacent pharyngeal endoderm (Six1/2, Six4/5, Eya, FoxI, Pax2/5/8, Tbx1/10, SoxB1) endowed the pharyngeal ectoderm (ALNNE) with a specific regional identity allowing the individualization of sensory neurons derived from this region (Fig. 6.4). Pharynx-associated muscles expressing Tbx1/10 and Islet may subsequently have evolved into the atrial siphon muscles in tunicates and the branchiomeric muscles ventilating the pharynx in vertebrates, a homology supported by recent studies (Stolfi et al. 2010; Diogo et al. 2015). Endodermally derived neurosecretory cells located in the pharynx, such as the serotonergic cells identified in both tunicates and vertebrates (Fritsch,

Van Noorden, and Pearse 1982; Georges 1985; Jonz et al. 2016; Hockman et al. 2017) may have contributed to modulate ciliary beating in a paracrine or possibly endocrine fashion – I will come back to this when discussing the evolution of viscerosensory cells below.

Transcription factors that are involved in early anteroposterior regionalization of the ectoderm in response to Wnt, RA, and FGF signals (e.g. anterior Otx, Six3/6, posterior Gbx, Irx) probably also acquired new roles in defining the boundary between ALNNE and ANNE and to properly position them along the anteroposterior axis.

6.4.3   Third Step: Formation of Proper Placodes Producing Large Sensory Organs

The third step in placode evolution, resulting in the origin of proper placodes that formed large and complex sensory and neurosecretory organs, happened only in the lineage leading to vertebrates (Fig. 6.4). In this lineage, a switch from ciliary to muscular ventilation of the pharynx increased the efficiency in feeding and gas exchange. Together with a further elaboration of sensory and neurosecretory organs at the front end and the formation of paired eyes, this enabled these animals to increase in size and adopt an increasingly active and predatory life style with macrophagous feeding (e.g. suction feeding powered by branchiomeric muscles) replacing microphagous feeding (filter feeding driven by cilia) along the way (Northcutt and Gans 1983; Purnell 2001; Northcutt 2005).

This step required the evolution of new regulatory linkages downstream of Six1/2, Six4/5, and Eya, which were already expressed in two contiguous protoplacodal regions in the dorsal non-neural ectoderm, the ANNE associated with the mouth opening at the front of the neural plate and the ALNNE overlying the pharyngeal region. These regulatory linkages enabled Six1/2, Six4/5, and Eya to activate new transcription factors driving progenitor expansion, differentiation, and morphogenesis (Fig. 6.3), thereby adopting a central role for the development of cranial sensory and neurosecretory organs. Consequently, ANNE and ALNNE were transformed into a novel type of embryonic tissues, the first cranial placodes (Fig. 6.4). However, some of the mucus producing cells that were present in the ANNE of the last common ancestor of tunicates and vertebrates (see Chapter 5), may have separated early in development from the developing placodes and clustered in separate cement glands. The latter were specified independently of Six1/2, Six4/5, and Eya but remained under regulatory control of anteriorly restricted transcription factors such as Otx and Pitx (Gammill and Sive 2001; Dickinson and Sive 2007; Jin and Weinstein 2018). Other mucus-secreting cells may have been retained in the anterior placode (and later in the olfactory placode).

Importantly, because Six1/2, Six4/5, and Eya were expressed in both ANNE and ALNNE, these regulatory changes may have affected both regions simultaneously in an example of concerted evolution (see Chapter 2). Thus, the generic placodal functions of Six1/2, Six4/5, and Eya may have evolved, when the dorsal non-neural ectoderm was already partitioned into an anterior and posterior part. At the same time, sensory neurons that were specified independently of Six1/2, Six4/5, and Eya disappeared from the non-neural ectoderm, possibly because other transcription factors that could initiate sensory differentiation in earlier chordate lost this ability or became dependent on Six1/2, Six4/5, and Eya themselves.

The stem lineage of vertebrates not only saw the evolution of proper placodes from the non-neural ectoderm, but also the evolution of another novel embryonic tissue, the migratory and multipotential neural crest, from the lateral border of the neural plate. However, we currently do not know, in which order of events proper placodes or neural crest cells originated during this period. Evolution of the neural crest involved the recruitment of transcription factors (e.g. FoxD, SoxE, TFAP2, Twist, Ets, Myc, Id) controlling epithelial-mesenchymal transitions, cell migration, and cytodifferentiation from the mesoderm and other embryonic tissues. These cells provided skeletal support elements for the pharyngeal arches and contributed to the elaboration of cranial ganglia and other tissues.

The evolution of two proper placodes (anterior and posterior placode) in the stem lineage of vertebrates was followed by the further subdivision of these placodes in steps 4–6 of placode evolution. The functional specializations of the different placodes together with the elaboration of other head structures probably allowed a more and more sophisticated analysis of environmental sensory stimuli in support of an increasingly complex exploratory behavior.

6.4.4   FOURTH STEP: DIVERSIFICATION OF THE ANTERIOR GROUP OF PLACODES

In the fourth step of placode evolution, the cell types arising from the anterior placode developing from the ANNE diversified further and the placode subdivided into distinct adenohypophyseal, olfactory and lens placodes (Fig. 6.4). In particular, differential expression of transcription factors led to the evolution of different subtypes of non-neuronal neurosecretory cells, specialized for the secretion of neuropeptides (Tbx19), glycoprotein hormones (FoxL2, GATA2), and four-helix cytokine-like hormones (Prop1, POU1f1). These became concentrated in the adenohypophyseal placode, which evolved in the medial part of the ANNE under the control of locally restricted transcription factors (e.g. Pitx, Lhx3/4, Islet, Hesx1).

On the other hand, the chemosensory neurons expressing odorant receptors as well as the neurosecretory sensory neurons releasing GnRH became concentrated in the olfactory placode evolving in a more lateral part of the ANNE under control of other transcription factors (DMRT4/5, FoxG1, Sp8). At least some of these neuropeptidergic, GnRH secreting cells became migratory and moved towards and partly into the forebrain (Fig. 6.6B, C). Some of these cells still send their dendrite to the ANNE, receive sensory input and continue to modulate its sensory and neurosecretory cells via their axon; these ultimately evolved into the GnRH neurons of the terminal nerve (Mousley et al. 2006; Kawai et al. 2009). Others, which lost the connection to the ANNE but started to secrete GnRH into small vessels carrying blood to the adenohypophyseal region probably evolved into the GnRH producing cells of the vertebrate hypothalamus, which secrete GnRH into the portal vessels of the median eminence and modulate the release of glycoprotein hormones from the adenohypophysis (fig. 1.11 B in Schlosser 2021) (Sower, Freamat, and Kavanaugh 2009). Later, brain-derived neurosecretory neurons with projections to the same portal vessels acquired control functions for other adenohypophyseal cell types.

The lens placode originated from the most posterior part of the ANNE. This may have been linked to the origin of paired eyes from an originally unpaired eye of chordates. However, we currently cannot rule out that paired eyes may already have been present in the last common ancestors of tunicates and vertebrates or even of chordates (see Chapter 4). Alternatively, lenses may have evolved in parallel to an increasing complexity of retinal organization in stem vertebrates, which made image-forming vision possible. The lens placode became distinct from other anterior placodes by differential maintenance of Pax6 and recruitment of additional transcription factors from the endoderm (FoxE) or the adjacent retina (large Mafs).

The recruitment of large Mafs, which already played a role in the specification of ciliary photoreceptors in the retina, may have facilitated the formation of a new specialized placodal cell type, the lens fiber cell. As discussed in Chapter 7 in Volume 1 (Schlosser 2021), ciliary photoreceptors in vertebrates use α-, β-, and γ-crystallins as chaperones. Moreover, in tunicates, $\beta\gamma$-crystallins were found to be expressed in the otolith, the pigmented sister cell of the photoreceptor in the neural tube (Shimeld et al. 2005), indicating that the association of $\beta\gamma$-crystallins with ciliary photoreceptors dates back at least to the last common-ancestor of vertebrates and tunicates. This suggests that in stem vertebrates $\beta\gamma$-crystallins were also expressed downstream of large Mafs in the ciliary photoreceptors of the retina. After becoming genetically distinct from other photoreceptors by expression of particular Maf paralogs (MafA, c-Maf), which drive the expression of high levels of crystallin, transparent lens fiber cells were subsequently recruited together with the large Mafs to the lens placode.

Many of the transcription factors that acquired differential expression in the adenohypophyseal, olfactory, and lens placodes in the stem lineage of vertebrates, remain co-expressed during early vertebrate development in the extended anterior placodal area before they become restricted to these individual placodes (remember that I use "extended anterior placodal area" to refer to the precursor of adenohypophyseal, olfactory and lens placodes, and "anterior placodal area" to refer to the precursor of only adenohypopyseal and olfactory placodes; see Chapter 4 in Schlosser 2021). This endows the entire extended anterior placodal area with a bias to develop into one or the other of the anterior placodes until signaling molecules from adjacent tissues consolidate a particular fate.

6.4.5   Fifth Step: Diversification of the Posterior Group of Placodes

The fifth step of placode evolution saw a similar diversification of cell types arising from the posterior placode developing from the ALNNE and a subdivision of this placode into distinct otic, lateral line, and epibranchial placodes (Fig. 6.4). With the evolution of a more efficient pharyngeal apparatus powered by branchiomeric muscles, the secondary sensory cells (Atoh1, POU4f3, Gfi) and their associated neurons (Neurog, POU4f1, Islet) located near the gill slits further diversified. Some sensory neurons kept their associated secondary sensory cells, which specialized as mechanoreceptors for responding to water movements, thereby evolving into hair cells, while the sensory neurons became the SSNs of vertebrates (Fig. 6.5C).

However, other sensory neurons lost their associated secondary sensory cell while their free nerve endings responded to various mechanical and chemical stimuli

(Fig. 6.5C). In one subpopulation of these sensory neurons, distinguished by expression of DRG11, dendrites maintained contact with ectodermal cells, while the axons predominantly kept projecting to the motor center innervating locomotory muscles of the trunk, thus evolving into GSNs. In another subpopulation, distinguished by Phox2 expression, dendrites branched out to form contacts with endodermal cells, while the axons predominantly kept projecting to motor neurons innervating the muscles ventilating the pharynx (branchiomeric muscles), thus evolving into VSNs. Some of the latter may have developed connections to neurosecretory (serotonergic) cells in the pharynx that monitor O_2 and/or CO_2 levels similar to the endodermal serotonergic hypoxia sensitive cells – so-called "neuroepithelial cells" (NECs) – found in fish gills (Jonz et al. 2016; Hockman et al. 2017). This may have allowed VSNs to mediate protective reflexes controlling gill ventilation in response to oxygen levels (Baker et al. 2008; Hockman et al. 2017). Subsequently the visceromotor neurons in the efferent arm of these reflexes gained additional functions by connecting to neural crest derived parasympathetic and enteric ganglia (Fig. 6.5C).

After evolution of the neural crest, GSNs were probably recruited to this new cell population which invaded the pharyngeal region. Some recent studies suggest that neural crest cells in lampreys contribute mostly glia rather than sensory neurons to the ganglia of cranial nerves (Modrell et al. 2014; Yuan, York, and McCauley 2020). Thus, recruitment of GSNs into the neural crest may have happened only in the stem lineage of gnathostomes. However, the evidence so far is not conclusive and needs to be corroborated in further studies of lampreys and hagfishes. While both a placode-derived and a neural crest-derived population of GSNs was maintained in the profundal and trigeminal placodes (see below), placodally derived GSNs were subsequently lost from the pharyngeal region, possibly because they were no longer required due to the presence of functionally redundant neural crest-derived GSNs (it remains unclear, however, why both populations were maintained in the profundal and trigeminal placodes despite significant functional redundancy).

Following the diversification of sensory cells, differential expression of transcription factors led to the concentration of hair cells and SSNs in the dorsal part of the ALNNE (Sox9/10, Hmx2/3), and of VSNs in the ventral part (FoxI2, Phox2a,2b). The ventral part then gave rise to epibranchial placodes, while the dorsal part of the ALNNE divided further into otic (Lmx1b, Sox9/10, Irx3, Gbx2) and lateral line placodes (Tbx3). The subdivision of the dorsal part allowed the spatially well-defined invagination of its central portion, which forms a vesicle giving rise to the inner ear, a statocyst-like organ with hair cells and sensory neurons dedicated to the perception of balance and body movement (see Chapters 1 and 6 in Schlosser 2021). It is currently not clear, whether this invagination of the otic vesicle evolved only in the vertebrate lineage, or whether part of the ALLNE may have already invaginated in the last common ancestor of tunicates and vertebrates, subsequently giving rise to the invaginated atrial primordia of tunicates and the otic vesicle in vertebrates (see above) (Katz 1983; Jefferies 1986; Kourakis et al. 2010).

Similar to the anterior group of placodes, many of the transcription factors that acquired differential expression in otic, lateral line, and epibranchial placodes in the stem lineage of vertebrates, remain co-expressed during early vertebrate development in the posterior placodal area (the common precursor of the otic, lateral line and epibranchial placodes; see Chapter 4 in Schlosser 2021), before they become restricted to these individual placodes. Consequently, the entire posterior placodal area has a bias to develop into one or the other of the posterior placodes until signaling molecules from adjacent tissues consolidate a particular fate.

6.4.6 Sixth Step: Origin of Profundal and Trigeminal Placodes

In the sixth and last step of placode evolution, the profundal and trigeminal placodes containing GSNs emerged as distinct entities (Fig. 6.4). It is not entirely clear, whether they branched off from the anterior or posterior group of placodes, since they share some transcription factors (e.g. Otx) with the former and others (Irx, FoxI1/3) with the latter. However, if the scenario presented above is correct that pairs of secondary sensory cells and their associated sensory neurons were initially concentrated near the pharyngeal arches and that GSNs originated, when sensory neurons lost their secondary sensory cells (Fig. 6.5), an origin by specialization from a posterior group of placodes appears more likely. Differential expression of transcription factors in the profundal (Pax3) and trigeminal placodes (unknown transcription factors) subsequently resulted in the generation of distinct profundal and trigeminal placodes.

In summary, the evolution of cranial placodes, like the evolution of sensory or neurosecretory cell types, illustrates how conservation and novelty intertwine during the evolutionary origin of these vertebrate specific structures. Many of the CoRNs driving the specification of placodal cell types have a long evolutionary history (even though they further diversified in the vertebrate lineage) and so do the general mechanisms patterning the ectoderm along the dorsoventral and anteroposterior axis, as well as general mechanisms regulating the differentiation of neurons from proliferative progenitors. However, the establishment of new regulatory linkages between these different processes allowed the origin of cranial placodes as novel structures in vertebrates. During this process, the transcription factor Six1/2, its paralog Six4/5 and their Eya cofactors, which had an evolutionarily ancient role in the specification of subsets of sensory cells (dating back to the last common ancestor of eumetazoans), acquired new upstream links to patterning mechanisms and downstream links to regulators of progenitor proliferation and differentiation. Due to these changes in the gene regulatory network (GRN) up- and downstream of Six1/2, Six4/5, and Eya, these transcriptional regulators adopted a central role as key coordinators of large and complex sensory and neurosecretory organs in the vertebrate "New Head".

6.5 CONCLUSION AND OUTLOOK

We have come to the end of our long journey tracking the evolutionary origins of cell types developing from cranial placodes and trying to understand how evolutionarily old sensory and neurosecretory cell types were redeployed to build these novel and vertebrate-specific structures. Hopefully, this survey of placode evolution has helped to illustrate some of the general principles of conservation and novelty laid out in Chapter 2 of this book. We have been able to elucidate the evolutionary history of placodes, because conserved networks of regulatory genes, in particular those encoding transcription factors, allow us to identify the same (homologous) placode in different vertebrates. For example, the posterior placodal area in different vertebrates is defined by a GRN among *Six1/2, Six4/5, Eya, FoxI, Pax2/5/8, Tbx1/10*, and *SoxB1* genes, while additional interactions with *Lmx1b, Sox9/10, Irx3*, and *Gbx2* define the otic placode. Similarly, we have been able to trace the evolutionary history of sensory cell types, because of conserved CoRNs of genes specifying their identity, such as *POU4, Islet, Pax2/5/8, Gfi, Prox*, and *BarH* for an evolutionarily old mechano- or chemoreceptor cell type.

In spite of the conservation of these GRNs defining character or cell type identity, we also encountered evidence for the appearance of new structures or cell types during evolution. We saw novelty arising at different levels, for example the origin of novel placodes or of novel cell types. Moreover, at each level we saw novelty arising in different ways, either by duplication and divergence of pre-existing structures or cell types, or by recombination and re-deployment of some of their components. For example, the evolution of placodes as novel structures in stem vertebrates involved significant recombination and rewiring of GRNs upstream and downstream of Six1/2. In contrast, the origin of adenohypophyseal and olfactory placodes from an originally single anterior placode can be largely understood as a case of duplication and divergence (with differential regional distribution of transcription factors helping to create different identities between the medial and lateral part of a common anterior placodal precursor). Similarly, the origin of new types of neurosecretory cells and neurons in the ANNE in the stem lineage of tunicates and vertebrates probably required the recombination and rewiring of GRNs up-and downstream of many transcription factors expressed in different germ layers, thereby forging new CoRNs. However, the subsequent diversification of one of these neurosecretory cell types into the various adenohypophyseal cell types of vertebrates probably proceeded largely by duplication and divergence.

While the survey presented in this book points to deep evolutionary links between sensory cell types in different animal groups, one important caveat needs to be added here. The ability to respond to various stimuli either from the environment or from adjacent cells is one of the defining features of a living cell. It is, therefore, perhaps not surprising that this fundamental property has been exploited in evolution many times independently to provide cells with enhanced capacities to respond to one or the other class of stimuli. In multicellular organisms, in particular, even cells that are not specialized as sensory cells need to communicate with each other by chemical and mechanical stimuli. Therefore, chemoreceptive (e.g. neurons responding to neurotransmitters, immune cells) and mechanoreceptive cells (e.g. chondrocytes, epithelial cells) that are distinct from the specialized sensory cell types discussed in this book, are found throughout the animal kingdom. These express a variety of chemo- (e.g. neurotransmitter gated ion channels, receptors of various signaling pathways) and mechanoreceptive proteins (e.g. Enac channels, some potassium channels, Piezo channels) and a diverse array of transcription factors, indicating that they belong to cell types, unrelated (non-homologous) to the mechano- and chemoreceptors discussed here (reviewed in Whitfield 2008; Chalfie 2009). Similarly, extraocular photoreception based on either opsins or other photopigments (e.g. cryptochromes) has evolved repeatedly in cell types (e.g. epidermal cells, gonads, skin chromatophores in fishes, amphibians and cephalopods) that are unrelated to the photoreceptors discussed in this book (reviewed in Porter 2016; Ramirez et al. 2011).

Recent advances in single-cell RNA sequencing and related genomic technologies have not only given us exciting new tools for elucidating the mechanisms of cell fate decisions during development, but have also opened up the door for exploring the evolutionary relatedness (homology) of cell types. In this book, I have drawn on recent single cell RNA-seq studies wherever possible, but since the techniques are still new, we can expect a flood of additional studies in the next couple of years. While these studies promise to provide us with an unprecedented repository of data that can be mined for addressing the evolutionary origin of sensory or other cell types, they will also pose significant challenges.

First, claims of "single cell" resolution may give us an unjustified sense of confidence in the completeness of these data, even though the shallowness of many single-cell sequencing studies poses difficult problems of interpretations. In particular, the low level expression of many transcription factors that may be important components of CoRNs may be missed, distorting our view of which genes may be important regulators of cell fate decisions. Second, differences in the methodology between different studies (e.g. methods of filtering data and applying thresholds for gene expression) still impede comparisons between different studies and hamper evolutionary interpretations of these data. And finally, but most importantly, RNA-seq studies per se only give us insights into the temporal (and to a limited extent spatial) pattern of gene expression. We have to be very careful in the interpretation of any correlations identified in these studies, which may or may not point to direct causative links. Ultimately, functional studies will be needed to confirm causative links, allowing us to identify core regulators in the GRNs underlying cell fate decisions. Keeping these challenges in mind, should hopefully help us to overcome them. With our increased understanding of cell type evolution, the story of how old cells make new senses, as told in this book, will likely turn out to be oversimplified or even wrong in parts. I told the story based on current evidence, which I tried to present objectively, even though necessarily colored by my own ideas and research experience. I am looking forward to see this story being embellished, enriched and retold in the light of new findings in the years to come.

References

Abitua, P. B., T. B. Gainous, A. N. Kaczmarczyk, C. J. Winchell, C. Hudson, K. Kamata, M. Nakagawa, M. Tsuda, T. G. Kusakabe, and M. Levine. 2015. The pre-vertebrate origins of neurogenic placodes. *Nature* 524:462–465.

Abitua, P. B., E. Wagner, I. A. Navarrete, and M. Levine. 2012. Identification of a rudimentary neural crest in a non-vertebrate chordate. *Nature* 492:104–107.

Acampora, D., M. P. Postiglione, V. Avantaggiato, M. Di Bonito, F. M. Vaccarino, J. Michaud, and A. Simeone. 1999. Progressive impairment of developing neuroendocrine cell lineages in the hypothalamus of mice lacking the orthopedia gene. *Genes Dev* 13:2787–2800.

Acar, M., H. Jafar-Nejad, N. Giagtzoglou, S. Yallampalli, G. David, Y. He, C. Delidakis, and H. J. Bellen. 2006. Senseless physically interacts with proneural proteins and functions as a transcriptional co-activator. *Development* 133:1979–1989.

Achim, K., N. Eling, H. M. Vergara, P. Y. Bertucci, J. Musser, P. Vopalensky, T. Brunet, P. Collier, V. Benes, J. C. Marioni, and D. Arendt. 2018. Whole-body single-cell sequencing reveals transcriptional domains in the annelid larval body. *Mol Biol Evol* 35:1047–1062.

Achim, K., J. B. Pettit, L. R. Saraiva, D. Gavriouchkina, T. Larsson, D. Arendt, and J. C. Marioni. 2015. High-throughput spatial mapping of single-cell RNA-seq data to tissue of origin. *Nat Biotechnol* 33:503–509.

Adams, B. A., J. A. Tello, J. Erchegyi, C. Warby, D. J. Hong, K. O. Akinsanya, G. O. Mackie, W. Vale, J. E. Rivier, and N. M. Sherwood. 2003. Six novel gonadotropin-releasing hormones are encoded as triplets on each of two genes in the protochordate, Ciona intestinalis. *Endocrinology* 144:1907–1919.

Agca, C., M. C. Elhajj, W. H. Klein, and J. M. Venuti. 2011. Neurosensory and neuromuscular organization in tube feet of the sea urchin *Strongylocentrotus purpuratus*. *J Comp Neurol* 519:3566–3579.

Aguinaldo, A. M. A., J. M. Turbeville, L. S. Linfold, M. C. Rivera, J. R. Garey, R. A. Raff, and J. A. Lake. 1997. Evidence for a clade of nematodes, arthropods and other moulting animals. *Nature* 387:489–493.

Ahmed, M., E. Y. Wong, J. Sun, J. Xu, F. Wang, and P. X. Xu. 2012. Eya1-Six1 interaction is sufficient to induce hair cell fate in the cochlea by activating atoh1 expression in cooperation with Sox2. *Dev Cell* 22:377–390.

Ahmed, M., J. Xu, and P. X. Xu. 2012. EYA1 and SIX1 drive the neuronal developmental program in cooperation with the SWI/SNF chromatin-remodeling complex and SOX2 in the mammalian inner ear. *Development* 139:1965–1977.

Albertin, C. B., O. Simakov, T. Mitros, Z. Y. Wang, J. R. Pungor, E. Edsinger-Gonzales, S. Brenner, C. W. Ragsdale, and D. S. Rokhsar. 2015. The octopus genome and the evolution of cephalopod neural and morphological novelties. *Nature* 524:220–224.

Albuixech-Crespo, B., C. Herrera-Ubeda, G. Marfany, M. Irimia, and J. Garcia-Fernàndez. 2017. Origin and evolution of the chordate central nervous system: insights from amphioxus genoarchitecture. *Int J Dev Biol* 61:655–664.

Albuixech-Crespo, B., L. López-Blanch, D. Burguera, I. Maeso, L. Sánchez-Arrones, J. A. Moreno-Bravo, I. Somorjai, J. Pascual-Anaya, E. Puelles, P. Bovolenta, J. Garcia-Fernàndez, L. Puelles, M. Irimia, and J. L. Ferran. 2017. Molecular regionalization of the developing amphioxus neural tube challenges major partitions of the vertebrate brain. *PLoS Biol* 15:e2001573.

Aldridge, R. J., and M. A. Purnell. 1996. The conodont controversies. *Trends Ecol Evol* 11:463–468.

Allan, D. W., and S. Thor. 2003. Together at last: bHLH and LIM-HD regulators cooperate to specify motor neurons. *Neuron* 38:675–677.

Allan, D. W., and S. Thor. 2015. Transcriptional selectors, masters, and combinatorial codes: regulatory principles of neural subtype specification. *Wiley Interdiscip Rev Dev Biol* 4:505–528.

Alvarez-Bolado, G. 2019. Development of neuroendocrine neurons in the mammalian hypothalamus. *Cell Tissue Res.* 375:23–39.

Alvarez, C. E. 2008. On the origins of arrestin and rhodopsin. *BMC Evol Biol* 8:222.

Amcheslavsky, A., W. Song, Q. Li, Y. Nie, I. Bragatto, D. Ferrandon, N. Perrimon, and Y. T. Ip. 2014. Enteroendocrine cells support intestinal stem-cell-mediated homeostasis in *Drosophila. Cell Rep* 9:32–39.

Anctil, M. 2009. Chemical transmission in the sea anemone *Nematostella vectensis*: a genomic perspective. *Comp Biochem Physiol Part D Genomics Proteomics* 4:268–289.

Anderson, P. A. V., and C. Bouchard. 2009. The regulation of cnidocyte discharge. *Toxicon* 54:1046–1053.

Ando, M., M. Goto, M. Hojo, A. Kita, M. Kitagawa, T. Ohtsuka, R. Kageyama, and S. Miyamoto. 2018. The proneural bHLH genes Mash1, Math3 and NeuroD are required for pituitary development. *J Mol Endocrinol* 61:127–138.

Angueyra, J. M., C. Pulido, G. Malagón, E. Nasi, and P. Gómez Mdel. 2012. Melanopsin-expressing amphioxus photoreceptors transduce light via a phospholipase C signaling cascade. *PLoS One* 7:e29813.

Aniello, F., A. Locascio, M. G. Villani, A. Di Gregorio, L. Fucci, and M. Branno. 1999. Identification and developmental expression of Ci-msxb: a novel homologue of *Drosophila* msh gene in *Ciona intestinalis. Mech Dev* 88:123–126.

Annona, G., N. D. Holland, and S. D'Aniello. 2015. Evolution of the notochord. *Evodevo* 6:30.

Appel, T. A. 1987. *The Cuvier-Geoffroy debate: French biology in the decades before Darwin*. New York: Oxford University Press.

Arenas-Mena, C., and K. S. Wong. 2007. HeOtx expression in an indirectly developing polychaete correlates with gastrulation by invagination. *Dev Genes Evol* 217:373–384.

Arendt, D. 2003. Evolution of eyes and photoreceptor cell types. *Int J Dev Biol* 47:563–571.

Arendt, D. 2008. The evolution of cell types in animals: emerging principles from molecular studies. *Nat Rev Genet* 9:868–882.

Arendt, D. 2018. Animal evolution: convergent nerve cords? *Curr Biol* 28:R225–R227.

Arendt, D., E. Benito-Gutiérrez, T. Brunet, and H. Marlow. 2015. Gastric pouches and the mucociliary sole: setting the stage for nervous system evolution. *Philos Trans R Soc Lond B Biol Sci* 370:20150286.

Arendt, D., P. Y. Bertucci, K. Achim, and J. M. Musser. 2019. Evolution of neuronal types and families. *Curr Opin Neurobiol* 56:144–152.

Arendt, D., A. S. Denes, G. Jékely, and K. Tessmar-Raible. 2008. The evolution of nervous system centralization. *Philos Trans R Soc Lond B Biol Sci* 363:1523–1528.

Arendt, D., J. M. Musser, C. V. Baker, A. Bergman, C. Cepko, D. H. Erwin, M. Pavlicev, G. Schlosser, S. Widder, M. D. Laubichler, and G. P. Wagner. 2016. The origin and evolution of cell types. *Nat Rev Genet* 17:744–757.

Arendt, D., and K. Nübler-Jung. 1994. Inversion of dorsoventral axis? *Nature* 371:26.

Arendt, D., and K. Nübler-Jung. 1996. Common ground plans in early brain development in mice and flies. *Bioessays* 18:255–259.

Arendt, D., and K. Nübler-Jung. 1997. Dorsal or ventral: similarities in fatemaps and gastrulation patterns in annelids, arthropods and chordates. *Mech Dev* 61:7–21.

Arendt, D., and K. Nübler-Jung. 1999. Comparison of early nerve cord development in insects and vertebrates. *Development* 126:2309–2325.

Arendt, D., K. Tessmar-Raible, H. Snyman, A. W. Dorresteijn, and J. Wittbrodt. 2004. Ciliary photoreceptors with a vertebrate-type opsin in an invertebrate brain. *Science* 306:869–871.

Arendt, D., K. Tessmar, M. I. Campos-Baptista, A. Dorresteijn, and J. Wittbrodt. 2002. Development of pigment-cup eyes in the polychaete *Platynereis dumerilii* and evolutionary conservation of larval eyes in Bilateria. *Development* 129:1143–1154.

Arendt, D., M. A. Tosches, and H. Marlow. 2016. From nerve net to nerve ring, nerve cord and brain–evolution of the nervous system. *Nat Rev Neurosci* 17:61–72.

Arendt, D., and J. Wittbrodt. 2001. Reconstructing the eyes of Urbilateria. *Philos Trans R Soc Lond B Biol Sci* 356:1545–1563.

Arkett, S. A., and A. N. Spencer. 1986. Neuronal mechanism of a hydromedusan shadow reflex. II. Graded responses of reflex components, possible mechanism of photic integration, and functional significance. *J Comp Physiol A* 159:215–225.

Aronowicz, J., and C. J. Lowe. 2006. Hox gene expression in the hemichordate *Saccoglossus kowalevskii* and the evolution of deuterostome nervous systems. *Integr Comp Biol* 46:890–901.

Arshavsky, V. Y., and M. E. Burns. 2012. Photoreceptor signaling: supporting vision across a wide range of light intensities. *J Biol Chem* 287:1620–1626.

Auger, H., Y. Sasakura, J. S. Joly, and W. R. Jeffery. 2010. Regeneration of oral siphon pigment organs in the ascidian *Ciona intestinalis*. *Dev Biol* 339:374–389.

Ayers, T., H. Tsukamoto, M. Gühmann, V. B. Veedin Rajan, and K. Tessmar-Raible. 2018. A Go-type opsin mediates the shadow reflex in the annelid *Platynereis dumerilii*. *BMC Biol* 16:41.

Baatrup, E. 1981. Primary sensory cells in the skin of amphioxus (*Branchiostoma lancelatum*). *Acta Zool* 62:147–157.

Baatrup, E. 1982. On the structure of the corpuscles of de Quatrefages (*Branchiostoma lanceolatum*). *Acta Zool* 63:39–44.

Babonis, L. S., M. B. Debiasse, W. H. Francis, L. M. Christianson, S. H. D. Haddock, M. Q. Martindale, and J. F. Ryan. 2018. Not your mama's tentacle: molecular characterization of ctenophore colloblasts. *Integr Comp Biol* 58:E271–E271.

Babonis, L. S., and M. Q. Martindale. 2014. Old cell, new trick? Cnidocytes as a model for the evolution of novelty. *Integr Comp Biol* 54:714–722.

Babonis, L. S., and M. Q. Martindale. 2017. PaxA, but not PaxC, is required for cnidocyte development in the sea anemone *Nematostella vectensis*. *Evodevo* 8:14.

Backfisch, B., Rajan V. Veedin, R. M. Fischer, C. Lohs, E. Arboleda, K. Tessmar-Raible, and F. Raible. 2013. Stable transgenesis in the marine annelid *Platynereis dumerilii* sheds new light on photoreceptor evolution. *Proc Natl Acad Sci USA* 110:193–198.

Baker, C. V. H., and M. Bronner-Fraser. 1997. The origins of the neural crest. Part II: an evolutionary perspective. *Mech Dev* 69:13–29.

Baker, C. V. H., and M. S. Modrell. 2018. Insights into electroreceptor development and evolution from molecular comparisons with hair cells. *Integr Comp Biol* 58:329–340.

Baker, C. V., M. S. Modrell, and J. A. Gillis. 2013. The evolution and development of vertebrate lateral line electroreceptors. *J Exp Biol* 216:2515–2522.

Baker, C. V., P. O'Neill, and R. B. McCole. 2008. Lateral line, otic and epibranchial placodes: developmental and evolutionary links? *J Exp Zoolog B Mol Dev Evol* 310:370–383.

Baker, C. V., and G. Schlosser. 2005. Editorial: the evolutionary origin of neural crest and placodes. *J Exp Zoolog B Mol Dev Evol* 304B:269–273.

Barber, V. C., E. M. Evans, and M. F. Land. 1967. The fine structure of the eye of the mollusc *Pecten maximus*. *Z Zellforsch Mikrosk Anat* 76:25–312.

Bargmann, C. I. 2006. Chemosensation in *C. elegans*. *WormBook* ed. The C. elegans Research Community, WormBook, doi/10.1895/wormbook.1.123.1.:1–29.

Bargmann, C. I. 2012. Beyond the connectome: how neuromodulators shape neural circuits. *Bioessays* 34:458–465.

Barresi, M. J., and S. F. Gilbert. 2019. *Developmental Biology*. 12th ed. New York: Oxford University Press.

Barton-Owen, T. B., D. E. K. Ferrier, and I. M. L. Somorjai. 2018. Pax3/7 duplicated and diverged independently in amphioxus, the basal chordate lineage. *Sci Rep*8:9414.

Bassham, S., and J. H. Postlethwait. 2005. The evolutionary history of placodes: a molecular genetic investigation of the larvacean urochordate *Oikopleura dioica*. *Development* 132:4259–4272.

Bateson, W. 1886. The ancestry of the Chordata. *Quart J Microscop Sci* 26:535–571.

Bear, D. M., J. M. Lassance, H. E. Hoekstra, and S. R. Datta. 2016. The evolving neural and genetic architecture of vertebrate olfaction. *Curr Biol* 26:R1039–R1049.

Beaster-Jones, L., S. L. Kaltenbach, D. Koop, S. Yuan, R. Chastain, and L. Z. Holland. 2008. Expression of somite segmentation genes in amphioxus: a clock without a wavefront? *Dev Genes Evol* 218:599–611.

Bebenek, I. G., R. D. Gates, J. Morris, V. Hartenstein, and D. K. Jacobs. 2004. Sine oculis in basal Metazoa. *Dev Genes Evol* 214:342–351.

Beckmann, H., L. Hering, M. J. Henze, A. Kelber, P. A. Stevenson, and G. Mayer. 2015. Spectral sensitivity in Onychophora (velvet worms) revealed by electroretinograms, phototactic behaviour and opsin gene expression. *J Exp Biol* 218:915–922.

Belecky-Adams, T., S. Tomarev, H. S. Li, L. Ploder, R. R. McInnes, O. Sundin, and R. Adler. 1997. Pax-6, Prox 1, and Chx10 homeobox gene expression correlates with phenotypic fate of retinal precursor cells. *Invest Ophthalmol Vis Sci* 38:1293–1303.

Bellefroid, E. J., L. Leclère, A. Saulnier, M. Keruzore, M. Sirakov, M. Vervoort, and S. De Clercq. 2013. Expanding roles for the evolutionarily conserved Dmrt sex transcriptional regulators during embryogenesis. *Cell Mol Life Sci* 70:3829–3845.

Benito-Gutiérrez, E., M. Illas, J. X. Comella, and J. Garcia-Fernàndez. 2005. Outlining the nascent nervous system of *Branchiostoma floridae* (amphioxus) by the pan-neural marker AmphiElav. *Brain Res Bull* 66:518–521.

Benito-Gutiérrez, E., C. Nake, M. Llovera, J. X. Comella, and J. Garcia-Fernàndez. 2005. The single AmphiTrk receptor highlights increased complexity of neurotrophin signalling in vertebrates and suggests an early role in developing sensory neuroepidermal cells. *Development* 132:2191–2202.

Benito, J., and F. Pardos. 1997. Hemichordata. In *Microscopic anatomy of invertebrates, Vol. 15: Hemichordata, Chaetognatha, and the invertebrate chordates*, edited by F. W. Harrison and E. E. Ruppert, 15–101. New York: Wiley-Liss.

Benton, R. 2006. On the ORigin of smell: odorant receptors in insects. *Cell Mol Life Sci* 63:1579–1585.

Bermingham, N. A., B. A. Hassan, S. D. Price, M. A. Vollrath, N. Ben Arie, R. A. Eatock, H. J. Bellen, A. Lysakowski, and H. Y. Zoghbi. 1999. Math1: an essential gene for the generation of inner ear hair cells. *Science* 284:1837–1841.

Berná, L., and F. Alvarez-Valin. 2014. Evolutionary genomics of fast evolving tunicates. *Genome Biol Evol* 6:1724–1738.

Berrill, N. J. 1955. *The origin of vertebrates*. Oxford: Oxford University Press.

Bertho, S., J. Pasquier, Q. Pan, G. Le Trionnaire, J. Bobe, J. H. Postlethwait, E. Pailhoux, M. Schartl, A. Herpin, and Y. Guiguen. 2016. Foxl2 and its relatives are evolutionary conserved players in gonadal sex differentiation. *Sex Dev* 10:111–129.

Bertrand, N., D. S. Castro, and F. Guillemot. 2002. Proneural genes and the specification of neural cell types. *Nat Rev Neurosci* 3:517–530.

Bertrand, S., A. Camasses, I. Somorjai, M. R. Belgacem, O. Chabrol, M. L. Escande, P. Pontarotti, and H. Escriva. 2011. Amphioxus FGF signaling predicts the acquisition of vertebrate morphological traits. *Proc Natl Acad Sci USA* 108:9160–9165.

Bertrand, S., and H. Escriva. 2011. Evolutionary crossroads in developmental biology: amphioxus. *Development* 138:4819–4830.

Bertrand, V., C. Hudson, D. Caillol, C. Popovici, and P. Lemaire. 2003. Neural tissue in ascidian embryos is induced by FGF9/16/20, acting via a combination of maternal GATA and Ets transcription factors. *Cell* 115:615–627.

Besnard, V., S. E. Wert, W. M. Hull, and J. A. Whitsett. 2004. Immunohistochemical localization of Foxa1 and Foxa2 in mouse embryos and adult tissues. *Gene Expr Patterns* 5:193–208.

Bezares-Calderón, L. A., J. Berger, S. Jasek, C. Veraszto, S. Mendes, M. Gühmann, R. Almeda, R. Shahidi, and G. Jékely. 2018. Neural circuitry of a polycystin-mediated hydrodynamic startle response for predator avoidance. *Elife* 7:e36262.

Bhati, M., E. Llamosas, D. A. Jacques, C. M. Jeffries, S. Dastmalchi, N. Ripin, H. R. Nicholas, and J. M. Matthews. 2017. Interactions between LHX3- and ISL1-family LIM-homeodomain transcription factors are conserved in *Caenorhabditis elegans. Sci Rep*7:4579.

Biryukova, I., and P. Heitzler. 2005. The *Drosophila* LIM-homeo domain protein Islet antagonizes pro-neural cell specification in the peripheral nervous system. *Dev Biol* 288:559–570.

Blackburn, D. C., K. W. Conley, D. C. Plachetzki, K. Kempler, B. A. Battelle, and N. L. Brown. 2008. Isolation and expression of Pax6 and atonal homologues in the American horseshoe crab, *Limulus polyphemus. Dev Dyn*237:2209–2219.

Bleckmann, H. 2008. Peripheral and central processing of lateral line information. *J Comp Physiol A Neuroethol Sens Neural Behav Physiol* 194:145–158.

Blochlinger, K., L. Y. Jan, and Y. N. Jan. 1991. Transformation of sensory organ identity by ectopic expression of Cut in *Drosophila. Genes Dev*5:1124–1135.

Blum, M., K. Feistel, T. Thumberger, and A. Schweickert. 2014. The evolution and conservation of left-right patterning mechanisms. *Development* 141:1603–1613.

Bodmer, R., S. Barbel, S. Sheperd, J. W. Jack, L. Y. Jan, and Y. N. Jan. 1987. Transformation of sensory organs by mutations of the cut locus of *D. melanogaster. Cell* 51:293–307.

Bodznick, D., and J. C. Montgomery. 2005. The physiology of low-frequency electrosensory systems. In *Electroreception*, edited by T. H. Bullock, C. D. Hopkins, A. N. Popper and R. R. Fay, 132–153. New York: Springer.

Boekhoff-Falk, G., and D. F. Eberl. 2014. The *Drosophila* auditory system. *Wiley Interdiscip Rev Dev Biol* 3:179–191.

Bohorquez, D. V., R. A. Shahid, A. Erdmann, A. M. Kreger, Y. Wang, N. Calakos, F. Wang, and R. A. Liddle. 2015. Neuroepithelial circuit formed by innervation of sensory enteroendocrine cells. *J Clin Invest* 125:782–786.

Bollner, T., P. W. Beesley, and M. C. Thorndyke. 1992. Pattern of substance P- and cholecystokinin-like immunoreactivity during regeneration of the neural complex in the ascidian *Ciona intestinalis. J Comp Neurol* 325:572–580.

Bollner, T., K. Holmberg, and R. Olsson. 1986. A rostral sensory mechanism in *Oikopleura dioica* (Appendicularia). *Acta Zool* 67:235–241.

Bone, Q. 1961. The organization of the atrial nervous system of amphioxus (*Branchiostoma lanceolatum* (Pallas)). *Phil Trans Roy Soc B* 243:241–269.

Bone, Q., and A. C. G. Best. 1978. Ciliated sensory cell in amphioxus (*Branchiostoma*). *J Mar Biol Ass UK* 58:479–486.

Bone, Q., and K. P. Ryan. 1978. Cupular sense organs in *Ciona* (Tunicata: Ascidiacea). *J Zool Lond* 186:417–429.

Bone, Q., and K. P. Ryan. 1979. Langerhans receptor of *Oikopleura* (Tunicata Larvacea). *J Mar Biol Assoc UK* 59:69–75.

Bonini, N. M., Q. T. Bui, G. L. Gray-Board, and J. M. Warrick. 1997. The *Drosophila* eyes absent gene directs ectopic eye formation in a pathway conserved between flies and vertebrates. *Development* 124:4819–4826.

Bonini, N. M., W. M. Leiserson, and S. Benzer. 1993. The eyes absent gene: genetic control of cell survival and differentiation in the developing *Drosophila* eye. *Cell* 72:379–395.

Boorman, C. J., and S. M. Shimeld. 2002. Pitx homeobox genes in *Ciona* and amphioxus show left-right asymmetry is a conserved chordate character and define the ascidian adenohypophysis *Evol Dev* 4:354–365.

Bouillon, J., and M. Nielsen. 1974. Etude de quelques organes sensoriels de cnidaires. *Arch Biol Bruxelles* 85:307–328.

Bourlat, S. J., T. Juliusdottir, C. J. Lowe, R. Freeman, J. Aronowicz, M. Kirschner, E. S. Lander, M. Thorndyke, H. Nakano, A. B. Kohn, A. Heyland, L. L. Moroz, R. R. Copley, and M. J. Telford. 2006. Deuterostome phylogeny reveals monophyletic chordates and the new phylum Xenoturbellida. *Nature* 444:85–88.

Bowler, P. J. 1996. *Life's splendid drama*. Chicago: University of Chicago Press.

Brandenburger, J. L., R. M. Woollacott, and R. M. Eakin. 1973. Fine structure of eyespots in tornarian larvae (phylum: Hemichordata). *Z Zellforsch Mikrosk Anat* 142:89–102.

Braun, K., S. Kaul-Strehlow, E. Ullrich-Lüter, and T. Stach. 2015. Structure and ultrastructure of eyes of tornaria larvae of *Glossobalanus marginatus*. *Org Divers Evol* 15:423–428.

Braun, K., and T. Stach. 2017. Structure and ultrastructure of eyes and brains of *Thalia democratica* (Thaliacea, Tunicata, Chordata). *J Morphol* 278:1421–1437.

Brigandt, I., and A. C. Love. 2012. Conceptualizing evolutionary novelty: moving beyond definitional debates. *J Exp Zool B Mol Dev Evol* 318:417–427.

Briggs, D. E. G., E. N. K. Clarkson, and R. J. Aldridge. 1983. The conodont animal. *Lethaia* 16:1–14.

Brooke, N. M., J. Garcia-Fernàndez, and P. W. H. Holland. 1998. The ParaHox cluster is an evolutionary sister of the Hox gene cluster. *Nature* 392:920–922.

Brown, F. D., A. Prendergast, and B. J. Swalla. 2008. Man is but a worm: chordate origins. *Genesis* 46:605–613.

Brown, F. D., and B. J. Swalla. 2012. Evolution and development of budding by stem cells: ascidian coloniality as a case study. *Dev Biol* 369:151–162.

Brown, S., and J. Castelli-Gair Hombria. 2000. *Drosophila* grain encodes a GATA transcription factor required for cell rearrangement during morphogenesis. *Development* 127:4867–4876.

Brunet, J. F., and A. Pattyn. 2002. Phox2 genes - from patterning to connectivity. *Curr Opin Genet Dev* 12:435–440.

Brusca, R. C., and G. J. Brusca. 2003. *Invertebrates*. 2nd ed. Sunderland: Sinauer.

Budelmann, B. U. 1989. Hydrodynamic receptor systems in invertebrates. In *The mechanosensory lateral line*, edited by S. Coombs, P. Görner and H. Münz, 607–631. New York: Springer.

Budelmann, B. U., and H. Bleckmann. 1988. A lateral line analog in cephalopods - waterwaves generate microphonic potentials in the epidermal head lines of *Sepia* and *Lolliguncula*. *J Comp Physiol A* 164:1–5.

Budelmann, B. U., M. Sachse, and M. Staudigl. 1987. The angular-acceleration receptor system of the statocyst of *Octopus vulgaris:* morphometry, ultrastructure, and neuronal and synaptic organization. *Philos Trans R Soc Lond B Biol Sci* 315:305–343.

Bullock, T. H. 1945. The anatomical organization of the nervous system of Enteropneusta. *Q J Microsc Sci* 86:55–111.

Burgoyne, R. D., and A. Morgan. 2003. Secretory granule exocytosis. *Physiol Rev* 83:581–632.

Burighel, P., F. Caicci, and L. Manni. 2011. Hair cells in non-vertebrate models: lower chordates and molluscs. *Hear Res* 273:14–24.

Burighel, P., N. J. Lane, G. Fabio, T. Stefano, G. Zaniolo, M. D. Carnevali, and L. Manni. 2003. Novel, secondary sensory cell organ in ascidians: in search of the ancestor of the vertebrate lateral line. *J Comp Neurol* 461:236–249.

Burighel, P., N. J. Lane, G. Zaniolo, and L. Manni. 1998. Neurogenic role of the neural gland in the development of the ascidian, *Botryllus schlosseri* (Tunicata, Urochordata). *J Comp Neurol* 394:230–241.

Burke, R. D. 1978. Structure of nervous system of pluteus larva of *Strongylocentrotus purpuratus*. *Cell Tissue Res*191:233–247.

Burke, R. D. 1980. Podial sensory receptors and the induction of metamorphosis in echinoids. *J Exp Mar Bio Ecol* 47:223–234.

Burke, R. D., L. M. Angerer, M. R. Elphick, G. W. Humphrey, S. Yaguchi, T. Kiyama, S. Liang, X. Mu, C. Agca, W. H. Klein, B. P. Brandhorst, M. Rowe, K. Wilson, A. M. Churcher, J. S. Taylor, N. Chen, G. Murray, D. Wang, D. Mellott, R. Olinski, F. Hallbook, and M. C. Thorndyke. 2006. A genomic view of the sea urchin nervous system. *Dev Biol* 300:434–460.

Burkhardt, P., and S. G. Sprecher. 2017. Evolutionary origin of synapses and neurons: bridging the gap. *Bioessays* 39.

Byrne, M., Y. Nakajima, F. C. Chee, and R. D. Burke. 2007. Apical organs in echinoderm larvae: insights into larval evolution in the Ambulacraria. *Evol Dev* 9:432–445.

Cai, T., and A. K. Groves. 2015. The role of atonal factors in mechanosensory cell specification and function. *Mol Neurobiol* 52:1315–1329.

Cai, X., X. Wang, S. Patel, and D. E. Clapham. 2015. Insights into the early evolution of animal calcium signaling machinery: a unicellular point of view. *Cell Calcium* 57:166–173.

Caicci, F., P. Burighel, and L. Manni. 2007. Hair cells in an ascidian (Tunicata) and their evolution in chordates. *Hear Res* 231:63–72.

Caicci, F., F. Gasparini, F. Rigon, G. Zaniolo, P. Burighel, and L. Manni. 2013. The oral sensory structures of Thaliacea (Tunicata) and consideration of the evolution of hair cells in Chordata. *J Comp Neurol* 521:2756–2771.

Caicci, F., G. Zaniolo, P. Burighel, V. Degasperi, F. Gasparini, and L. Manni. 2010. Differentiation of papillae and rostral sensory neurons in the larva of the ascidian *Botryllus schlosseri* (Tunicata). *J Comp Neurol* 518:547–566.

Callaerts, P., A. M. Muñoz-Marmol, S. Glardon, E. Castillo, H. Sun, W. H. Li, W. J. Gehring, and E. Salo. 1999. Isolation and expression of a Pax-6 gene in the regenerating and intact planarian *Dugesia(G)tigrina*. *Proc Natl Acad Sci USA* 96:558–563.

Cameron, C. B., J. R. Garey, and B. J. Swalla. 2000. Evolution of the chordate body plan: new insights from phylogenetic analyses of deuterostome phyla. *Proc Nat Acad Sci USA* 97:4469–4474.

Cameron, C. B., G. O. Mackie, J. F. Powell, D. W. Lescheid, and N. M. Sherwood. 1999. Gonadotropin-releasing hormone in mulberry cells of *Saccoglossus* and *Ptychodera* (Hemichordata: Enteropneusta). *Gen Comp Endocrinol* 114:2–10.

Campbell, R. K., N. Satoh, and B. M. Degnan. 2004. Piecing together evolution of the vertebrate endocrine system. *Trends Genet* 20:359–366.

Campo-Paysaa, F., F. Marlétaz, V. Laudet, and M. Schubert. 2008. Retinoic acid signaling in development: tissue-specific functions and evolutionary origins. *Genesis* 46:640–656.

Candiani, S., N. D. Holland, D. Oliveri, M. Parodi, and M. Pestarino. 2008. Expression of the amphioxus Pit-1 gene (AmphiPOU1F1/Pit-1) exclusively in the developing pre-oral organ, a putative homolog of the vertebrate adenohypophysis. *Brain Res Bull* 75:324–330.

Candiani, S., D. Oliveri, M. Parodi, E. Bertini, and M. Pestarino. 2006. Expression of AmphiPOU-IV in the developing neural tube and epidermal sensory neural precursors in amphioxus supports a conserved role of class IV POU genes in the sensory cells development. *Dev Genes Evol* 216:623–633.

Candiani, S., R. Pennati, D. Oliveri, A. Locascio, M. Branno, P. Castagnola, M. Pestarino, and F. de Bernardi. 2005. Ci-POU-IV expression identifies PNS neurons in embryos and larvae of the ascidian *Ciona intestinalis*. *Dev Genes Evol* 215:41–45.

Canestro, C., and J. H. Postlethwait. 2007. Development of a chordate anterior-posterior axis without classical retinoic acid signaling. *Dev Biol* 305:522–538.

Cannon, J. T., K. M. Kocot, D. S. Waits, D. A. Weese, B. J. Swalla, S. R. Santos, and K. M. Halanych. 2014. Phylogenomic resolution of the hemichordate and echinoderm clade. *Curr Biol* 24:2827–2832.

Cannon, J. T., A. L. Rychel, H. Eccleston, K. M. Halanych, and B. J. Swalla. 2009. Molecular phylogeny of hemichordata, with updated status of deep-sea enteropneusts. *Mol Phylogenet Evol* 52:17–24.

Cannon, J. T., B. C. Vellutini, J. Smith, 3rd, F. Ronquist, U. Jondelius, and A. Hejnol. 2016. Xenacoelomorpha is the sister group to Nephrozoa. *Nature* 530:89–93.

Cao, C., L. A. Lemaire, W. Wang, P. H. Yoon, Y. A. Choi, L. R. Parsons, J. C. Matese, W. Wang, M. Levine, and K. Chen. 2019. Comprehensive single-cell transcriptome lineages of a proto-vertebrate. *Nature* 571:349–354.

Cao, J., J. S. Packer, V. Ramani, D. A. Cusanovich, C. Huynh, R. Daza, X. Qiu, C. Lee, S. N. Furlan, F. J. Steemers, A. Adey, R. H. Waterston, C. Trapnell, and J. Shendure. 2017. Comprehensive single-cell transcriptional profiling of a multicellular organism. *Science* 357:661–667.

Castro, A., M. Becerra, M. J. Manso, and R. Anadón. 2015. Neuronal organization of the brain in the adult amphioxus (*Branchiostoma lanceolatum*): a study with acetylated tubulin immunohistochemistry. *J Comp Neurol* 523:2211–2232.

Castro, A., M. Becerra, M. J. Manso, N. M. Sherwood, and R. Anadón. 2006. Anatomy of the Hesse photoreceptor cell axonal system in the central nervous system of amphioxus. *J Comp Neurol* 494:54–62.

Castro, A., M. J. Manso, and R. Anadón. 2003. Distribution of neuropeptide Y immunoreactivity in the central and peripheral nervous systems of amphioxus (*Branchiostoma lanceolatum* Pallas). *J Comp Neurol* 461:350–361.

Castro, L. F., S. L. Rasmussen, P. W. Holland, N. D. Holland, and L. Z. Holland. 2006. A Gbx homeobox gene in amphioxus: insights into ancestry of the ANTP class and evolution of the midbrain/hindbrain boundary. *Dev Biol* 295:40–51.

Cattell, M. V., A. T. Garnett, M. W. Klymkowsky, and D. M. Medeiros. 2012. A maternally established SoxB1/SoxF axis is a conserved feature of chordate germ layer patterning. *Evol Dev* 14:104–115.

Certel, S. J., P. J. Clyne, J. R. Carlson, and W. A. Johnson. 2000. Regulation of central neuron synaptic targeting by the *Drosophila* POU protein, Acj6. *Development* 127:2395–2405.

Chabry, L. 1887. *Embryologie normale et tératologique des Ascidie*. Paris: Felix Alcan Editeur.

Chagnaud, B. P., J. Engelmann, B. Fritzsch, J. C. Glover, and H. Straka. 2017. Sensing external and self-motion with hair cells: a comparison of the lateral line and vestibular systems from a developmental and evolutionary perspective. *Brain Behav Evol* 90:98–116.

Chalfie, M. 2009. Neurosensory mechanotransduction. *Nat Rev Mol Cell Biol* 10:44–52.

Chalfie, M., and J. Sulston. 1981. Developmental genetics of the mechanosensory neurons of *Caenorhabditis elegans*. *Dev Biol* 82:358–370.

Chambery, A., A. Parente, E. Topo, J. Garcia-Fernàndez, and S. D'Aniello. 2009. Characterization and putative role of a type I gonadotropin-releasing hormone in the cephalochordate amphioxus. *Endocrinology* 150:812–820.

Chang, C., Y. Chu, and D. Chen. 1982. Immunocytochemical demonstration of luteinizing hormone (LH) in Hatschek's pit of amphioxus (*Branchiostoma belcheri* Gray). *Kexue Tongbao* 27:1233–1234.

Chang, C. Y., Y. X. Liu, Y. Zhu, and H. Zhu. 1984. The reproductive endocrinology of amphioxus. In *Frontiers in physiological research*, edited by D. G. Carlick and P. I. Korner, 79–86. Cambridge: Cambridge University Press.

Chang, I., and M. Parrilla. 2016. Expression patterns of homeobox genes in the mouse vomeronasal organ at postnatal stages. *Gene Expr Patterns* 21:69–80.

Chapman, D. M. 1999. Microanatomy of the bell rim of *Aurelia aurita* (Cnidaria: Scyphozoa). *Can J Zool* 77:34–46.

Chen, B., E. H. Kim, and P. X. Xu. 2009. Initiation of olfactory placode development and neurogenesis is blocked in mice lacking both Six1 and Six4. *Dev Biol* 326:75–85.

Chen, G., T. R. Korfhagen, Y. Xu, J. Kitzmiller, S. E. Wert, Y. Maeda, A. Gregorieff, H. Clevers, and J. A. Whitsett. 2009. SPDEF is required for mouse pulmonary goblet cell differentiation and regulates a network of genes associated with mucus production. *J Clin Invest* 119:2914–2924.

Chen, G., A. S. Volmer, K. J. Wilkinson, Y. Deng, L. C. Jones, D. Yu, X. M. Bustamante-Marin, K. A. Burns, B. R. Grubb, W. K. O'Neal, A. Livraghi-Butrico, and R. C. Boucher. 2018. Role of Spdef in the regulation of Muc5b expression in the airways of naive and mucoobstructed mice. *Am J Respir Cell Mol Biol* 59:383–396.

Chen, J.-Y., J. Dzik, G. D. Edgecombe, L. Ramsköld, and G.-Q. Zhou. 1995. A possible Early Cambrian chordate. *Nature* 377:720–722.

Chen, J., L. A. Jacox, F. Saldanha, and H. Sive. 2017. Mouth development. *Wiley Interdiscip Rev Dev Biol* 6:e275.

Chen, S., Q. L. Wang, Z. Nie, H. Sun, G. Lennon, N. G. Copeland, D. J. Gilbert, N. A. Jenkins, and D. J. Zack. 1997. Crx, a novel Otx-like paired-homeodomain protein, binds to and transactivates photoreceptor cell-specific genes. *Neuron* 19:1017–1030.

Cheng, L., C. L. Chen, P. Luo, M. Tan, M. Qiu, R. Johnson, and Q. Ma. 2003. Lmx1b, Pet-1, and Nkx2.2 coordinately specify serotonergic neurotransmitter phenotype. *J Neurosci* 23:9961–9967.

Cheung, M., A. Tai, P. J. Lu, and K. S. Cheah. 2019. Acquisition of multipotent and migratory neural crest cells in vertebrate evolution. *Curr Opin Genet Dev* 57:84–90.

Cheyette, B. N., P. J. Green, K. Martin, H. Garren, V. Hartenstein, and S. L. Zipursky. 1994. The *Drosophila* sine oculis locus encodes a homeodomain-containing protein required for the development of the entire visual system. *Neuron* 12:977–996.

Chia, F. S., and R. Koss. 1979. Fine-structural studies of the nervous system and the apical organ in the planula larva of the sea-anemone *Anthopleura elegantissima*. *J Morphol* 160:275–298.

Chiba, S., A. Sasaki, A. Nakayama, K. Takamura, and N. Satoh. 2004. Development of *Ciona intestinalis* juveniles (through 2nd ascidian stage). *Zool Sci* 21:285–298.

Chiodin, M., A. Børve, E. Berezikov, P. Ladurner, P. Martinez, and A. Hejnol. 2013. Mesodermal gene expression in the acoel *Isodiametra pulchra* indicates a low number of mesodermal cell types and the endomesodermal origin of the gonads. *PLoS One* 8:e55499.

Chipman, A. D. 2010. Parallel evolution of segmentation by co-option of ancestral gene regulatory networks. *Bioessays* 32:60–70.

Christiaen, L., P. Burighel, W. C. Smith, P. Vernier, F. Bourrat, and J. S. Joly. 2002. Pitx genes in tunicates provide new molecular insight into the evolutionary origin of pituitary. *Gene* 287:107–113.

Christodoulou, F., F. Raible, R. Tomer, O. Simakov, K. Trachana, S. Klaus, H. Snyman, G. J. Hannon, P. Bork, and D. Arendt. 2010. Ancient animal microRNAs and the evolution of tissue identity. *Nature* 463:1084–1088.

Churcher, A. M., and J. S. Taylor. 2009. Amphioxus (*Branchiostoma floridae*) has orthologs of vertebrate odorant receptors. *BMC Evol Biol* 9:242.

Churcher, A. M., and J. S. Taylor. 2011. The antiquity of chordate odorant receptors is revealed by the discovery of orthologs in the cnidarian *Nematostella vectensis*. *Genome Biol Evol* 3:36–43.

Cirillo, L. A., F. R. Lin, I. Cuesta, D. Friedman, M. Jarnik, and K. S. Zaret. 2002. Opening of compacted chromatin by early developmental transcription factors HNF3 (FoxA) and GATA-4. *Mol Cell* 9:279–289.

Cloney, R. A. 1977. Larval adhesive organs and metamorphosis in ascidians. I. Fine structure of the everting papillae of *Distaplia occidentalis*. *Cell Tissue Res* 183:423–444.

Clyne, P. J., S. J. Certel, M. de Bruyne, L. Zaslavsky, W. A. Johnson, and J. R. Carlson. 1999. The odor specificities of a subset of olfactory receptor neurons are governed by Acj6, a POU-domain transcription factor. *Neuron* 22:339–347.

Clyne, P. J., C. G. Warr, M. R. Freeman, D. Lessing, J. Kim, and J. R. Carlson. 1999. A novel family of divergent seven-transmembrane proteins: candidate odorant receptors in *Drosophila*. *Neuron* 22:327–338.

Cobb, J. L. S. 1968. The fine structure of the pedicellariae of *Echinus esculentus* (L.). II. The sensory system. *J R Microsc Soc* 88:223–233.

Cong, P.-Y., X.-G. Hou, R. J. Aldridge, M. A. Purnell, and Y.-Z. Li. 2014. New data on the palaeobiology of the enigmatic yunnanozoans from the Chengjiang Biota, Lower Cambrian, China. *Palaeontology* 58:45–70.

Conklin, E. 1905. The organization and cell lineage of the ascidian egg. *J Acad Nat Sci Phila* 13 13:1–119.

Conzelmann, M., E. A. Williams, S. Tunaru, N. Randel, R. Shahidi, A. Asadulina, J. Berger, S. Offermanns, and G. Jékely. 2013. Conserved MIP receptor-ligand pair regulates *Platynereis* larval settlement. *Proc Natl Acad Sci USA* 110:8224–8229.

Cook, T., F. Pichaud, R. Sonneville, D. Papatsenko, and C. Desplan. 2003. Distinction between color photoreceptor cell fates is controlled by Prospero in *Drosophila*. *Dev Cell* 4:853–864.

Coombs, S., J. Janssen, and J. F. Webb. 1988. Diversity of lateral line systems: evolutionary and functional considerations. In *Sensory biology of aquatic animals*, edited by J. Atema, R. R. Fay, A. N. Popper and W. N. Tavolga, 553–593. New York: Springer.

Corbo, J. C., A. Erives, A. DiGregorio, A. Chang, and M. Levine. 1997. Dorsoventral patterning of the vertebrate neural tue is conserved in a protochordate. *Development* 124:2335–2344.

Costa, A., L. M. Powell, S. Lowell, and A. P. Jarman. 2017. Atoh1 in sensory hair cell development: constraints and cofactors. *Semin Cell Dev Biol* 65:60–68.

Costa, A., L. M. Powell, A. Soufi, S. Lowell, and A. P. Jarman. 2019. Atoh1 is repurposed from neuronal to hair cell determinant by Gfi1 acting as a coactivator without redistributing Atoh1's genomic binding sites. *Biorxiv* doi: https://doi.org/10.1101/767574.

Costa, A., L. Sánchez-Guardado, S. Juniat, J. E. Gale, N. Daudet, and D. Henrique. 2015. Generation of sensory hair cells by genetic programming with a combination of transcription factors. *Development* 142:1948–1959.

Crisp, M. 1971. Structure and abundance of receptors of unspecialized external epithelium of *Nassarius reticulatus* [Gastropoda, Prosobranchia]. *J Mar Biol Assoc UK* 51:865.

Croll, R. P., and A. J. G. Dickinson. 2004. Form and function of the larval nervous system in molluscs. *Invertebr Reprod Dev* 46:173–187.

Cronin, T. W., and M. L. Porter. 2014. The evolution of invertebrate photopigments and photoreceptors. In *Evolution of visual and non-visual pigments*, edited by D. M. Hunt, M. W. Hankins, S. P. Collin and N. J. Marshall, 105–135. New York: Springer.

Cummins, S. F., L. Leblanc, B. M. Degnan, and G. T. Nagle. 2009. Molecular identification of candidate chemoreceptor genes and signal transduction components in the sensory epithelium of *Aplysia*. *J Exp Biol* 212:2037–2044.

Cunningham, D., and E. S. Casey. 2014. Spatiotemporal development of the embryonic nervous system of *Saccoglossus kowalevskii*. *Dev Biol* 386:252–263.

Czech-Damal, N. U., G. Dehnhardt, P. Manger, and W. Hanke. 2013. Passive electroreception in aquatic mammals. *J Comp Physiol A Neuroethol Sens Neural Behav Physiol* 199:555–563.

Czech-Damal, N. U., A. Liebschner, L. Miersch, G. Klauer, F. D. Hanke, C. Marshall, G. Dehnhardt, and W. Hanke. 2012. Electroreception in the Guiana dolphin (*Sotalia guianensis*). *Proc Biol Sci* 279:663–668.

Czerny, T., M. Bouchard, Z. Kozmik, and M. Busslinger. 1997. The characterization of Pax genes of the sea urchin and *Drosophila* reveal an ancient evolutionary origin of the Pax2/5/8 subfamily. *Mech Dev* 67:179–192.

D'Alessio, M., and M. Frasch. 1996. msh may play a conserved role in dorsoventral patterning of the neuroectoderm and mesoderm. *Mech Dev* 58:217–231.

D'Aniello, S., E. D'Aniello, A. Locascio, A. Memoli, M. Corrado, M. T. Russo, F. Aniello, L. Fucci, E. R. Brown, and M. Branno. 2006. The ascidian homolog of the vertebrate homeobox gene Rx is essential for ocellus development and function. *Differentiation* 74:222–234.

D'Aniello, S., J. Delroisse, A. Valero-Gracia, E. K. Lowe, M. Byrne, J. T. Cannon, K. M. Halanych, M. R. Elphick, J. Mallefet, S. Kaul-Strehlow, C. J. Lowe, P. Flammang, E. Ullrich-Lüter, A. Wanninger, and M. I. Arnone. 2015. Opsin evolution in the Ambulacraria. *Mar Genomics* 24 Pt 2:177–183.

D'Autreaux, F., E. Coppola, M. R. Hirsch, C. Birchmeier, and J. F. Brunet. 2011. Homeoprotein Phox2b commands a somatic-to-visceral switch in cranial sensory pathways. *Proc Natl Acad Sci USA* 108:20018–20023.

Dabdoub, A., C. Puligilla, J. M. Jones, B. Fritzsch, K. S. Cheah, L. H. Pevny, and M. W. Kelley. 2008. Sox2 signaling in prosensory domain specification and subsequent hair cell differentiation in the developing cochlea. *Proc Natl Acad Sci USA* 105:18396–18401.

Dailey, S. C., I. Kozmikova, and I. M. L. Somorjai. 2017. Amphioxus Sp5 is a member of a conserved Specificity Protein complement and is modulated by Wnt/beta-catenin signalling. *Int J Dev Biol* 61:723–732.

Dalton, R. P., D. B. Lyons, and S. Lomvardas. 2013. Co-opting the unfolded protein response to elicit olfactory receptor feedback. *Cell* 155:321–332.

Daniel, A., K. Dumstrei, J. A. Lengyel, and V. Hartenstein. 1999. The control of cell fate in the embryonic visual system by atonal, tailless and EGFR signaling. *Development* 126:2945–2954.

Darras, S., J. Gerhart, M. Terasaki, M. Kirschner, and C. J. Lowe. 2011. Beta-catenin specifies the endomesoderm and defines the posterior organizer of the hemichordate *Saccoglossus kowalevskii*. *Development* 138:959–970.

Darras, S., and H. Nishida. 2001a. The BMP signaling pathway is required together with the FGF pathway for notochord induction in the ascidian embryo. *Development* 128:2629–2638.

Darras, S., and H. Nishida. 2001b. The BMP/CHORDIN antagonism controls sensory pigment cell specification and differentiation in the ascidian embryo. *Dev Biol* 236:271–288.

Davidson, E. H., and D. H. Erwin. 2006. Gene regulatory networks and the evolution of animal body plans. *Science* 311:796–800.

Davidson, E. H., J. P. Rast, P. Oliveri, A. Ransick, C. Calestani, C. H. Yuh, T. Minokawa, G. Amore, V. Hinman, C. Arenas-Mena, O. Otim, C. T. Brown, C. B. Livi, P. Y. Lee, R. Revilla, M. J. Schilstra, P. J. Clarke, A. G. Rust, Z. Pan, M. I. Arnone, L. Rowen, R. A. Cameron, D. R. McClay, L. Hood, and H. Bolouri. 2002. A provisional regulatory gene network for specification of endomesoderm in the sea urchin embryo. *Dev Biol* 246:162–190.

Davies, W. I., L. Zheng, S. Hughes, T. K. Tamai, M. Turton, S. Halford, R. G. Foster, D. Whitmore, and M. W. Hankins. 2011. Functional diversity of melanopsins and their global expression in the teleost retina. *Cell Mol Life Sci* 68:4115–4132.

Davis, R. J., B. C. Tavsanli, C. Dittrich, U. Walldorf, and G. Mardon. 2003. *Drosophila* retinal homeobox (drx) is not required for establishment of the visual system, but is required for brain and clypeus development. *Dev Biol* 259:272–287.

Davis, R. L., H. Weintraub, and A. B. Lassar. 1987. Expression of a single transfected cDNA converts fibroblasts to myoblasts. *Cell* 51:987–1000.

Dawkins, R. 2004. *The ancestor's tale*. London: Weidenfeld and Nicholson.

de-Leon, S. B. 2011. The conserved role and divergent regulation of foxa, a pan-eumetazoan developmental regulatory gene. *Dev Biol* 357:21–26.

de Navascues, J., and J. Modolell. 2010. The pronotum LIM-HD gene tailup is both a positive and a negative regulator of the proneural genes achaete and scute of *Drosophila*. *Mech Dev* 127:393–406.

de Quatrefages, A. 1845. Mémoire sur le système nerveux et sur l'histologie du branchio-stome ou amphioxus. *Ann Sci Nat Zool Sér* 3:197–248.

De Robertis, E. M., and Y. Sasai. 1996. A common plan for dorsoventral patterning in Bilateria. *Nature* 380:37–40.

De Velasco, B., T. Erclik, D. Shy, J. Sclafani, H. Lipshitz, R. McInnes, and V. Hartenstein. 2007. Specification and development of the pars intercerebralis and pars lateralis, neuroendocrine command centers in the *Drosophila* brain. *Dev Biol* 302:309–323.

De Velasco, B., L. Mandal, M. Mkrtchyan, and V. Hartenstein. 2006. Subdivision and developmental fate of the head mesoderm in *Drosophila melanogaster*. *Dev Genes Evol* 216:39–51.

De Velasco, B., J. Shen, S. Go, and V. Hartenstein. 2004. Embryonic development of the *Drosophila* corpus cardiacum, a neuroendocrine gland with similarity to the vertebrate pituitary, is controlled by sine oculis and glass. *Dev Biol* 274:280–294.

Degl'Innocenti, A., and A. D'Errico. 2017. Regulatory features for odorant receptor genes in the mouse genome. *Front Genet* 8:19.

Dehal, P., Y. Satou, R. K. Campbell, J. Chapman, B. Degnan, A. De Tomaso, B. Davidson, A. Di Gregorio, M. Gelpke, D. M. Goodstein, N. Harafuji, K. E. Hastings, I. Ho, K. Hotta, W. Huang, T. Kawashima, P. Lemaire, D. Martinez, I. A. Meinertzhagen, S. Necula, M. Nonaka, N. Putnam, S. Rash, H. Saiga, M. Satake, A. Terry, L. Yamada, H. G. Wang, S. Awazu, K. Azumi, J. Boore, M. Branno, S. Chin-Bow, R. DeSantis, S. Doyle, P. Francino, D. N. Keys, S. Haga, H. Hayashi, K. Hino, K. S. Imai, K. Inaba, S. Kano, K. Kobayashi, M. Kobayashi, B. I. Lee, K. W. Makabe, C. Manohar, G. Matassi, M. Medina, Y. Mochizuki, S. Mount, T. Morishita, S. Miura, A. Nakayama, S. Nishizaka, H. Nomoto, F. Ohta, K. Oishi, I. Rigoutsos, M. Sano, A. Sasaki, Y. Sasakura, E. Shoguchi, T. Shin-i, A. Spagnuolo, D. Stainier, M. M. Suzuki, O. Tassy, N. Takatori, M. Tokuoka, K. Yagi, F. Yoshizaki, S. Wada, C. Zhang, P. D. Hyatt, F. Larimer, C. Detter, N. Doggett, T. Glavina, T. Hawkins, P. Richardson, S. Lucas, Y. Kohara, M. Levine, N. Satoh, and D. S. Rokhsar. 2002. The draft genome of *Ciona intestinalis*: insights into chordate and vertebrate origins. *Science* 298:2157–2167.

Del Giacco, L., A. Pistocchi, F. Cotelli, A. E. Fortunato, and P. Sordino. 2008. A peek inside the neurosecretory brain through orthopedia lenses. *Dev Dyn* 237:2295–2303.

Delsuc, F., H. Brinkmann, D. Chourrout, and H. Philippe. 2006. Tunicates and not cephalochordates are the closest living relatives of vertebrates. *Nature* 439:965–968.

Delsuc, F., H. Philippe, G. Tsagkogeorga, P. Simion, M. K. Tilak, X. Turon, S. López-Legentil, J. Piette, P. Lemaire, and E. J. P. Douzery. 2018. A phylogenomic framework and timescale for comparative studies of tunicates. *BMC Biol* 16:39.

Deneris, E. S., and S. C. Wyler. 2012. Serotonergic transcriptional networks and potential importance to mental health. *Nat Neurosci* 15:519–527.

Denes, A. S., G. Jékely, P. R. Steinmetz, F. Raible, H. Snyman, B. Prud'homme, D. E. Ferrier, G. Balavoine, and D. Arendt. 2007. Molecular architecture of annelid nerve cord supports common origin of nervous system centralization in bilateria. *Cell* 129:277–288.

Denker, E., E. Bapteste, H. Le Guyader, M. Manuel, and N. Rabet. 2008. Horizontal gene transfer and the evolution of cnidarian stinging cells. *Curr Biol* 18:R858–R859.

Derelle, R., and M. Manuel. 2007. Ancient connection between NKL genes and the mesoderm? Insights from Tlx expression in a ctenophore. *Dev Genes Evol* 217:253–261.

Díaz, C., N. Morales-Delgado, and L. Puelles. 2015. Ontogenesis of peptidergic neurons within the genoarchitectonic map of the mouse hypothalamus. *Front Neuroanat* 8:162.

Dickinson, A., and H. Sive. 2007. Positioning the extreme anterior in *Xenopus*: cement gland, primary mouth and anterior pituitary. *Semin Cell Dev Biol* 18:525–533.

Dijkgraaf, S. 1962. The functioning and significance of the lateral-line organs. *Biol Rev* 38:51–105.

Dilly, D.N., Wolken, J.J. 1973. Studies on the receptors in *Ciona intestinalis*. IV. The ocellus in the adult. *Micron* 4:11–29.

Dilly, P. N. 1972. The structures of the tentacles of *Rhabdopleura compacta* (Hemichordata) with special reference to neurociliary control. *Z Zellforsch Mikrosk Anat* 129:20–39.

Diogo, R., R. G. Kelly, L. Christiaen, M. Levine, J. M. Ziermann, J. L. Molnar, D. M. Noden, and E. Tzahor. 2015. A new heart for a new head in vertebrate cardiopharyngeal evolution. *Nature* 520:466–473.

Dohrmann, M., and G. Wörheide. 2013. Novel scenarios of early animal evolution–is it time to rewrite textbooks? *Integr Comp Biol* 53:503–511.

Dohrn, A. 1875. *Der Ursprung der Wirbelthiere und das Princip des Funktionswechsels.* Leipzig Engelmann.

Dolcemascolo, G., R. Pennati, F. De Bernardi, F. Damiani, and M. Gianguzza. 2009. Ultrastructural comparative analysis on the adhesive papillae of the swimming larvae of three ascidian species. *Invertebr Surv J* 6:S77–S86.

Donoghue, P. C. J., P. L. Forey, and R. J. Aldridge. 2000. Conodont affinity and chordate phylogeny. *Biol Rev Cambridge Philos Soc* 75:191–251.

Donoghue, P. C., and M. A. Purnell. 2005. Genome duplication, extinction and vertebrate evolution. *Trends Ecol Evol* 20:312–319.

Donoghue, P. C., and M. A. Purnell. 2009. Distinguishing heat from light in debate over controversial fossils. *Bioessays* 31:178–189.

Donoghue, P. C., and I. J. Sansom. 2002. Origin and early evolution of vertebrate skeletonization. *Microsc Res Tech* 59:352–372.

Dores, R. M., and A. J. Baron. 2011. Evolution of POMC: origin, phylogeny, posttranslational processing, and the melanocortins. *Ann N Y Acad Sci* 1220:34–48.

Dorsett, D. A. 1986. Brains to cells: the neuroanatomy of selected gastropod species. In *The Mollusca, Vol. 9. Neurobiology and behavior, Part 2*, edited by A. O. D. Willows, 101–187. Orlando: Academic Press.

Dos Santos, S., C. Bardet, S. Bertrand, H. Escriva, D. Habert, and B. Querat. 2009. Distinct expression patterns of glycoprotein hormone-alpha2 and -beta5 in a basal chordate suggest independent developmental functions. *Endocrinology* 150:3815–3822.

Dove, H., and A. Stollewerk. 2003. Comparative analysis of neurogenesis in the myriapod *Glomeris marginata* (Diplopoda) suggests more similarities to chelicerates than to insects. *Development* 130:2161–2171.

Dubaissi, E., K. Rousseau, R. Lea, X. Soto, S. Nardeosingh, A. Schweickert, E. Amaya, D. J. Thornton, and N. Papalopulu. 2014. A secretory cell type develops alongside multiciliated cells, ionocytes and goblet cells, and provides a protective, anti-infective function in the frog embryonic mucociliary epidermis. *Development* 141:1514–1525.

Duboc, V., E. Röttinger, F. Lapraz, L. Besnardeau, and T. Lepage. 2005. Left-right asymmetry in the sea urchin embryo is regulated by nodal signaling on the right side. *Dev Cell* 9:147–158.

Duffy, D. J., G. Plickert, T. Kuenzel, W. Tilmann, and U. Frank. 2010. Wnt signaling promotes oral but suppresses aboral structures in *Hydractinia* metamorphosis and regeneration. *Development* 137:3057–3066.

Dufour, H. D., Z. Chettouh, C. Deyts, R. de Rosa, C. Goridis, J. S. Joly, and J. F. Brunet. 2006. Precraniate origin of cranial motoneurons. *Proc Natl Acad Sci USA* 103:8727–8732.

Dunn, E. F., V. N. Moy, L. M. Angerer, R. C. Angerer, R. L. Morris, and K. J. Peterson. 2007. Molecular paleoecology: using gene regulatory analysis to address the origins of complex life cycles in the late Precambrian. *Evol Dev* 9:10–24.

Duvaux-Miret, O., C. Dissous, J. P. Gautron, E. Pattou, C. Kordon, and A. Capron. 1990. The helminth *Schistosoma mansoni* expresses a peptide similar to human beta-endorphin and possesses a proopiomelanocortin-related gene. *New Biol* 2:93–99.

Eakin, R. M. 1972. Structure of invertebrate photoreceptors. In *Photochemistry of vision*, edited by E. W. E. W. Abrahamson and H. J. A. Dartnall, 625–684. Berlin: Springer.

Eakin, R. M., and A. Kuda. 1971. Ultrastructure of sensory receptors in ascidian tadpoles. *Z Zellforsch Mikrosk Anat* 112:287–312.

Eakin, R. M., and J. A. Westfall. 1962. Fine structure of photoreceptors in Amphioxus. *J Ultrastruct Res* 6:531–539.

Eakin, R.M. 1963. Lines of evolution of photoreceptors. In *General physiology of cell specialization*, edited by D. Mazia and A. Tyler, 393–425. New York: McGraw-Hill.

Eakin, R.M. 1979. Evolutionary significance of photoreceptors: in retrospect. *Am Zool* 19:647–653.

Eames, B. F., D. M. Medeiros, and I. Adameyko. 2020. *Evolving neural crest cells*. Boca Raton: CRC Press.

Eeckhaut, I., P. Flammang, C. LoBue, and M. Jangoux. 1997. Functional morphology of the tentacles and tentilla of *Coeloplana bannworthi* (Ctenophora, Platyctenida), an ectosymbiont of *Diadema setosum* (Echinodermata, Echinoida). *Zoomorphology* 117:165–174.

Eggert, T., B. Hauck, N. Hildebrandt, W. J. Gehring, and U. Walldorf. 1998. Isolation of a Drosophila homolog of the vertebrate homeobox gene Rx and its possible role in brain and eye development. *Proc Natl Acad Sci USA* 95:2343–2348.

Ehlers, U., and B. Ehlers. 1977. Monociliary receptors in interstitial Proseriata and Neorhabdocoela (Turbellaria Neoophora). *Zoomorphologie* 86:197–222.

Eijkelkamp, N., K. Quick, and J. N. Wood. 2013. Transient receptor potential channels and mechanosensation. *Annu Rev Neurosci* 36:519–546.

Eisenhoffer, G. T., and J. Rosenblatt. 2013. Bringing balance by force: live cell extrusion controls epithelial cell numbers. *Trends Cell Biol* 23:185–192.

El Hashash, A. H., D. Al Alam, G. Turcatel, O. Rogers, X. Li, S. Bellusci, and D. Warburton. 2011. Six1 transcription factor is critical for coordination of epithelial, mesenchymal and vascular morphogenesis in the mammalian lung. *Dev Biol* 353:242–258.

Elphick, M. R., O. Mirabeau, and D. Larhammar. 2018. Evolution of neuropeptide signalling systems. *J Exp Biol* 221:jeb151092.

Ema, M., M. Morita, S. Ikawa, M. Tanaka, Y. Matsuda, O. Gotoh, Y. Saijoh, H. Fujii, H. Hamada, Y. Kikuchi, and Y. Fujii-Kuriyama. 1996. Two new members of the murine Sim gene family are transcriptional repressors and show different expression patterns during mouse embryogenesis. *Mol Cell Biol* 16:5865–5875.

Emery, D. G. 1992. Fine structure of olfactory epithelia of gastropod molluscs. *Microsc Res Tech* 22:307–324.

Erclik, T., V. Hartenstein, H. D. Lipshitz, and R. R. McInnes. 2008. Conserved role of the Vsx genes supports a monophyletic origin for bilaterian visual systems. *Curr Biol* 18:1278–1287.

Erclik, T., V. Hartenstein, R. R. McInnes, and H. D. Lipshitz. 2009. Eye evolution at high resolution: the neuron as a unit of homology. *Dev Biol* 332:70–79.

Eriksson, B. J., D. Fredman, G. Steiner, and A. Schmid. 2013. Characterisation and localisation of the opsin protein repertoire in the brain and retinas of a spider and an onychophoran. *BMC Evol Biol* 13:186.

Erzurumlu, R. S., Y. Murakami, and F. M. Rijli. 2010. Mapping the face in the somatosensory brainstem. *Nat Rev Neurosci* 11:252–263.

Escriva, H., N. D. Holland, H. Gronemeyer, V. Laudet, and L. Z. Holland. 2002. The retinoic acid signaling pathway regulates anterior/posterior patterning in the nerve cord and pharynx of amphioxus, a chordate lacking neural crest. *Development* 129:2905–2916.

Esposito, R., C. Racioppi, M. R. Pezzotti, M. Branno, A. Locascio, F. Ristoratore, and A. Spagnuolo. 2015. The ascidian pigmented sensory organs: structures and developmental programs. *Genesis* 53:15–33.

Fain, G. L., R. Hardie, and S. B. Laughlin. 2010. Phototransduction and the evolution of photoreceptors. *Curr Biol* 20:R114–R124.

Fang, Y., W. Huang, and L. Chen. 1999. Immunohistochemical localization of gonadotropin-releasing hormone receptors (GnRHR) in the nervous system, Hatschek's pit and gonads of amphioxus, *Branchiostoma belcheri*. *Chin Sci Bull* 44:908–912.

Feinberg, S., A. Roure, J. Piron, and S. Darras. 2019. Antero-posterior ectoderm patterning by canonical Wnt signaling during ascidian development. *PLoS Genet* 15:e1008054.

Fekete, D. M., and D. K. Wu. 2002. Revisiting cell fate specification in the inner ear. *Curr Opin Neurobiol* 12:35–42.

Ferrer, C., G. Malagón, P. Gómez Mdel, and E. Nasi. 2012. Dissecting the determinants of light sensitivity in amphioxus microvillar photoreceptors: possible evolutionary implications for melanopsin signaling. *J Neurosci* 32:17977–17987.

Feuda, R., M. Dohrmann, W. Pett, H. Philippe, O. Rota-Stabelli, N. Lartillot, G. Wörheide, and D. Pisani. 2017. Improved modeling of compositional heterogeneity supports sponges as sister to all other animals. *Curr Biol* 27:3864–3870e4.

Feuda, R., S. C. Hamilton, J. O. McInerney, and D. Pisani. 2012. Metazoan opsin evolution reveals a simple route to animal vision. *Proc Natl Acad Sci USA* 109:18868–18872.

Feuda, R., O. Rota-Stabelli, T. H. Oakley, and D. Pisani. 2014. The comb jelly opsins and the origins of animal phototransduction. *Genome Biol Evol* 6:1964–1971.

Fitch, W. M. 1970. Distinguishing homologous from analogous proteins. *Syst Zool* 19:99–113.

Flasse, L. C., D. G. Stern, J. L. Pirson, I. Manfroid, B. Peers, and M. L. Voz. 2013. The bHLH transcription factor Ascl1a is essential for the specification of the intestinal secretory cells and mediates Notch signaling in the zebrafish intestine. *Dev Biol* 376:187–197.

Fleming, J. F., R. Feuda, N. W. Roberts, and D. Pisani. 2020. A novel approach to investigate the effect of tree reconstruction artifacts in single-gene analysis clarifies opsin evolution in nonbilaterian metazoans. *Genome Biol Evol* 12:3906–3916.

Focareta, L., S. Sesso, and A. G. Cole. 2014. Characterization of homeobox genes reveals sophisticated regionalization of the central nervous system in the European cuttlefish *Sepia officinalis*. *PLoS One* 9:e109627.

Fortunato, S. A., M. Adamski, O. M. Ramos, S. Leininger, J. Liu, D. E. Ferrier, and M. Adamska. 2014. Calcisponges have a ParaHox gene and dynamic expression of dispersed NK homeobox genes. *Nature* 514:620–623.

Fortunato, S. A., S. Leininger, and M. Adamska. 2014. Evolution of the Pax-Six-Eya-Dach network: the calcisponge case study. *Evodevo* 5:23.

Fortunato, S. A. V., M. Vervoort, M. Adamski, and M. Adamska. 2016. Conservation and divergence of bHLH genes in the calcisponge *Sycon ciliatum*. *Evodevo* 7:23.

Franc, J. M. 1978. Organization and function of ctenophore colloblasts: an ultrastructural study. *Biol Bull* 155:527–541.

Frank, C. A., P. D. Baum, and G. Garriga. 2003. HLH-14 is a *C. elegans* achaete-scute protein that promotes neurogenesis through asymmetric cell division. *Development* 130:6507–6518.

Frankfort, B. J., and G. Mardon. 2002. R8 development in the *Drosophila* eye: a paradigm for neural selection and differentiation. *Development* 129:1295–1306.

Frankfort, B. J., R. Nolo, Z. Zhang, H. Bellen, and G. Mardon. 2001. senseless repression of rough is required for R8 photoreceptor differentiation in the developing *Drosophila* eye. *Neuron* 32:403–414.

Fredriksson, R., and H. B. Schiöth. 2005. The repertoire of G-protein-coupled receptors in fully sequenced genomes. *Mol Pharmacol* 67:1414–1425.

Friedman, J. R., and K. H. Kaestner. 2006. The Foxa family of transcription factors in development and metabolism. *Cell Mol Life Sci* 63:2317–2328.

Friedrich, M. 2013. Development and evolution of the *Drosophila* Bolwig's organ: a compound eye relict. In *Molecular genetics of axial patterning. Growth and disease in the Drosophila eye*, edited by A. Singh and M. Kango-Singh, 329–357. New York: Springer.

Fritsch, H. A., S. Van Noorden, and A. G. Pearse. 1982. Gastro-intestinal and neurohormonal peptides in the alimentary tract and cerebral complex of *Ciona intestinalis* (Ascidiaceae). *Cell Tissue Res* 223:369–402.

Fritzenwanker, J. H., J. Gerhart, R. M. Freeman, Jr., and C. J. Lowe. 2014. The Fox/Forkhead transcription factor family of the hemichordate *Saccoglossus kowalevskii*. *Evodevo* 5:17.

Fritzsch, B., K. W. Beisel, and N. A. Bermingham. 2000. Developmental evolutionary biology of the vertebrate ear: conserving mechanoelectric transduction and developmental pathways in diverging morphologies. *Neuroreport* 11:R35–R44.

Fritzsch, B., K. W. Beisel, S. Pauley, and G. Soukup. 2007. Molecular evolution of the vertebrate mechanosensory cell and ear. *Int J Dev Biol* 51:663–678.

Fritzsch, B., D. F. Eberl, and K. W. Beisel. 2010. The role of bHLH genes in ear development and evolution: revisiting a 10-year-old hypothesis. *Cell Mol Life Sci* 67:3089–3099.

Fritzsch, B., and K. L. Elliott. 2017. Gene, cell, and organ multiplication drives inner ear evolution. *Dev Biol* 431:3–15.

Fritzsch, B., N. Pan, I. Jahan, J. S. Duncan, B. J. Kopecky, K. L. Elliott, J. Kersigo, and T. Yang. 2013. Evolution and development of the tetrapod auditory system: an organ of Corti-centric perspective. *Evol Dev* 15:63–79.

Fritzsch, B., J. Piatigorsky, K. Tessmar-Raible, G. Jékely, K. Guy, F. Raible, J. Wittbrodt, and D. Arendt. 2005. Ancestry of photic and mechanic sensation? *Science* 308:1113–1114.

Fu, W., H. Duan, E. Frei, and M. Noll. 1998. Shaven and sparkling are mutations in separate enhancers of the *Drosophila* Pax2 homolog. *Development* 125:2943–2950.

Fung, S., F. Wang, M. Chase, D. Godt, and V. Hartenstein. 2008. Expression profile of the cadherin family in the developing *Drosophila* brain. *J Comp Neurol* 506:469–488.

Furlong, R. F., and P. W. Holland. 2002. Bayesian phylogenetic analysis supports monophyly of ambulacraria and of cyclostomes. *Zoolog Sci* 19:593–599.

Furukawa, T., E. M. Morrow, and C. L. Cepko. 1997. Crx, a novel otx-like homeobox gene, shows photoreceptor-specific expression and regulates photoreceptor differentiation. *Cell* 91:531–541.

Galliot, B., M. Quiquand, L. Ghila, R. de Rosa, M. Miljkovic-Licina, and S. Chera. 2009. Origins of neurogenesis, a cnidarian view. *Dev Biol* 332:2–24.

Gammill, L. S., and H. Sive. 2001. Otx2 expression in the ectoderm activates anterior neural determination and is required for *Xenopus* cement gland formation *Dev Biol* 240:223–236.

Gans, C., and R. G. Northcutt. 1983. Neural crest and the origin of vertebrates: a new head. *Science* 220:268–274.

Garcia-Fernàndez, J., and E. Benito-Gutiérrez. 2009. It's a long way from amphioxus: descendants of the earliest chordate. *Bioessays* 31:665–675.

Garner, S., I. Zysk, G. Byrne, M. Kramer, D. Moller, V. Taylor, and R. D. Burke. 2016. Neurogenesis in sea urchin embryos and the diversity of deuterostome neurogenic mechanisms. *Development* 143:286–297.

Garstang, W. 1894. Preliminary note on a new theory of the phylogeny of the chordata. *Zool Anz* 22:122–125.

Garstang, W. 1922. The theory of recapitulation. *J Linn Soc Lond (Zool)* 35:81–101.

Garstang, W. 1928. The morphology of the Tunicata. *Quart J Microsc Sci* 72:51–189.

Gasparini, F., V. Degasperi, S. M. Shimeld, P. Burighel, and L. Manni. 2013. Evolutionary conservation of the placodal transcriptional network during sexual and asexual development in chordates. *Dev Dyn* 242:752–766.

Gauchat, D., S. Kreger, T. Holstein, and B. Galliot. 1998. Prdl-a, a gene marker for *Hydra* apical differentiation related to triploblastic paired-like head-specific genes. *Development* 125:1637–1645.

Gazave, E., P. Lapebie, E. Renard, C. Bezac, N. Boury-Esnault, J. Vacelet, T. Perez, M. Manuel, and C. Borchiellini. 2008. NK homeobox genes with choanocyte-specific expression in homoscleromorph sponges. *Dev Genes Evol* 218:479–489.

Gee, H. 1996. *Before the backbone: views on the origin of the vertebrates.* London: Chapman and Hall.

Gee, H. 2018. *Across the bridge. Understanding the origin of vertebrates.* Chicago: University of Chicago Press.

Geeta, R. 2003. Structure trees and species trees: what they say about morphological development and evolution. *Evol Dev* 5:609–621.

Gegenbaur, C. 1859. *Grundzüge der vergleichenden Anatomie.* Leipzig: Wilhelm Engelmann.

Gehring, W. J., and K. Ikeo. 1999. Pax 6: mastering eye morphogenesis and eye evolution. *Trends Genet* 15:371–377.

Georges, D. 1985. Presence of cells resembling serotonergic elements in four species of tunicates. *Cell Tissue Res* 242:341–348.

Gerhart, J. 2006. The deuterostome ancestor. *J Cell Physiol* 209:677–685.

Gerhart, J., C. Lowe, and M. Kirschner. 2005. Hemichordates and the origin of chordates. *Curr Opin Genet Dev* 15:461–467.

Gerkema, M. P., W. I. Davies, R. G. Foster, M. Menaker, and R. A. Hut. 2013. The nocturnal bottleneck and the evolution of activity patterns in mammals. *Proc Biol Sci* 280:20130508.

Ghysen, A., and C. Dambly-Chaudière. 1989. Genesis of the *Drosophila* peripheral nervous system. *Trends Genet* 5:251–255.

Gillis, J. A., J. H. Fritzenwanker, and C. J. Lowe. 2012. A stem-deuterostome origin of the vertebrate pharyngeal transcriptional network. *Proc Biol Sci* 279:237–246.

Gillis, W. J., B. Bowerman, and S. Q. Schneider. 2007. Ectoderm- and endomesoderm-specific GATA transcription factors in the marine annelid *Platynereis dumerilli. Evol Dev* 9:39–50.

Giri, D., M. L. Vignola, A. Gualtieri, V. Scagliotti, P. McNamara, M. Peak, M. Didi, C. Gaston-Massuet, and S. Senniappan. 2017. Novel FOXA2 mutation causes hyperinsulinism, hypopituitarism with craniofacial and endoderm-derived organ abnormalities. *Hum Mol Genet* 26:4315–4326.

Gislén, T. 1930. Affinities between the Echinodermata, Enteropneusta and Chordonia. *Zool Bidrag Uppsala* 12:199–304.

Giuliano, P., R. Marino, M. R. Pinto, and R. De Santis. 1998. Identification and developmental expression of Ci-isl, a homologue of vertebrate islet genes, in the ascidian *Ciona intestinalis. Mech Dev* 78:199–202.

Glardon, S., L. Z. Holland, W. J. Gehring, and N. D. Holland. 1998. Isolation and developmental expression of the amphioxus pax-6 gene (amphipax-6): insights into eye and photoreceptor evolution. *Development* 125:2701–2710.

Gold, D. A., R. D. Gates, and D. K. Jacobs. 2014. The early expansion and evolutionary dynamics of POU class genes. *Mol Biol Evol* 31:3136–3147.

Gómez-Díaz, C., F. Martin, J. M. Garcia-Fernàndez, and E. Alcorta. 2018. The two main olfactory receptor families in *Drosophila*, ORs and IRs: a comparative approach. *Front Cell Neurosci* 12:253.

Gómez, M. P., and E. Nasi. 2000. Light transduction in invertebrate hyperpolarizing photoreceptors: possible involvement of a Go-regulated guanylate cyclase. *J Neurosci* 20:5254–5263.

Gómez, M. P., J. M. Angueyra, and E. Nasi. 2009. Light-transduction in melanopsin-expressing photoreceptors of Amphioxus. *Proc Natl Acad Sci USA* 106:9081–9086.

Gong, J., Y. Yuan, A. Ward, L. Kang, B. Zhang, Z. Wu, J. Peng, Z. Feng, J. Liu, and X. Z. S. Xu. 2016. The *C. elegans* taste receptor homolog LITE-1 is a photoreceptor. *Cell* 167:1252–1263e10.

Gonzalez, P., K. R. Uhlinger, and C. J. Lowe. 2017. The adult body plan of indirect developing hemichordates develops by adding a Hox-patterned trunk to an anterior larval territory. *Curr Biol* 27:87–95.

Goodman, M. B. 2006. Mechanosensation. ed. The C. elegans Research Community, WormBook, doi/10.1895/wormbook.1.62.1. *WormBook*:1–14.

Goodman, M. B., and P. Sengupta. 2019. How *Caenorhabditis elegans* senses mechanical stress, temperature, and other physical stimuli. *Genetics* 212:25–51.

Goodrich, E. S. 1917. "Proboscis pores" in craniate vertebrates, a suggestion concerning the premandibular somites and hypophysis. *Quart J Microsc Sci* 62:539–553.

Gorbman, A., M. Nozaki, and K. Kubokawa. 1999. A brain-Hatschek's pit connection in amphioxus. *Gen comp Endocrinol* 113:251–254.

Gorbman, A. 1999. Brain - Hatschek's pit relationships in amphioxus species. *Acta-Zoologica* 80:301–305.

Gordon, P. M., and O. Hobert. 2015. A competition mechanism for a homeotic neuron identity transformation in *C. elegans*. *Dev Cell* 34:206–219.

Gorman, A. L., J. S. McReynolds, and S. N. Barnes. 1971. Photoreceptors in primitive chordates: fine structure, hyperpolarizing receptor potentials, and evolution. *Science* 172:1052–1054.

Gostling, N. J., and S. M. Shimeld. 2003. Protochordate Zic genes define primitive somite compartments and highlight molecular changes underlying neural crest evolution. *Evol Dev* 5:136–144.

Goulding, S. E., P. zur Lage, and A. P. Jarman. 2000. amos, a proneural gene for *Drosophila* olfactory sense organs that is regulated by lozenge. *Neuron* 25:69–78.

Gracheva, E. O., N. T. Ingolia, Y. M. Kelly, J. F. Cordero-Morales, G. Hollopeter, A. T. Chesler, E. E. Sánchez, J. C. Perez, J. S. Weissman, and D. Julius. 2010. Molecular basis of infrared detection by snakes. *Nature* 464:1006–1011.

Graf, T., and T. Enver. 2009. Forcing cells to change lineages. *Nature* 462:587–594.

Graham, A., T. Butts, A. Lumsden, and C. Kiecker. 2014. What can vertebrates tell us about segmentation? *Evodevo* 5:24.

Graham, A., and S. M. Shimeld. 2013. The origin and evolution of the ectodermal placodes. *J Anat* 222:32–40.

Grande, C., and N. H. Patel. 2009a. Lophotrochozoa get into the game: the nodal pathway and left/right asymmetry in bilateria. *Cold Spring Harb Symp Quant Biol* 74:281–287.

Grande, C., and N. H. Patel. 2009b. Nodal signalling is involved in left-right asymmetry in snails. *Nature* 457:1007–1011.

Grave, C., and G. Riley. 1935. Development of the sense organs of the larva of *Botryllus schlosseri*. *J Morph Physiol* 57:185–211.

Graziussi, D. F., H. Suga, V. Schmid, and W. J. Gehring. 2012. The "eyes absent" (eya) gene in the eye-bearing hydrozoan jellyfish *Cladonema radiatum*: conservation of the retinal determination network. *J Exp Zool B Mol Dev Evol* 318:257–267.

Green, S. A., M. Simoes-Costa, and M. E. Bronner. 2015. Evolution of vertebrates as viewed from the crest. *Nature* 520:474–482.

Grens, A., E. Mason, J. L. Marsh, and H. R. Bode. 1995. Evolutionary conservation of a cell fate specification gene: the *Hydra* achaete-scute homolog has proneural activity in *Drosophila*. *Development* 121:4027–4035.

Gribble, F. M., and F. Reimann. 2016. Enteroendocrine cells: chemosensors in the intestinal epithelium. *Annu Rev Physiol* 78:277–299.

Grimmelikhuijzen, C. J. P., M. Williamson, and G. N. Hansen. 2004. Neuropeptides in cnidarians. In *Cell signalling in prokaryotes and lower metazoans*, edited by I. Fairweather, 115–139. Dordrecht: Kluwer Academic Publ.

Grimmelikhuijzen, C. J., and J. A. Westfall. 1995. The nervous systems of cnidarians. *EXS* 72:7–24.

Grobben, K. 1908. Die systematische Einteilung des Tierreichs. *Verh Zool Bot Ges Wien* 58:491–510.

Gröger, H., P. Callaerts, W. J. Gehring, and V. Schmid. 2000. Characterization and expression analysis of an ancestor-type Pax gene in the hydrozoan jellyfish *Podocoryne carnea*. *Mech Dev* 94:157–169.

Grossniklaus, U., R. K. Pearson, and W. J. Gehring. 1992. The *Drosophila* sloppy paired locus encodes two proteins involved in segmentation that show homology to mammalian transcription factors. *Genes Dev* 6:1030–1051.

Groves, A. K., and D. M. Fekete. 2012. Shaping sound in space: the regulation of inner ear patterning. *Development* 139:245–257.

Groves, A. K., and C. LaBonne. 2014. Setting appropriate boundaries: fate, patterning and competence at the neural plate border. *Dev Biol* 389:2–12.

Grus, W. E., and J. Zhang. 2009. Origin of the genetic components of the vomeronasal system in the common ancestor of all extant vertebrates. *Mol Biol Evol* 26:407–419.

Gühmann, M., H. Jia, N. Randel, C. Veraszto, L. A. Bezares-Calderón, N. K. Michiels, S. Yokoyama, and G. Jékely. 2015. Spectral tuning of phototaxis by a Go-opsin in the rhabdomeric eyes of *Platynereis*. *Curr Biol* 25:2265–2271.

Guijarro-Clarke, C., P. W. H. Holland, and J. Paps. 2020. Widespread patterns of gene loss in the evolution of the animal kingdom. *Nat Ecol Evol* 4:519–523.

Gupta, B. P., and V. Rodrigues. 1997. Atonal is a proneural gene for a subset of olfactory sense organs in *Drosophila*. *Genes Cells* 2:225–233.

Gupta, D., S. A. Harvey, N. Kaminski, and S. K. Swamynathan. 2011. Mouse conjunctival forniceal gene expression during postnatal development and its regulation by Kruppel-like factor 4. *Invest Ophthalmol Vis Sci* 52:4951–4962.

Guthrie, D. M. 1975. The physiology and structure of the nervous system of amphioxus, *Branchiostoma lanceolatum*. In *Symposia of the zoological society of London*, edited by E. J. W. Barrington and R. P. S. Jefferies, 43–80. London: Academic Press.

Guthrie, S. 2007. Patterning and axon guidance of cranial motor neurons. *Nat Rev Neurosci* 8:859–871.

Hadfield, M. G., E. A. Meleshkevitch, and D. Y. Boudko. 2000. The apical sensory organ of a gastropod veliger is a receptor for settlement cues. *Biol Bull* 198:67–76.

Haeckel, E. 1866. *Generelle Morphologie der Organismen*. Berlin: Reimer.

Halanych, K. M. 2004. The new view of animal phylogeny. *Annu Rev Ecol Evol Syst* 35:229–256.

Halanych, K. M., J. D. Bacheller, A. M. A. Aguinaldo, S. M. Liva, D. M. Hillis, and J. A. Lake. 1995. Evidence from 18S ribosomal DNA that the lophophorates are protostome animals. *Science* 267:1641–1643.

Halder, G., P. Callaerts, S. Flister, U. Walldorf, U. Kloter, and W. J. Gehring. 1998. Eyeless initiates the expression of both sine oculis and eyes absent during *Drosophila* compound eye development. *Development* 125:2181–2191.

Halder, G., P. Callaerts, and W. J. Gehring. 1995. Induction of ectopic eyes by targeted expression of the eyeless gene in *Drosophila*. *Science* 267:1788–1792.

Hall, B. K. 2013. Homology, homoplasy, novelty, and behavior. *Dev Psychobiol* 55:4–12.

Hall, B. K., and R. Kerney. 2012. Levels of biological organization and the origin of novelty. *J Exp Zool B Mol Dev Evol* 318:428–437.

Hallberg, E., and B. S. Hansson. 1999. Arthropod sensilla: morphology and phylogenetic considerations. *Microsc Res Tech* 47:428–439.

Hallgrimsson, B., H. A. Jamniczky, N. M. Young, C. Rolian, U. Schmidt-Ott, and R. S. Marcucio. 2012. The generation of variation and the developmental basis for evolutionary novelty. *J Exp Zool B Mol Dev Evol* 318:501–517.

Hamada, M., N. Shimozono, N. Ohta, Y. Satou, T. Horie, T. Kawada, H. Satake, Y. Sasakura, and N. Satoh. 2011. Expression of neuropeptide- and hormone-encoding genes in the *Ciona intestinalis* larval brain. *Dev Biol* 352:202–214.

Hansen, T. F. 2013. Why epistasis is important for selection and adaptation. *Evolution* 67:3501–3511.

Hardie, R. C., and K. Franze. 2012. Photomechanical responses in *Drosophila* photoreceptors. *Science* 338:260–263.

Hardie, R. C., and M. Juusola. 2015. Phototransduction in *Drosophila*. *Curr Opin Neurobiol* 34:37–45.

Hardie, R. C., and M. Postma. 2008. Phototransduction in microvillar photoreceptors of *Drosophila* and other invertebrates. In *The senses: a comprehensive reference. Vol. 1: Vision*, edited by A. I. Basbaum, A. Kaneko, G. M. Shepherd and G. Westheimer, 77–130. San Diego: Academic Press.

Hartenstein, V. 1997. Development of the insect stomatogastric nervous system. *Trends Neurosci* 20:421–7.

Hartenstein, V. 2005. Development of insect sensilla. In *Comprehensive molecular insect science*, edited by L. I. Gilbert, K. Iatrou and S. S. Gill, 379–419. Amsterdam: Elsevier.

Hartenstein, V. 2006. The neuroendocrine system of invertebrates: a developmental and evolutionary perspective. *J Endocrinol* 190:555–570.

Hartenstein, V., S. Takashima, P. Hartenstein, S. Asanad, and K. Asanad. 2017. bHLH proneural genes as cell fate determinants of entero-endocrine cells, an evolutionarily conserved lineage sharing a common root with sensory neurons. *Dev Biol* 431:36–47.

Hartmann, B., P. N. Lee, Y. Y. Kang, S. Tomarev, H. G. de Couet, and P. Callaerts. 2003. Pax6 in the sepiolid squid *Euprymna scolopes*: evidence for a role in eye, sensory organ and brain development. *Mech Dev* 120:177–183.

Haszprunar, G. 1985. The fine morphology of the osphradial sense-organs of the Mollusca .1. Gastropoda, Prosobranchia. *Philos Trans R Soc Lond B Biol Sci* 307:457–496.

Haszprunar, G. 1992. Ultrastructure of the osphradium of the Tertiary relict snail, *Campanile symbolicum* Iredale (Mollusca, Streptoneura). *Philos Trans R Soc Lond B-Biol Sci* 337:457–469.

Hatschek, B. 1881. Studien über Entwickelung des Amphioxus. *Arbeiten Zool Inst Wien* 4:1–88.

Hatschek, B. 1884. Mittheilungen über Amphioxus. *Zool Anz* 7:517–520.

Hausken, K. N., B. Tizon, M. Shpilman, S. Barton, W. Decatur, D. Plachetzki, S. Kavanaugh, S. Ul-Hasan, B. Levavi-Sivan, and S. A. Sower. 2018. Cloning and characterization of a second lamprey pituitary glycoprotein hormone, thyrostimulin (GpA2/GpB5). *Gen Comp Endocrinol* 264:16–27.

Hayakawa, E., C. Fujisawa, and T. Fujisawa. 2004. Involvement of *Hydra* achaete-scute gene CnASH in the differentiation pathway of sensory neurons in the tentacles. *Dev Genes Evol* 214:486–492.

Hayashi, T., T. Kojima, and K. Saigo. 1998. Specification of primary pigment cell and outer photoreceptor fates by BarH1 homeobox gene in the developing *Drosophila* eye. *Dev Biol* 200:131–145.

Heimberg, A. M., R. Cowper-Sal-lari, M. Semon, P. C. Donoghue, and K. J. Peterson. 2010. microRNAs reveal the interrelationships of hagfish, lampreys, and gnathostomes and the nature of the ancestral vertebrate. *Proc Natl Acad Sci USA* 107:19379–19383.

Henderson, S. R., H. Reuss, and R. C. Hardie. 2000. Single photon responses in *Drosophila* photoreceptors and their regulation by Ca^2+. *J Physiol* 524 Pt 1:179–194.

Hennig, A. K., G. H. Peng, and S. Chen. 2008. Regulation of photoreceptor gene expression by Crx-associated transcription factor network. *Brain Res* 1192:114–133.

Hennig, W. 1966. *Phylogenetic systematics*. Urbana: University of Illinois Press.

Hering, L., and G. Mayer. 2014. Analysis of the opsin repertoire in the tardigrade *Hypsibius dujardini* provides insights into the evolution of opsin genes in Panarthropoda. *Genome Biol Evol* 6:2380–2391.

Hernandez-Nicaise, M.-L. 1984. Ctenophora. In *Biology of the integument. 1: Invertebrates*, edited by J. Bereiter-Hahn, A. G. Matoltsy and K. S. Richards, 96–111. Heidelberg: Springer.

Heyland, A., D. Plachetzki, E. Donelly, D. Gunaratne, Y. Bobkova, J. Jacobson, A. B. Kohn, and L. L. Moroz. 2012. Distinct expression patterns of glycoprotein hormone subunits in the lophotrochozoan *Aplysia*: implications for the evolution of neuroendocrine systems in animals. *Endocrinology* 153:5440–5451.

Hibino, T., A. Nishino, and S. Amemiya. 2006. Phylogenetic correspondence of the body axes in bilaterians is revealed by the right-sided expression of Pitx genes in echinoderm larvae. *Dev Growth Differ* 48:587–595.

Higashijima, S., T. Michiue, Y. Emori, and K. Saigo. 1992. Subtype determination of *Drosophila* embryonic external sensory organs by redundant homeo box genes BarH1 and BarH2. *Genes Dev* 6:1005–1018.

Hill, A., W. Boll, C. Ries, L. Warner, M. Osswalt, M. Hill, and M. Noll. 2010. Origin of Pax and Six gene families in sponges: single PaxB and Six1/2 orthologs in *Chalinula loosanoffi*. *Dev Biol* 343:106–123.

Hirth, F., L. Kammermeier, E. Frei, U. Walldorf, M. Noll, and H. Reichert. 2003. An urbilaterian origin of the tripartite brain: developmental genetic insights from *Drosophila*. *Development* 130:2365–2373.

Hirth, F., and H. Reichert. 2007. Basic nervous system types: one or many. In *Evolution of nervous systems*, Vol. 1, edited by J. H. Kaas, 55–72. Amsterdam: Academic Press.

Hisatomi, O., and F. Tokunaga. 2002. Molecular evolution of proteins involved in vertebrate phototransduction. *Comp Biochem Physiol B Biochem Mol Biol* 133:509–522.

Hobert, O. 2008. Regulatory logic of neuronal diversity: terminal selector genes and selector motifs. *Proc Natl Acad Sci USA* 105:20067–20071.

Hobert, O. 2011. Regulation of terminal differentiation programs in the nervous system. *Annu Rev Cell Dev Biol* 27:681–696.

Hobert, O. 2016. A map of terminal regulators of neuronal identity in *Caenorhabditis elegans*. *Wiley Interdiscip Rev Dev Biol* 5:474–498.

Hobert, O., and H. Westphal. 2000. Functions of LIM-homeobox genes. *Trends Genet* 16:75–83.

Hobmayer, E., T. W. Holstein, and C. N. David. 1990. Tentacle morphogenesis in *Hydra*. 2. Formation of a complex between a sensory nerve-cell and a battery cell. *Development* 109:897–904.

Hockman, D., A. J. Burns, G. Schlosser, K. P. Gates, B. Jevans, A. Mongera, S. Fisher, G. Unlu, E. W. Knapik, C. K. Kaufman, C. Mosimann, L. I. Zon, J. J. Lancman, P. D. S. Dong, H. Lickert, A. S. Tucker, and C. V. Baker. 2017. Evolution of the hypoxia-sensitive cells involved in amniote respiratory reflexes. *Elife* 6:e21231.

Holland, L. Z. 2002. Heads or tails? Amphioxus and the evolution of anterior-posterior patterning in deuterostomes. *Dev Biol* 241:209–228.

Holland, L. Z. 2005a. Non-neural ectoderm is really neural: evolution of developmental patterning mechanisms in the non-neural ectoderm of chordates and the problem of sensory cell homologies. *J Exp Zoolog B Mol Dev Evol* 304B:304–323.

Holland, L. Z. 2015. Genomics, evolution and development of amphioxus and tunicates: the Goldilocks principle. *J Exp Zool B Mol Dev Evol* 324:342–352.

Holland, L. Z. 2016a. Tunicates. *Curr Biol* 26:R146–R152.

Holland, L. Z., R. Albalat, K. Azumi, E. Benito-Gutiérrez, M. J. Blow, M. Bronner-Fraser, F. Brunet, T. Butts, S. Candiani, L. J. Dishaw, D. E. Ferrier, J. Garcia-Fernàndez, J. J. Gibson-Brown, C. Gissi, A. Godzik, F. Hallbook, D. Hirose, K. Hosomichi, T. Ikuta, H. Inoko, M. Kasahara, J. Kasamatsu, T. Kawashima, A. Kimura, M. Kobayashi, Z. Kozmik, K. Kubokawa, V. Laudet, G. W. Litman, A. C. McHardy, D. Meulemans, M. Nonaka, R. P. Olinski, Z. Pancer, L. A. Pennacchio, M. Pestarino, J. P. Rast, I. Rigoutsos, M. Robinson-Rechavi, G. Roch, H. Saiga, Y. Sasakura, M. Satake, Y. Satou, M. Schubert, N. Sherwood, T. Shiina, N. Takatori, J. Tello, P. Vopalensky, S. Wada, A. Xu, Y. Ye, K. Yoshida, F. Yoshizaki, J. K. Yu, Q. Zhang, C. M. Zmasek, P. J. de Jong, K. Osoegawa, N. H. Putnam, D. S. Rokhsar, N. Satoh, and P. W. Holland. 2008. The amphioxus genome illuminates vertebrate origins and cephalochordate biology. *Genome Res* 18:1100–1111.

Holland, L. Z., J. E. Carvalho, H. Escriva, V. Laudet, M. Schubert, S. M. Shimeld, and J. K. Yu. 2013. Evolution of bilaterian central nervous systems: a single origin? *Evodevo* 4:27.

Holland, L. Z., G. Gorsky, and R. Fenaux. 1988. Fertilization in *Oikopleura dioica* (Tunicata, Appendicularia): acrosome reaction, cortical reaction and sperm-egg fusion. *Zoomorphol* 108:229–243.

Holland, L. Z., and N. D. Holland. 2001. Evolution of neural crest and placodes: amphioxus as a model for the ancestral vertebrate? *J Anat* 199:85–98.

Holland, L. Z., and T. Onai. 2012. Early development of cephalochordates (amphioxus). *Wiley Interdiscip Rev Dev Biol* 1:167–183.

Holland, L. Z., K. A. Panfilio, R. Chastain, M. Schubert, and N. D. Holland. 2005. Nuclear beta-catenin promotes non-neural ectoderm and posterior cell fates in amphioxus embryos. *Dev Dyn* 233:1430–1443.

Holland, L. Z., M. Schubert, N. D. Holland, and T. Neuman. 2000. Evolutionary conservation of the presumptive neural plate markers AmphiSox1/2/3 and AmphiNeurogenin in the invertebrate chordate amphioxus. *Dev Biol* 226:18–33.

Holland, L. Z., M. Schubert, Z. Kozmik, and N. D. Holland. 1999. AmphiPax3/7, an amphioxus paired box gene: insights into chordate myogenesis, neurogenesis, and the possible evolutionary precursor of definitive vertebrate neural crest. *Evol Dev* 1:153–165.

Holland, N. D. 2003. Early central nervous system evolution: an era of skin brains? *Nat Rev Neurosci* 4:617–627.

Holland, N. D. 2005b. Chordates. *Curr Biol* 15:R911–R914.

Holland, N. D. 2016b. Nervous systems and scenarios for the invertebrate-to-vertebrate transition. *Philos Trans R Soc Lond B Biol Sci* 371:20150047.

Holland, N. D. 2018. Formation of the initial kidney and mouth opening in larval amphioxus studied with serial blockface scanning electron microscopy (SBSEM). *Evodevo* 9:16.

Holland, N. D., and L. Z. Holland. 2017. The ups and downs of amphioxus biology: a history. *Int J Dev Biol* 61:575–583.

Holland, N. D., L. Z. Holland, and P. W. Holland. 2015. Scenarios for the making of vertebrates. *Nature* 520:450–455.

Holland, N. D., G. Panganiban, E. L. Henyey, and L. Z. Holland. 1996. Sequence and developmental expression of AmphiDll, an amphioxus distalless gene trabscribed in the ectoderm, epidermis and nervous system: insights into evolution of craniate forebrain and neural crest. *Development* 122:2911–2920.

Holland, N. D., M. Paris, and D. Koop. 2009. The club-shaped gland of amphioxus: export of secretion to the pharynx in pre-metamorphic larvae and apoptosis during metamorphosis. *Acta Zool* 90:372–379.

Holmberg, K. 1986. The neural connection between the Langerhans receptor-cells and the central nervous system in *Oikopleura dioica* (Appendicularia). *Zoomorphology* 106:31–34.

Holstein, T. W. 2012. A view to kill. *BMC Biol* 10:18.

Holstein, T. W., and C. N. David. 1990. Putative intermediates in the nerve cell differentiation pathway in *Hydra* have properties of multipotent stem cells. *Dev Biol* 142:401–405.

Holstein, T. W., H. Watanabe, and S. Ozbek. 2011. Signaling pathways and axis formation in the lower metazoa. *Curr Top Dev Biol* 97:137–177.

Honoré, E., J. R. Martins, D. Penton, A. Patel, and S. Demolombe. 2015. The Piezo mechanosensitive ion channels: may the force be with you! *Rev. Physiol Biochem Pharmacol* 169:25–41.

Horie, R., A. Hazbun, K. Chen, C. Cao, M. Levine, and T. Horie. 2018. Shared evolutionary origin of vertebrate neural crest and cranial placodes. *Nature* 560:228–232.

Horie, T., R. Horie, K. Chen, C. Cao, M. Nakagawa, T. G. Kusakabe, N. Satoh, Y. Sasakura, and M. Levine. 2018. Regulatory cocktail for dopaminergic neurons in a protovertebrate identified by whole-embryo single-cell transcriptomics. *Genes Dev* 32:1297–1302.

Horie, T., D. Sakurai, H. Ohtsuki, A. Terakita, Y. Shichida, J. Usukura, T. Kusakabe, and M. Tsuda. 2008. Pigmented and nonpigmented ocelli in the brain vesicle of the ascidian larva. *J Comp Neurol* 509:88–102.

Horie, T., R. Shinki, Y. Ogura, T. G. Kusakabe, N. Satoh, and Y. Sasakura. 2011. Ependymal cells of chordate larvae are stem-like cells that form the adult nervous system. *Nature* 469:525–528.

Horridge, G. A. 1964. Presumed photoreceptive cilia in ctenophore. *Q J Microsc Sci* 105:311–317.

Hosp, J., Y. Sagane, G. Danks, and E. M. Thompson. 2012. The evolving proteome of a complex extracellular matrix, the *Oikopleura* house. *PLoS One* 7:e40172.

Hozumi, A., S. Matsunobu, K. Mita, N. Treen, T. Sugihara, T. Horie, T. Sakuma, T. Yamamoto, A. Shiraishi, M. Hamada, N. Satoh, K. Sakurai, H. Satake, and Y. Sasakura. 2020. GABA-induced GnRH release triggers chordate metamorphosis. *Curr Biol* 30:1555–1561e4.

Hroudova, M., P. Vojta, H. Strnad, Z. Krejcik, J. Ridl, J. Paces, C. Vlcek, and V. Paces. 2012. Diversity, phylogeny and expression patterns of Pou and Six homeodomain transcription factors in hydrozoan jellyfish *Craspedacusta sowerbyi*. *PLoS One* 7:e36420.

Hsiung, F., and K. Moses. 2002. Retinal development in *Drosophila*: specifying the first neuron. *Hum Mol Genet* 11:1207–1214.

Huang, C., J. A. Chan, and C. Schuurmans. 2014. Proneural bHLH genes in development and disease. *Curr Top Dev Biol* 110:75–127.

Huang, J., C. H. Liu, S. A. Hughes, M. Postma, C. J. Schwiening, and R. C. Hardie. 2010. Activation of TRP channels by protons and phosphoinositide depletion in *Drosophila* photoreceptors. *Curr Biol* 20:189–197.

Hudson, C., S. Darras, D. Caillol, H. Yasuo, and P. Lemaire. 2003. A conserved role for the MEK signalling pathway in neural tissue specification and posteriorisation in the invertebrate chordate, the ascidian *Ciona intestinalis*. *Development* 130:147–159.

Hudson, C., and P. Lemaire. 2001. Induction of anterior neural fates in the ascidian *Ciona intestinalis*. *Mech Dev* 100:189–203.

Hudson, C., S. Lotito, and H. Yasuo. 2007. Sequential and combinatorial inputs from Nodal, Delta2/Notch and FGF/MEK/ERK signalling pathways establish a grid-like organisation of distinct cell identities in the ascidian neural plate. *Development* 134:3527–3537.

Huising, M. O., C. P. Kruiswijk, and G. Flik. 2006. Phylogeny and evolution of class-I helical cytokines. *J Endocrinol* 189:1–25.

Hündgen, M., and C. Biela. 1982. Fine-structure of touch-plates in the scyphomedusan *Aurelia aurita. J Ultrastruct Res* 80:178–184.

Hunnekuhl, V. S., and M. Akam. 2014. An anterior medial cell population with an apical-organ-like transcriptional profile that pioneers the central nervous system in the centipede *Strigamia maritima. Dev Biol* 396:136–149.

Ikuta, T., and H. Saiga. 2007. Dynamic change in the expression of developmental genes in the ascidian central nervous system: revisit to the tripartite model and the origin of the midbrain-hindbrain boundary region. *Dev Biol* 312:631–643.

Ikuta, T., N. Satoh, and H. Saiga. 2010. Limited functions of Hox genes in the larval development of the ascidian *Ciona intestinalis. Development* 137:1505–1513.

Ikuta, T., N. Yoshida, N. Satoh, and H. Saiga. 2004. *Ciona intestinalis* Hox gene cluster: its dispersed structure and residual colinear expression in development. *Proc Natl Acad Sci USA* 101:15118–15123.

Imai, J. H., and I. A. Meinertzhagen. 2007. Neurons of the ascidian larval nervous system in *Ciona intestinalis*: II. Peripheral nervous system. *J Comp Neurol* 501:335–352.

Imai, K. S., H. Hikawa, K. Kobayashi, and Y. Satou. 2017. Tfap2 and Sox1/2/3 cooperatively specify ectodermal fates in ascidian embryos. *Development* 144:33–37.

Imai, K. S., K. Hino, K. Yagi, N. Satoh, and Y. Satou. 2004. Gene expression profiles of transcription factors and signaling molecules in the ascidian embryo: towards a comprehensive understanding of gene networks. *Development* 131:4047–4058.

Imai, K. S., M. Levine, N. Satoh, and Y. Satou. 2006. Regulatory blueprint for a chordate embryo. *Science* 312:1183–1187.

Imai, K. S., N. Satoh, and Y. Satou. 2003. A Twist-like bHLH gene is a downstream factor of an endogenous FGF and determines mesenchymal fate in the ascidian embryos. *Development* 130:4461–4472.

Inoue, J., Y. Yasuoka, H. Takahashi, and N. Satoh. 2017. The chordate ancestor possessed a single copy of the Brachyury gene for notochord acquisition. *Zool Lett* 3:4.

Irimia, M., C. Pineiro, I. Maeso, J. L. Gómez-Skarmeta, F. Casares, and J. Garcia-Fernàndez. 2010. Conserved developmental expression of Fezf in chordates and *Drosophila* and the origin of the Zona Limitans Intrathalamica (ZLI) brain organizer. *Evodevo* 1:7.

Irvine, S. Q., M. C. Cangiano, B. J. Millette, and E. S. Gutter. 2007. Non-overlapping expression patterns of the clustered Dll-A/B genes in the ascidian *Ciona intestinalis. J Exp Zoolog B Mol Dev Evol* 308:428–441.

Irvine, S. Q., V. C. Fonseca, M. A. Zompa, and R. Antony. 2008. Cis-regulatory organization of the Pax6 gene in the ascidian *Ciona intestinalis. Dev Biol* 317:649–659.

Irvine, S. Q., D. A. Vierra, B. J. Millette, M. D. Blanchette, and R. E. Holbert. 2011. Expression of the Distalless-B gene in *Ciona* is regulated by a pan-ectodermal enhancer module. *Dev Biol* 353:432–439.

Isshiki, T., M. Takeichi, and A. Nose. 1997. The role of the msh homeobox gene during *Drosophila* neurogenesis: implication for the dorsoventral specification of the neuroectoderm. *Development* 124:3099–3109.

Itoh, N., and D. M. Ornitz. 2011. Fibroblast growth factors: from molecular evolution to roles in development, metabolism and disease. *J Biochem* 149:121–130.

Ivashkin, E., and I. Adameyko. 2013. Progenitors of the protochordate ocellus as an evolutionary origin of the neural crest. *Evodevo* 4:12.

Iwafuchi-Doi, M. 2019. The mechanistic basis for chromatin regulation by pioneer transcription factors. *Wiley Interdiscip Rev Syst Biol Med* 11:e1427.

Iwafuchi-Doi, M., and K. S. Zaret. 2016. Cell fate control by pioneer transcription factors. *Development* 143:1833–1837.

Jackman, W. R., J. A. Langeland, and C. B. Kimmel. 2000. Islet reveals segmentation in the amphioxus hindbrain homolog. *Dev Biol* 220:16–26.

Jacob, F. 1977. Evolution and tinkering. *Science* 196:1161–1166.

Jacobs, D. K., and R. D. Gates. 2003. Developmental genes and the reconstruction of meta-zoan evolution–implications of evolutionary loss, limits on inference of ancestry and type 2 errors. *Integr Comp Biol* 43:11–18.

Jacobs, D. K., N. Nakanishi, D. Yuan, A. Camara, S. A. Nichols, and V. Hartenstein. 2007. Evolution of sensory structures in basal metazoa. *Integr Comp Biol* 47:712–723.

Jafar-Nejad, H., M. Acar, R. Nolo, H. Lacin, H. Pan, S. M. Parkhurst, and H. J. Bellen. 2003. Senseless acts as a binary switch during sensory organ precursor selection. *Genes Dev* 17:2966–2978.

Jafar-Nejad, H., and H. J. Bellen. 2004. Gfi/Pag-3/senseless zinc finger proteins: a unifying theme? *Mol Cell Biol* 24:8803–8812.

Jafari, S., L. Alkhori, A. Schleiffer, A. Brochtrup, T. Hummel, and M. Alenius. 2012. Combinatorial activation and repression by seven transcription factors specify *Drosophila* odorant receptor expression. *PLoS Biol* 10:e1001280.

Janvier, P. 1996. *Early vertebrates*. Oxford: Clarendon Press.

Janvier, P. 2015. Facts and fancies about early fossil chordates and vertebrates. *Nature* 520:483–489.

Jarman, A. P. 2014. Development of the auditory organ (Johnston's organ) in *Drosophila*. In *Development of auditory and vestibular systems*, edited by R. Romand and I. Varela-Nieto, 31–61. Oxford: Academic Press.

Jarman, A. P., Y. Grau, L. Y. Jan, and Y. N. Jan. 1993. atonal is a proneural gene that directs chor-dotonal organ formation in the *Drosophila* peripheral nervous system. *Cell* 73:1307–1321.

Jarman, A. P., E. H. Grell, L. Ackerman, L. Y. Jan, and Y. N. Jan. 1994. Atonal is the proneu-ral gene for *Drosophila* photoreceptors. *Nature* 369:398–400.

Jefferies, R. P. S. 1986. *The ancestry of vertebrates*. London: British Museum (Natural History).

Jefferies, R. P. S. 1991. Two types of bilateral symmetry in the Metazoa: chordate and bilat-erian. In *Biological asymmetries and handedness*, edited by G. R. Bock and J. Marsh, 94–127. Chichester: Wiley.

Jeffery, W. R. 2001. Determinants of cell and positional fate in ascidian embryos. *Int Rev Cytol* 203:3–62.

Jeffery, W. R. 2015. Regeneration, stem cells, and aging in the tunicate *Ciona*: insights from the oral siphon. *Int Rev Cell Mol Biol* 319:255–282.

Jeffery, W. R., T. Chiba, F. R. Krajka, C. Deyts, N. Satoh, and J. S. Joly. 2008. Trunk lat-eral cells are neural crest-like cells in the ascidian *Ciona intestinalis*: insights into the ancestry and evolution of the neural crest. *Dev Biol* 324:152–160.

Jeffery, W. R., A. G. Strickler, and Y. Yamamoto. 2004. Migratory neural crest-like cells form body pigmentation in a urochordate embryo. *Nature* 431:696–699.

Jékely, G. 2013. Global view of the evolution and diversity of metazoan neuropeptide signal-ing. *Proc Natl Acad Sci USA* 110:8702–8707.

Jékely, G., J. Colombelli, H. Hausen, K. Guy, E. Stelzer, F. Nedelec, and D. Arendt. 2008. Mechanism of phototaxis in marine zooplankton. *Nature* 456:395–399.

Jenny, M., C. Uhl, C. Roche, I. Duluc, V. Guillermin, F. Guillemot, J. Jensen, M. Kedinger, and G. Gradwohl. 2002. Neurogenin3 is differentially required for endocrine cell fate specification in the intestinal and gastric epithelium. *EMBO J* 21:6338–6347.

Jin, Y., R. Hoskins, and H. R. Horvitz. 1994. Control of type-D GABAergic neuron differen-tiation by *C. elegans* UNC-30 homeodomain protein. *Nature* 372:780–783.

Jin, Y., and D. C. Weinstein. 2018. Pitx1 regulates cement gland development in *Xenopus laevis* through activation of transcriptional targets and inhibition of BMP signaling. *Dev Biol* 437:41–49.

Johnsen, S. 1997. Identification and localization of a possible rhodopsin in the echinoderms *Asterias forbesi* (Asteroidea) and *Ophioderma brevispinum* (Ophiuroidea). *Biol Bull* 193:97–105.

Joly, J. S., J. Osorio, A. Alunni, H. Auger, S. Kano, and S. Retaux. 2007. Windows of the brain: towards a developmental biology of circumventricular and other neurohemal organs. *Semin Cell Dev Biol* 18:512–524.

Jonz, M. G., L. T. Buck, S. F. Perry, T. Schwerte, and G. Zaccone. 2016. Sensing and surviving hypoxia in vertebrates. *Ann N Y Acad Sci* 1365:43–58.

Jørgensen, J. M. 1989. Evolution of octavolateralis cells. In *The mechanosensory lateral line*, edited by S. Coombs, P. Görner and H. Münz, 115–145. New York: Springer.

Jørgensen, J. M. 2005. Morphology of electroreceptive sensory organs. In *Electroreception*, edited by T. H. Bullock, C. D. Hopkins, A. N. Popper and R. R. Fay, 47–67. New York: Springer.

José-Edwards, D. S., I. Oda-Ishii, J. E. Kugler, Y. J. Passamaneck, L. Katikala, Y. Nibu, and A. Di Gregorio. 2015. Brachyury, Foxa2 and the cis-regulatory origins of the notochord. *PLoS Genet* 11:e1005730.

Joyce Tang, W., J. S. Chen, and R. W. Zeller. 2013. Transcriptional regulation of the peripheral nervous system in *Ciona intestinalis*. *Dev Biol* 378:183–193.

Julin, C. 1881. Recherche sur l'organisation des Ascides simples: sur l'hypophyse. *Arch de Biol* 2:59–126.

Kaelberer, M. M., K. L. Buchanan, M. E. Klein, B. B. Barth, M. M. Montoya, X. Shen, and D. V. Bohorquez. 2018. A gut-brain neural circuit for nutrient sensory transduction. *Science* 361:eaat5236.

Kage-Nakadai, E., A. Ohta, T. Ujisawa, S. Sun, Y. Nishikawa, A. Kuhara, and S. Mitani. 2016. *Caenorhabditis elegans* homologue of Prox1/Prospero is expressed in the glia and is required for sensory behavior and cold tolerance. *Genes Cells* 21:936–948.

Kaji, T., S. Aizawa, M. Uemura, and K. Yasui. 2001. Establishment of left-right asymmetric innervation in the lancelet oral region. *J Comp Neurol* 435:394–405.

Kaji, T., J. D. Reimer, A. R. Morov, S. Kuratani, and K. Yasui. 2016. Amphioxus mouth after dorso-ventral inversion. *Zoological Lett* 2:2. doi: 10.1186/s40851-016-0038-3.

Kaltenbach, S. L., L. Z. Holland, N. D. Holland, and D. Koop. 2009. Developmental expression of the three iroquois genes of amphioxus (BfIrxA, BfIrxB, and BfIrxC) with special attention to the gastrula organizer and anteroposterior boundaries in the central nervous system. *Gene Expr Patterns* 9:329–334.

Kaltenbach, S. L., J. K. Yu, and N. D. Holland. 2009. The origin and migration of the earliest-developing sensory neurons in the peripheral nervous system of amphioxus. *Evol Dev* 11:142–151.

Kamiya, C., N. Ohta, Y. Ogura, K. Yoshida, T. Horie, T. G. Kusakabe, H. Satake, and Y. Sasakura. 2014. Nonreproductive role of gonadotropin-releasing hormone in the control of ascidian metamorphosis. *Dev Dyn* 243:1524–1535.

Kanda, M., H. Wada, and S. Fujiwara. 2009. Epidermal expression of Hox1 is directly activated by retinoic acid in the *Ciona intestinalis* embryo. *Dev Biol* 335:454–463.

Kang, L., J. Gao, W. R. Schafer, Z. Xie, and X. Z. Xu. 2010. *C. elegans* TRP family protein TRP-4 is a pore-forming subunit of a native mechanotransduction channel. *Neuron* 67:381–391.

Kapsimali, M. 2017. Epithelial cell behaviours during neurosensory organ formation. *Development* 144:1926–1936.

Kardong, K. V. 2009. *Vertebrates. Comparative anatomy, function, evolution.* 5th ed. New York: McGraw Hill.

Kardong, K. V. 2018. *Vertebrates: comparative anatomy, function, evolution.* 8th ed. New York: McGraw Hill.

Kasahara, M. 2007. The 2R hypothesis: an update. *Curr Opin Immunol* 19:547–552.

Kasai, A., and N. Oshima. 2006. Light-sensitive motile iridophores and visual pigments in the neon tetra, *Paracheirodon innesi*. *Zoolog Sci* 23:815–819.

Kasberg, A. D., E. W. Brunskill, and S. Steven Potter. 2013. SP8 regulates signaling centers during craniofacial development. *Dev Biol* 381:312–323.

Kass-Simon, G., and L. A. Hufnagel. 1992. Suspected chemoreceptors in coelenterates and ctenophores. *Microsc Res Tech* 22:265–284.

Kassmer, S. H., S. Nourizadeh, and A. W. De Tomaso. 2019. Cellular and molecular mechanisms of regeneration in colonial and solitary ascidians. *Dev Biol* 448:271–278.

Kassmer, S. H., D. Rodriguez, and A. W. De Tomaso. 2016. Colonial ascidians as model organisms for the study of germ cells, fertility, whole body regeneration, vascular biology and aging. *Curr Opin Genet Dev* 39:101–106.

Katsuyama, Y., and H. Saiga. 1998. Retinoic acid affects patterning along the anterior-posterior axis of the ascidian embryo. *Dev Growth Differ* 40:413–422.

Katsuyama, Y., S. Wada, S. Yasugi, and H. Saiga. 1995. Expression of the labial group Hox gene HrHox-1 and its alteration induced by retinoic acid in development of the ascidian *Halocynthia roretzi*. *Development* 121:3197–3205.

Katz, M. J. 1983. Comparative anatomy of the tunicate tadpole, *Ciona intestinalis*. *Biol Bull* 164:1–27.

Kaul-Strehlow, S., M. Urata, T. Minokawa, T. Stach, and A. Wanninger. 2015. Neurogenesis in directly and indirectly developing enteropneusts: of nets and cords. *Org Divers Evol* 15:405–422.

Kaul-Strehlow, S., M. Urata, D. Praher, and A. Wanninger. 2017. Neuronal patterning of the tubular collar cord is highly conserved among enteropneusts but dissimilar to the chordate neural tube. *Sci Rep* 7:7003.

Kaul, S., and T. Stach. 2010. Ontogeny of the collar cord: neurulation in the hemichordate *Saccoglossus kowalevskii*. *J Morphol* 271:1240–1259.

Kaupp, U. B. 2010. Olfactory signalling in vertebrates and insects: differences and commonalities. *Nat Rev Neurosci* 11:188–200.

Kavaler, J., W. Fu, H. Duan, M. Noll, and J. W. Posakony. 1999. An essential role for the *Drosophila* Pax2 homolog in the differentiation of adult sensory organs. *Development* 126:2261–2272.

Kavanaugh, S. I., A. R. Root, and S. A. Sower. 2005. Distribution of gonadotropin-releasing hormone (GnRH) by in situ hybridization in the tunicate *Ciona intestinalis*. *Gen Comp Endocrinol* 141:76–83.

Kawada, T., M. Aoyama, I. Okada, T. Sakai, T. Sekiguchi, M. Ogasawara, and H. Satake. 2009. A novel inhibitory gonadotropin-releasing hormone-related neuropeptide in the ascidian, *Ciona intestinalis*. *Peptides* 30:2200–2205.

Kawai, T., Y. Oka, and H. Eisthen. 2009. The role of the terminal nerve and GnRH in olfactory system neuromodulation. *Zoolog Sci* 26:669–680.

Kawauchi, H., and S. A. Sower. 2006. The dawn and evolution of hormones in the adenohypophysis. *Gen Comp Endocrinol* 148:3–14.

Kazanskaya, O. V., E. A. Severtzova, K. A. Barth, G. V. Ermakova, S. A. Lukyanov, A. O. Benyumov, M. Pannese, E. Boncinelli, S. W. Wilson, and A. G. Zarais ky. 1997. Anf: a novel class of vertebrate homeobox genes expressed at the anterior end of the main embryonic axis. *Gene* 200:25–34.

Keil, T. A. 2012. Sensory cilia in arthropods. *Arthropod Struct Dev* 41:515–534.

Kelberman, D., K. Rizzoti, R. Lovell-Badge, I. C. Robinson, and M. T. Dattani. 2009. Genetic regulation of pituitary gland development in human and mouse. *Endocr Rev* 30:790–829.

Kernan, M. J. 2007. Mechanotransduction and auditory transduction in *Drosophila*. *Pflugers Arch* 454:703–720.

Kikkawa, S., N. Yoshida, M. Nakagawa, T. Iwasa, and M. Tsuda. 1998. A novel rhodopsin kinase in octopus photoreceptor possesses a pleckstrin homology domain and is activated by G protein betagamma-subunits. *J Biol Chem* 273:7441–7447.

Kim, K., R. Kim, and P. Sengupta. 2010. The HMX/NKX homeodomain protein MLS-2 specifies the identity of the AWC sensory neuron type via regulation of the ceh-36 Otx gene in *C. elegans*. *Development* 137:963–974.

Kim, T., M. C. Gondre-Lewis, I. Arnaoutova, and Y. P. Loh. 2006. Dense-core secretory granule biogenesis. *Physiology (Bethesda)* 21:124–133.

Kimura, S., Y. Hara, T. Pineau, P. Fernàndez-Salguero, C. H. Fox, J. M. Ward, and F. J. Gonzalez. 1996. The T/ebp null mouse: thyroid-specific enhancer-binding protein is essential for the organogenesis of the thyroid, lung, ventral forebrain, and pituitary. *Genes Dev* 10:60–69.

Kirjavainen, A., M. Sulg, F. Heyd, K. Alitalo, S. Yla-Herttuala, T. Moroy, T. V. Petrova, and U. Pirvola. 2008. Prox1 interacts with Atoh1 and Gfi1, and regulates cellular differentiation in the inner ear sensory epithelia. *Dev Biol* 322:33–45.

Klann, M., and A. Stollewerk. 2017. Evolutionary variation in neural gene expression in the developing sense organs of the crustacean *Daphnia magna*. *Dev Biol* 424:50–61.

Knight-Jones, E. W. 1952. On the nervous system of *Saccoglossus cambrensis* (Enteropneusta). *Philos Trans R Soc Lond B* 236:315–354.

Kobayashi, M., N. Takatori, Y. Nakajima, G. Kumano, H. Nishida, and H. Saiga. 2010. Spatial and temporal expression of two transcriptional isoforms of Lhx3, a LIM class homeobox gene, during embryogenesis of two phylogenetically remote ascidians, *Halocynthia roretzi* and *Ciona intestinalis*. *Gene Expr Patterns* 10:98–104.

Koizumi, O., N. Sato, and C. Goto. 2004. Chemical anatomy of *Hydra* nervous system using antibodies against *Hydra* neuropeptides: a review. *Hydrobiologia* 530:41–47.

Kojima, D., A. Terakita, T. Ishikawa, Y. Tsukahara, A. Maeda, and Y. Shichida. 1997. A novel Go-mediated phototransduction cascade in scallop visual cells. *J Biol Chem* 272:22979–22982.

Kojima, T., T. Tsuji, and K. Saigo. 2005. A concerted action of a paired-type homeobox gene, aristaless, and a homolog of Hox11/tlx homeobox gene, clawless, is essential for the distal tip development of the *Drosophila* leg. *Dev Biol* 279:434–445.

Konno, A., M. Kaizu, K. Hotta, T. Horie, Y. Sasakura, K. Ikeo, and K. Inaba. 2010. Distribution and structural diversity of cilia in tadpole larvae of the ascidian *Ciona intestinalis*. *Dev Biol* 337:42–62.

Koop, D., N. D. Holland, M. Semon, S. Alvarez, A. R. de Lera, V. Laudet, L. Z. Holland, and M. Schubert. 2010. Retinoic acid signaling targets Hox genes during the amphioxus gastrula stage: insights into early anterior-posterior patterning of the chordate body plan. *Dev Biol* 338:98–106.

Kourakis, M. J., C. Borba, A. Zhang, E. Newman-Smith, P. Salas, B. Manjunath, and W. C. Smith. 2019. Parallel visual circuitry in a basal chordate. *Elife* 8:e44753.

Kourakis, M. J., E. Newman-Smith, and W. C. Smith. 2010. Key steps in the morphogenesis of a cranial placode in an invertebrate chordate, the tunicate *Ciona savignyi*. *Dev Biol* 340:134–144.

Kourakis, M. J., and W. C. Smith. 2007. A conserved role for FGF signaling in chordate otic/atrial placode formation. *Dev Biol* 312:257.

Kourakis, M. J., and W. C. Smith. 2015. An organismal perspective on *C. intestinalis* development, origins and diversification. *Elife* 4:e06024.

Kowalevsky, A. 1866. Entwicklungsgeschichte der einfachen Ascidien. *Mem Acad Sci St Petersbourgh* 10 (7):1–119.

Kowalevsky, A. 1867. Entwicklungsgeschichte des *Amphioxus lanceolatus*. *Mem Acad Sci St Petersbourgh* 11 (7):1–117.

Koyanagi, M., K. Kubokawa, H. Tsukamoto, Y. Shichida, and A. Terakita. 2005. Cephalochordate melanopsin: evolutionary linkage between invertebrate visual cells and vertebrate photosensitive retinal ganglion cells. *Curr Biol* 15:1065–1069.

Koyanagi, M., K. Takano, H. Tsukamoto, K. Ohtsu, F. Tokunaga, and A. Terakita. 2008. Jellyfish vision starts with cAMP signaling mediated by opsin-G(s) cascade. *Proc Natl Acad Sci USA* 105:15576–15580.

Kozmik, Z., M. Daube, E. Frei, B. Norman, L. Kos, L. J. Dishaw, M. Noll, and J. Piatigorsky. 2003. Role of Pax genes in eye evolution: a cnidarian PaxB gene uniting Pax2 and Pax6 functions. *Dev Cell* 5:773–785.

Kozmik, Z., L. Z. Holland, M. Schubert, T. C. Lacalli, J. Kreslova, C. Vlcek, and N. D. Holland. 2001. Characterization of amphioxus AmphiVent, an evolutionarily conserved marker for chordate ventral mesoderm. *Genesis* 29:172–179.

Kozmik, Z., N. D. Holland, A. Kalousova, J. Paces, M. Schubert, and L. Z. Holland. 1999. Characterization of an amphioxus paired box gene, AmphiPax2/5/8: developmental expression patterns in optic support cells, nephridium, thyroid-like structures and pharyngeal gill slits, but not in the midbrain-hindbrain boundary region. *Development* 126:1295–1304.

Kozmik, Z., N. D. Holland, J. Kreslova, D. Oliveri, M. Schubert, K. Jonasova, L. Z. Holland, M. Pestarino, V. Benes, and S. Candiani. 2007. Pax-Six-Eya-Dach network during amphioxus development: conservation in vitro but context-specificity in vivo *Dev Biol* 306:143–159.

Kozmik, Z., P. Pfeffer, J. Kralova, J. Paces, V. Paces, A. Kalousova, and A. Cvekl. 1999. Molecular cloning and expression of the human and mouse homologues of the *Drosophila* dachshund gene. *Dev Genes Evol* 209:537–545.

Kozmik, Z., J. Ruzickova, K. Jonasova, Y. Matsumoto, P. Vopalensky, I. Kozmikova, H. Strnad, S. Kawamura, J. Piatigorsky, V. Paces, and C. Vlcek. 2008. Assembly of the cnidarian camera-type eye from vertebrate-like components. *Proc Natl Acad Sci USA* 105:8989–8993.

Kozmikova, I., S. Candiani, P. Fabian, D. Gurska, and Z. Kozmik. 2013. Essential role of Bmp signaling and its positive feedback loop in the early cell fate evolution of chordates. *Dev Biol* 382:538–554.

Kozmikova, I., J. Smolikova, C. Vlcek, and Z. Kozmik. 2011. Conservation and diversification of an ancestral chordate gene regulatory network for dorsoventral patterning. *PLoS One* 6:e14650.

Kozmikova, I., and J. K. Yu. 2017. Dorsal-ventral patterning in amphioxus: current understanding, unresolved issues, and future directions. *Int J Dev Biol* 61:601–610.

Krisch, B. 1973. Apical organ (statocyst) of *Pleurobrachia pileus. Z Zellforsch Mikrosk Anat.* 142:241–262.

Krishnan, A., M. S. Almen, R. Fredriksson, and H. B. Schiöth. 2013. Remarkable similarities between the hemichordate (*Saccoglossus kowalevskii*) and vertebrate GPCR repertoire. *Gene* 526:122–133.

Krishnan, A., and H. B. Schiöth. 2015. The role of G protein-coupled receptors in the early evolution of neurotransmission and the nervous system. *J Exp Biol* 218:562–571.

Kumamoto, T., and C. Hanashima. 2017. Evolutionary conservation and conversion of Foxg1 function in brain development. *Dev Growth Differ* 59:258–269.

Kumar, J. P. 2009. The sine oculis homeobox (SIX) family of transcription factors as regulators of development and disease. *Cell Mol Life Sci* 66:565–583.

Kumar, J. P., and K. Moses. 2001. Expression of evolutionarily conserved eye specification genes during *Drosophila* embryogenesis. *Dev Genes Evol* 211:406–414.

Kuraku, S., D. Hoshiyama, K. Katoh, H. Suga, and T. Miyata. 1999. Monophyly of lampreys and hagfishes supported by nuclear DNA-coded genes. *J Mol Evol* 49:729–735.

Kuratani, S., N. Adachi, N. Wada, Y. Oisi, and F. Sugahara. 2012. Developmental and evolutionary significance of the mandibular arch and prechordal/premandibular cranium in vertebrates: revising the heterotopy scenario of gnathostome jaw evolution. *J Anat* 222:41–55.

Kurn, U., S. Rendulic, S. Tiozzo, and R. J. Lauzon. 2011. Asexual propagation and regeneration in colonial ascidians. *Biol Bull* 221:43–61.

Kusakabe, T. G., T. Sakai, M. Aoyama, Y. Kitajima, Y. Miyamoto, T. Takigawa, Y. Daido, K. Fujiwara, Y. Terashima, Y. Sugiuchi, G. Matassi, H. Yagisawa, M. K. Park, H. Satake, and M. Tsuda. 2012. A conserved non-reproductive GnRH system in chordates. *PLoS One* 7:e41955.

Kusakabe, T., R. Kusakabe, I. Kawakami, Y. Satou, N. Satoh, and M. Tsuda. 2001. Ci-opsin1, a vertebrate-type opsin gene, expressed in the larval ocellus of the ascidian *Ciona intestinalis*. *FEBS Lett* 506:69–72.

Kusakabe, T., and M. Tsuda. 2007. Photoreceptive systems in ascidians. *Photochem Photobiol* 83:248–252.

Lacalli, T. 2005. Protochordate body plans and the evolutionary role of larvae: old controversies resolved? *Can J Zool* 83:216–224.

Lacalli, T. 2008a. Mucus secretion and transport in amphioxus larvae: organisation and ultrastructure of the food trapping system, and implications for head evolution. *Acta Zool* 88:219–230.

Lacalli, T. 2018. Amphioxus, motion detection, and the evolutionary origin of the vertebrate retinotectal map. *Evodevo* 9:6.

Lacalli, T. C. 1981. Structure and development of the apical organ in trochophores of *Spirobranchus polycerus*, *Phyllodoce maculata* and *Phyllodoce mucosa* (Polychaeta). *Proc R Soc Lond B-Biol Sci* 212:381.

Lacalli, T. C. 1996. Frontal eye circuitry, rostral sensory pathways and brain organization in amphioxus larvae: evidence from 3D reconstructions. *Philos Trans R Soc Lond B Biol Sci* 351:243–263.

Lacalli, T. C. 2001. New perspectives on the evolution of protochordate sensory and locomotory systems, and the origin of brains and heads. *Philos Trans R Soc Lond B Biol Sci* 356:1565–1572.

Lacalli, T. C. 2002a. The dorsal compartment locomotory control system in amphioxus larvae. *J Morphol* 252:227–237.

Lacalli, T. C. 2002b. Sensory pathways in amphioxus larvae. I. Constituent fibers of the rostral and anterodorsal nerves, their targets and evolutionary significance. *Acta Zool* 83:149–166.

Lacalli, T. C. 2004. Sensory systems in amphioxus: a window on the ancestral chordate condition. *Brain Behav Evol* 64:148–162.

Lacalli, T. C. 2008b. Basic features of the ancestral chordate brain: a protochordate perspective. *Brain Res Bull* 75:319–323.

Lacalli, T. C., T. H. J. Gilmour, and S. J. Kelly. 1999. The oral nerve plexus in amphioxus larvae: function, cell types and phylogenetic significance. *Proc R Soc Lond B Biol Sci* 266:1461–1470.

Lacalli, T. C., T. H. J. Gilmour, and J. E. West. 1990. Ciliary band innervation in the bipinnaria larva of *Pisaster ochraceus*. *Philos Trans R Soc B-Biol Sci* 330:371–390.

Lacalli, T. C., N. D. Holland, and J. E. West. 1994. Landmarks in the anterior central nervous system of amphioxus larvae. *Philos Trans R Soc Lond Biol Sci* 344:165–185.

Lacalli, T. C., and S. Hou. 1999. A reexamination of the epithelial sensory cells of amphioxus (*Branchiostoma*) *Acta Zool* 80:125–134.

Lacalli, T. C., and J. E. West. 1993. A distinctive nerve-cell type common to diverse deuterostome larvae: comparative data from echinoderms, hemichordates and amphioxus. *Acta Zool* 74:1–8.

Lagman, D., D. Ocampo Daza, J. Widmark, X. M. Abalo, G. Sundström, and D. Larhammar. 2013. The vertebrate ancestral repertoire of visual opsins, transducin alpha subunits and oxytocin/vasopressin receptors was established by duplication of their shared genomic region in the two rounds of early vertebrate genome duplications. *BMC Evol Biol* 13:238.

Lagman, D., G. Sundström, D. Ocampo Daza, X. M. Abalo, and D. Larhammar. 2012. Expansion of transducin subunit gene families in early vertebrate tetraploidizations. *Genomics* 100:203–211.

Lai, E. C., and V. Orgogozo. 2004. A hidden program in *Drosophila* peripheral neurogenesis revealed: fundamental principles underlying sensory organ diversity. *Dev Biol* 269:1–17.

Lamb, T. D. 2009. Evolution of vertebrate retinal photoreception. *Philos Trans R Soc Lond B Biol Sci* 364:2911–2924.

Lamb, T. D. 2013. Evolution of phototransduction, vertebrate photoreceptors and retina. *Prog Retin Eye Res* 36:52–119.

Lamb, T. D. 2020. Evolution of the genes mediating phototransduction in rod and cone photoreceptors. *Prog Retin Eye Res* 76:100823.

Lamb, T. D., S. P. Collin, and E. N. Pugh, Jr. 2007. Evolution of the vertebrate eye: opsins, photoreceptors, retina and eye cup. *Nat Rev Neurosci* 8:960–976.

Lamb, T. D., and D. M. Hunt. 2017. Evolution of the vertebrate phototransduction cascade activation steps. *Dev Biol* 431:77–92.

Lamb, T. D., H. R. Patel, A. Chuah, and D. M. Hunt. 2018. Evolution of the shut-off steps of vertebrate phototransduction. *Open Biol* 8:170232.

Lamy, C., U. Rothbächer, D. Caillol, and P. Lemaire. 2006. Ci-FoxA-a is the earliest zygotic determinant of the ascidian anterior ectoderm and directly activates Ci-sFRP1/5. *Development* 133:2835–2844.

Land, M. F. 1984. Molluscs. In *Photoreception and vision in invertebrates. NATO ASI Series (Series A: Life Sciences)*, Vol. 74, edited by M. A. Ali, 699–725. Boston: Springer.

Land, M. F., and D.-E. Nilsson. 2012. *Animal eyes.* 2nd ed. Oxford: Oxford University Press.

Land, M. F., and R. D. Fernald. 1992. The evolution of eyes. *Annu Rev Neurosci* 15:1–29.

Lange, K. 2011. Fundamental role of microvilli in the main functions of differentiated cells: outline of an universal regulating and signaling system at the cell periphery. *J Cell Physiol* 226:896–927.

Langeland, J. A., J. M. Tomsa, W. R. Jackman, and C. B. Kimmel. 1998. An amphioxus snail gene: expression in paraxial mesoderm and neural plate suggests a conserved role in patterning the chordate embryo. *Dev Genes Evol* 208:569–577.

Lankester, E. R. 1870. On the use of the term homology in modern zoology, and the distinction between homogenetic and homoplastic agreements. *J Nat His* 6:34–43.

Larroux, C., B. Fahey, S. M. Degnan, M. Adamski, D. S. Rokhsar, and B. M. Degnan. 2007. The NK homeobox gene cluster predates the origin of Hox genes. *Curr Biol* 17:706–710.

Larroux, C., G. N. Luke, P. Koopman, D. S. Rokhsar, S. M. Shimeld, and B. M. Degnan. 2008. Genesis and expansion of metazoan transcription factor gene classes. *Mol Biol Evol* 25:980–996.

Latorre, R., C. Sternini, R. De Giorgio, and B. Greenwood-Van Meerveld. 2016. Enteroendocrine cells: a review of their role in brain-gut communication. *Neurogastroenterol Motil* 28:620–630.

Laverack, M. S. 1988. The diversity of chemoreceptors. In *Sensory systems of aquatic animals*, edited by J. Atema, R. R. Fay, A. N. Popper and W. N. Tavolga, 287–312. New York: Springer.

Layden, M. J., M. Boekhout, and M. Q. Martindale. 2012. *Nematostella vectensis* achaete-scute homolog NvashA regulates embryonic ectodermal neurogenesis and represents an ancient component of the metazoan neural specification pathway. *Development* 139:1013–1022.

Le Petillon, Y., G. Luxardi, P. Scerbo, M. Cibois, A. Leon, L. Subirana, M. Irimia, L. Kodjabachian, H. Escriva, and S. Bertrand. 2017. Nodal/activin pathway is a conserved neural induction signal in chordates. *Nat Ecol Evol* 1:1192–1200.

Lee, M. H., and P. M. Salvaterra. 2002. Abnormal chemosensory jump 6 is a positive transcriptional regulator of the cholinergic gene locus in *Drosophila* olfactory neurons. *J Neurosci* 22:5291–5299.

Lee, S. J., H. Xu, and C. Montell. 2004. Rhodopsin kinase activity modulates the amplitude of the visual response in *Drosophila*. *Proc Natl Acad Sci USA* 101:11874–11879.

Lee, S., B. Lee, K. Joshi, S. L. Pfaff, J. W. Lee, and S. K. Lee. 2008. A regulatory network to segregate the identity of neuronal subtypes. *Dev Cell* 14:877–889.

Leininger, S., M. Adamski, B. Bergum, C. Guder, J. Liu, M. Laplante, J. Brate, F. Hoffmann, S. Fortunato, S. Jordal, H. T. Rapp, and M. Adamska. 2014. Developmental gene expression provides clues to relationships between sponge and eumetazoan body plans. *Nat Commun* 5:3905.

Lek, A., F. J. Evesson, R. B. Sutton, K. N. North, and S. T. Cooper. 2012. Ferlins: regulators of vesicle fusion for auditory neurotransmission, receptor trafficking and membrane repair. *Traffic* 13:185–194.

Lek, A., M. Lek, K. N. North, and S. T. Cooper. 2010. Phylogenetic analysis of ferlin genes reveals ancient eukaryotic origins. *BMC Evol Biol* 10:231.

Lemaire, P. 2009. Unfolding a chordate developmental program, one cell at a time: invariant cell lineages, short-range inductions and evolutionary plasticity in ascidians. *Dev Biol* 332:48–60.

Lemaire, P., and J. Piette. 2015. Tunicates: exploring the sea shores and roaming the open ocean. A tribute to Thomas Huxley. *Open Biol* 5:150053.

Lemaire, P., W. C. Smith, and H. Nishida. 2008. Ascidians and the plasticity of the chordate developmental program. *Curr Biol* 18:R620–R631.

Lemons, D., J. H. Fritzenwanker, J. Gerhart, C. J. Lowe, and W. McGinnis. 2010. Co-option of an anteroposterior head axis patterning system for proximodistal patterning of appendages in early bilaterian evolution. *Dev Biol* 344:358–362.

Lenhoff, H. M. 1961. Activation of feeding reflex in *Hydra littoralis*. 1. Role played by reduced glutathione, and quantitative assay of feeding reflex. *J Gen Physiol* 45:331–344.

Lentz, T. L., and R. J. Barrnett. 1965. Fine structure of the nervous system of *Hydra*. *Am Zool* 5:341–356.

Lesh-Laurie, G. E. 1988. Coelenterate endocrinology. In *Invertebrate endocrinology, Vol. 2: endocrinology of selected invertebrate types*, edited by H. Laufer and R. G. H. Downer, 3–29. New York: Alan R. Liss.

Lesser, M. P., K. L. Carleton, S. A. Bottger, T. M. Barry, and C. W. Walker. 2011. Sea urchin tube feet are photosensory organs that express a rhabdomeric-like opsin and PAX6. *Proc Biol Sci* 278:3371–3379.

Levy, O., L. Appelbaum, W. Leggat, Y. Gothlif, D. C. Hayward, D. J. Miller, and O. Hoegh-Guldberg. 2007. Light-responsive cryptochromes from a simple multicellular animal, the coral *Acropora millepora*. *Science* 318:467–470.

Leys, S. P., T. W. Cronin, B. M. Degnan, and J. N. Marshall. 2002. Spectral sensitivity in a sponge larva. *J Comp Physiol A Neuroethol Sens Neural Behav Physiol* 188:199–202.

Leys, S. P., and B. M. Degnan. 2001. Cytological basis of photoresponsive behavior in a sponge larva. *Biol Bull* 201:323–38.

Leys, S. P., J. L. Mah, P. R. McGill, L. Hamonic, F. C. De Leo, and A. S. Kahn. 2019. Sponge behavior and the chemical basis of responses: a post-genomic view. *Integr Comp Biol* 59:751–764.

Li, H., F. Horns, B. Wu, Q. Xie, J. Li, T. Li, D. J. Luginbuhl, S. R. Quake, and L. Luo. 2017. Classifying *Drosophila* olfactory projection neuron subtypes by single-cell RNA sequencing. *Cell* 171:1206–1220 e22.

Li, K. L., T. M. Lu, and J. K. Yu. 2014. Genome-wide survey and expression analysis of the bHLH-PAS genes in the amphioxus *Branchiostoma floridae* reveal both conserved and diverged expression patterns between cephalochordates and vertebrates. *Evodevo* 5:20.

Li, M., Z. Gao, D. Ji, and S. Zhang. 2014. Functional characterization of GH-like homolog in amphioxus reveals an ancient origin of GH/GH receptor system. *Endocrinology* 155:4818–4830.

Li, M., C. Jiang, Y. Zhang, and S. Zhang. 2017. Activities of amphioxus GH-like protein in osmoregulation: insight into origin of vertebrate GH family. *Int J Endocrinol* 2017:9538685.

Li, T., N. Giagtzoglou, D. F. Eberl, S. N. Jaiswal, T. Cai, D. Godt, A. K. Groves, and H. J. Bellen. 2016. The E3 ligase Ubr3 regulates Usher syndrome and MYH9 disorder proteins in the auditory organs of *Drosophila* and mammals. *Elife* 5:e15258.

Li, X., K. A. Oghi, J. Zhang, A. Krones, K. T. Bush, C. K. Glass, S. K. Nigam, A. K. Aggarwal, R. Maas, D. W. Rose, and M. G. Rosenfeld. 2003. Eya protein phosphatase activity regulates Six1-Dach-Eya transcriptional effects in mammalian organogenesis. *Nature* 426:247–254.

Lichtneckert, R., and H. Reichert. 2005. Insights into the urbilaterian brain: conserved genetic patterning mechanisms in insect and vertebrate brain development. *Heredity* 94:465–477.

Liegertova, M., J. Pergner, I. Kozmikova, P. Fabian, A. R. Pombinho, H. Strnad, J. Paces, C. Vlcek, P. Bartunek, and Z. Kozmik. 2015. Cubozoan genome illuminates functional diversification of opsins and photoreceptor evolution. *Sci Rep* 5:11885.

Light, V. 1930. Photoreceptors in *Mya arenaria*, with special reference to their distribution, structure, and function. *J Morphol* 49:1–43.

Liu, B., and Y. Satou. 2019. Foxg specifies sensory neurons in the anterior neural plate border of the ascidian embryo. *Nat Commun* 10:4911.

Liu, B., and Y. Satou. 2020. The genetic program to specify ectodermal cells in ascidian embryos. *Dev Growth Differ* 62:301–310.

Liu, C., and C. Montell. 2015. Forcing open TRP channels: mechanical gating as a unifying activation mechanism. *Biochem Biophys Res Commun* 460:22–25.

Liu, J., A. Ward, J. Gao, Y. Dong, N. Nishio, H. Inada, L. Kang, Y. Yu, D. Ma, T. Xu, I. Mori, Z. Xie, and X. Z. Xu. 2010. *C. elegans* phototransduction requires a G protein-dependent cGMP pathway and a taste receptor homolog. *Nat Neurosci* 13:715–722.

Lokits, A. D., H. Indrischek, J. Meiler, H. E. Hamm, and P. F. Stadler. 2018. Tracing the evolution of the heterotrimeric G protein alpha subunit in Metazoa. *BMC Evol Biol* 18:51.

Loomis, W. F. 1955. Glutathione control of the specific feeding reactions of *Hydra*. *Ann N Y Acad Sci* 62:211–227.

Loosli, F., M. Kmita-Cunisse, and W. J. Gehring. 1996. Isolation of a Pax-6 homolog from the ribbonworm *Lineus sanguineus*. *Proc Natl Acad Sci USA* 93:2658–2663.

López-Díaz, L., R. N. Jain, T. M. Keeley, K. L. VanDussen, C. S. Brunkan, D. L. Gumucio, and L. C. Samuelson. 2007. Intestinal Neurogenin 3 directs differentiation of a bipotential secretory progenitor to endocrine cell rather than goblet cell fate. *Dev Biol* 309:298–305.

Lowe, C. J., D. N. Clarke, D. M. Medeiros, D. S. Rokhsar, and J. Gerhart. 2015. The deuterostome context of chordate origins. *Nature* 520:456–465.

Lowe, C. J., M. Terasaki, M. Wu, R. M. Freeman, Jr., L. Runft, K. Kwan, S. Haigo, J. Aronowicz, E. Lander, C. Gruber, M. Smith, M. Kirschner, and J. Gerhart. 2006. Dorsoventral patterning in hemichordates: insights into early chordate evolution. *PLoS Biol* 4:1602–1619.

Lowe, C. J., M. Wu, A. Salic, L. Evans, E. Lander, N. Stange-Thomann, C. E. Gruber, J. Gerhart, and M. Kirschner. 2003. Anteroposterior patterning in hemichordates and the origins of the chordate nervous system. *Cell* 113:853–865.

Lu, T. M., Y. J. Luo, and J. K. Yu. 2012. BMP and Delta/Notch signaling control the development of amphioxus epidermal sensory neurons: insights into the evolution of the peripheral sensory system. *Development* 139:2020–2030.

Ludeman, D. A., N. Farrar, A. Riesgo, J. Paps, and S. P. Leys. 2014. Evolutionary origins of sensation in metazoans: functional evidence for a new sensory organ in sponges. *BMC Evol Biol* 14:3.

Ma, L., B. J. Swalla, J. Zhou, S. L. Dobias, Bell, J. Chen, R. E. Maxson, and W. R. Jeffery. 1996. Expression of an msx homeobox gene in ascidians: insights into the archetypal chordate expression pattern. *Dev Dyn* 205:308–318.

Macias-Muñoz, A., R. Murad, and A. Mortazavi. 2019. Molecular evolution and expression of opsin genes in *Hydra vulgaris*. *BMC Genomics* 20:992.

Mackie, G. O., and P. Burighel. 2005. The nervous system in adult tunicates: current research directions. *Can J Zool* 83:151–183.

Mackie, G. O., P. Burighel, F. Caicci, and L. Manni. 2006. Innervation of ascidian siphons and their responses to stimulation. *Can J Zool* 84:1146–1162.

Mackie, G. O., D. H. Paul, C. M. Singla, M. A. Sleigh, and D. E. Williams. 1974. Branchial innervation and ciliary control in the ascidian *Corella*. *Proc R Soc Lond B Biol Sci* 187:1–35.

Mackie, G. O., and C. L. Singla. 2003. The capsular organ of *Chelyosoma productum* (Ascidiacea: Corellidae): a new tunicate hydrodynamic sense organ. *Brain Behav Evol* 61:45–58.

Mackie, G. O., and C. L. Singla. 2004. Cupular organs in two species of *Corella* (Tunicata: Ascidiacea). *Invert Biol* 123:269–281.

Magklara, A., A. Yen, B. M. Colquitt, E. J. Clowney, W. Allen, E. Markenscoff-Papadimitriou, Z. A. Evans, P. Kheradpour, G. Mountoufaris, C. Carey, G. Barnea, M. Kellis, and S. Lomvardas. 2011. An epigenetic signature for monoallelic olfactory receptor expression. *Cell* 145:555–570.

Mah, J. L., and S. P. Leys. 2017. Think like a sponge: the genetic signal of sensory cells in sponges. *Dev Biol* 431:93–100.

Mahadevan, N. R., A. C. Horton, and J. J. Gibson-Brown. 2004. Developmental expression of the amphioxus Tbx1/10 gene illuminates the evolution of vertebrate branchial arches and sclerotome. *Dev Genes Evol* 214:559–566.

Mahoney, J. L., E. M. Graugnard, P. Mire, and G. M. Watson. 2011. Evidence for involvement of TRPA1 in the detection of vibrations by hair bundle mechanoreceptors in sea anemones. *J Comp Physiol A Neuroethol Sens Neural Behav Physiol* 197:729–742.

Maienschein, J. 1994. 'It's a long way From "amphioxus"'. Anton Dohrn and late nineteenth century debates about vertebrate origins. *His Philos Life Sci* 16:465–478.

Mallatt, J., and J. Y. Chen. 2003. Fossil sister group of craniates: predicted and found. *J Morphol* 258:1–31.

Mallatt, J., and N. Holland. 2013. *Pikaia gracilens* Walcott: stem chordate, or already specialized in the Cambrian? *J Exp Zool B Mol Dev Evol* 320:247–271.

Manni, L., A. Agnoletto, G. Zaniolo, and P. Burighel. 2005. Stomodeal and neurohypophysial placodes in *Ciona intestinalis*: insights into the origin of the pituitary gland. *J Exp Zoolog B Mol Dev Evol* 304B:324–339.

Manni, L., F. Caicci, F. Gasparini, G. Zaniolo, and P. Burighel. 2004. Hair cells in ascidians and the evolution of lateral line placodes. *Evol Dev* 6:379–381.

Manni, L., N. J. Lane, J. S. Joly, F. Gasparini, S. Tiozzo, F. Caicci, G. Zaniolo, and P. Burighel. 2004. Neurogenic and non-neurogenic placodes in ascidians. *J Exp Zoolog Part B Mol Dev Evol* 302B:483.

Manni, L., N. J. Lane, M. Sorrentino, G. Zaniolo, and P. Burighel. 1999. Mechanism of neurogenesis during the embryonic development of a tunicate. *J Comp Neurol* 412:527–541.

Manni, L., G. O. Mackie, F. Caicci, G. Zaniolo, and P. Burighel. 2006. Coronal organ of ascidians and the evolutionary significance of secondary sensory cells in chordates. *J Comp Neurol* 495:363–373.

Manni, L., G. Zaniolo, F. Cima, P. Burighel, and L. Ballarin. 2007. *Botryllus schlosseri*: a model ascidian for the study of asexual reproduction. *Dev Dyn* 236:335–352.

Manning, L., and C. Q. Doe. 1999. Prospero distinguishes sibling cell fate without asymmetric localization in the *Drosophila* adult external sense organ lineage. *Development* 126:2063–2071.

Mannini, L., P. Deri, J. Picchi, and R. Batistoni. 2008. Expression of a retinal homeobox (Rx) gene during planarian regeneration. *Int J Dev Biol* 52:1113–1117.

Mannini, L., L. Rossi, P. Deri, V. Gremigni, A. Salvetti, E. Salo, and R. Batistoni. 2004. Djeyes absent (Djeya) controls prototypic planarian eye regeneration by cooperating with the transcription factor Djsix-1. *Dev Biol* 269:346–359.

Manoli, M., and W. Driever. 2014. nkx2.1 and nkx2.4 genes function partially redundant during development of the zebrafish hypothalamus, preoptic region, and pallidum. *Front Neuroanat* 8:145.

Maor, R., T. Dayan, H. Ferguson-Gow, and K. E. Jones. 2017. Temporal niche expansion in mammals from a nocturnal ancestor after dinosaur extinction. *Nat Ecol Evol* 1:1889–1895.

Marasco, P. D., P. R. Tsuruda, D. M. Bautista, D. Julius, and K. C. Catania. 2006. Neuroanatomical evidence for segregation of nerve fibers conveying light touch and pain sensation in Eimer's organ of the mole. *Proc Natl Acad Sci USA* 103:9339–9344.

Marder, E. 2012. Neuromodulation of neuronal circuits: back to the future. *Neuron* 76:1–11.

Marlow, H., D. Q. Matus, and M. Q. Martindale. 2013. Ectopic activation of the canonical wnt signaling pathway affects ectodermal patterning along the primary axis during larval development in the anthozoan *Nematostella vectensis*. *Dev Biol* 380:324–334.

Marlow, H., M. A. Tosches, R. Tomer, P. R. Steinmetz, A. Lauri, T. Larsson, and D. Arendt. 2014. Larval body patterning and apical organs are conserved in animal evolution. *BMC Biol* 12:7.

Marois, R., and T. J. Carew. 1997. Fine structure of the apical ganglion and its serotonergic cells in the larva of *Aplysia californica*. *Biol Bull* 192:388–398.

Marquardt, T., R. Ashery-Padan, N. Andrejewski, R. Scardigli, F. Guillemot, and P. Gruss. 2001. Pax6 is required for the multipotent state of retinal progenitor cells. *Cell* 105:43–55.

Martin-Durán, J. M., F. Monjo, and R. Romero. 2012. Morphological and molecular development of the eyes during embryogenesis of the freshwater planarian *Schmidtea polychroa*. *Dev Genes Evol* 222:45–54.

Martin-Durán, J. M., K. Pang, A. Børve, H. S. Le, A. Furu, J. T. Cannon, U. Jondelius, and A. Hejnol. 2018. Convergent evolution of bilaterian nerve cords. *Nature* 553:45–50.

Martin, V. J. 2002. Photoreceptors of cnidarians. *Can J Zool* 80:1703–1722.

Martindale, M. Q., and A. Hejnol. 2009. A developmental perspective: changes in the position of the blastopore during bilaterian evolution. *Dev Cell* 17:162–174.

Martindale, M. Q., K. Pang, and J. R. Finnerty. 2004. Investigating the origins of triploblasty: 'mesodermal' gene expression in a diploblastic animal, the sea anemone *Nematostella vectensis* (phylum, Cnidaria; class, Anthozoa). *Development* 131:2463–2474.

Martínez, P., V. Hartenstein, and S. G. Sprecher. 2017. Xenacoelomorpha nervous systems. In *Oxford research encyclopedia of neuroscience*. Oxford Univ. Press, https://doi.org/10.1093/acrefore/9780190264086.013.203

März, M., F. Seebeck, and K. Bartscherer. 2013. A Pitx transcription factor controls the establishment and maintenance of the serotonergic lineage in planarians. *Development* 140:4499–4509.

Mason, B., M. Schmale, P. Gibbs, M. W. Miller, Q. Wang, K. Levay, V. Shestopalov, and V. Z. Slepak. 2012. Evidence for multiple phototransduction pathways in a reef-building coral. *PLoS One* 7:e50371.

Massari, M., S. Candiani, and M. Pestarino. 1999. Distribution and localization of immunore-active FMRFamide-like peptides in the lancelet. *Eur J Histochem* 43:63–69.

Masuda, M., D. Dulon, K. Pak, L. M. Mullen, Y. Li, L. Erkman, and A. F. Ryan. 2011. Regulation of POU4F3 gene expression in hair cells by 5' DNA in mice. *Neuroscience* 197:48–64.

Masuda, M., Y. Li, K. Pak, E. Chavez, L. Mullen, and A. F. Ryan. 2017. The promoter and multiple enhancers of the pou4f3 gene regulate expression in inner ear hair cells. *Mol Neurobiol* 54:5414–5426.

Materna, S. C., A. Ransick, E. Li, and E. H. Davidson. 2013. Diversification of oral and aboral mesodermal regulatory states in pregastrular sea urchin embryos. *Dev Biol* 375:92–104.

Mathger, L. M., S. B. Roberts, and R. T. Hanlon. 2010. Evidence for distributed light sensing in the skin of cuttlefish, *Sepia officinalis*. *Biol Lett* 6:600–603.

Matsubara, S., T. Kawada, T. Sakai, M. Aoyama, T. Osugi, A. Shiraishi, and H. Satake. 2016. The significance of *Ciona intestinalis* as a stem organism in integrative studies of func-tional evolution of the chordate endocrine, neuroendocrine, and nervous systems. *Gen Comp Endocrinol* 227:101–108.

Matthews, G., and P. Fuchs. 2010. The diverse roles of ribbon synapses in sensory neurotrans-mission. *Nat Rev Neurosci* 11:812–822.

Matthysse, A. G., K. Deschet, M. Williams, M. Marry, A. R. White, and W. C. Smith. 2004. A functional cellulose synthase from ascidian epidermis. *Proc Natl Acad Sci USA* 101:986–991.

Matus, D. Q., R. R. Copley, C. W. Dunn, A. Hejnol, H. Eccleston, K. M. Halanych, M. Q. Martindale, and M. J. Telford. 2006. Broad taxon and gene sampling indicate that chae-tognaths are protostomes. *Curr Biol* 16:R575–R576.

Matus, D. Q., K. Pang, M. Daly, and M. Q. Martindale. 2007. Expression of Pax gene family members in the anthozoan cnidarian, *Nematostella vectensis*. *Evol Dev* 9:25–38.

Mayorova, T. D., C. L. Smith, K. Hammar, C. A. Winters, N. B. Pivovarova, M. A. Aronova, R. D. Leapman, and T. S. Reese. 2018. Cells containing aragonite crystals mediate responses to gravity in *Trichoplax adhaerens* (Placozoa), an animal lacking neurons and synapses. *PLoS One* 13:e0190905.

Mayorova, T. D., S. Tian, W. Cai, D. C. Semmens, E. A. Odekunle, M. Zandawala, Y. Badi, M. L. Rowe, M. Egertova, and M. R. Elphick. 2016. Localization of neuropeptide gene expression in larvae of an echinoderm, the starfish *Asterias rubens*. *Front Neurosci* 10:553.

Mazet, F., J. A. Hutt, J. Millard, and S. M. Shimeld. 2003. Pax gene expression in the develop-ing central nervous system of *Ciona intestinalis*. *Gene Expr Patterns* 3:743–745.

Mazet, F., J. A. Hutt, J. Milloz, J. Millard, A. Graham, and S. M. Shimeld. 2005. Molecular evidence from *Ciona intestinalis* for the evolutionary origin of vertebrate sensory plac-odes. *Dev Biol* 282:494–508.

Mazet, F., S. Masood, G. N. Luke, N. D. Holland, and S. M. Shimeld. 2004. Expression of AmphiCoe, an amphioxus COE/EBF gene, in the developing central nervous system and epidermal sensory neurons. *Genesis* 38:58–65.

Mazet, F., J. K. Yu, D. A. Liberles, L. Z. Holland, and S. M. Shimeld. 2003. Phylogenetic relationships of the Fox (Forkhead) gene family in the Bilateria. *Gene* 316:79–89.

Mazza, M. E., K. Pang, A. M. Reitzel, M. Q. Martindale, and J. R. Finnerty. 2010. A con-served cluster of three PRD-class homeobox genes (homeobrain, rx and orthopedia) in the Cnidaria and Protostomia. *Evodevo* 1:3.

McCauley, H. A., and G. Guasch. 2015. Three cheers for the goblet cell: maintaining homeo-stasis in mucosal epithelia. *Trends Mol Med* 21:492–503.

McEvilly, R. J., L. Erkman, L. Luo, P. E. Sawchenko, A. F. Ryan, and M. G. Rosenfeld. 1996. Requirement for brn-3.0 in differentiation and survival of sensory and motor neurons. *Nature* 384:574–577.

McGregor, A. P., M. Pechmann, E. E. Schwager, N. M. Feitosa, S. Kruck, M. Aranda, and W. G. Damen. 2008. Wnt8 is required for growth-zone establishment and development of opisthosomal segments in a spider. *Curr Biol* 18:1619–1623.

McLaughlin, S. 2017. Evidence that polycystins are involved in Hydra cnidocyte discharge. *Invert Neurosci* 17:1.

McReynolds, J. S., and A. L. Gorman. 1970. Photoreceptor potentials of opposite polarity in the eye of the scallop, *Pecten irradians*. *J Gen Physiol* 56:376–391.

Medeiros, D. M. 2013. The evolution of the neural crest: new perspectives from lamprey and invertebrate neural crest-like cells. *Wiley Interdiscip Rev Dev Biol* 2:1–15.

Merritt, D. J. 2007. The organule concept of insect sense organs: sensory transduction and organule evolution. *Adv Insect Physiol* 33:192–241.

Metschnikoff, V. E. 1881. Über die systematische Stellung von *Balanoglossus*. *Zool Anz* 4:139–157.

Meulemans, D., and M. Bronner-Fraser. 2002. Amphioxus and lamprey AP-2 genes: implications for neural crest evolution and migration patterns. *Development* 129:4953–4962.

Meulemans, D., and M. Bronner-Fraser. 2004. Gene-regulatory interactions in neural crest evolution and development. *Dev Cell* 7:291–299.

Meulemans, D., and M. Bronner-Fraser. 2007. The amphioxus SoxB family: implications for the evolution of vertebrate placodes. *Int J Biol Sci*3:356–364.

Meyer, A., and P. Y. Van de Peer. 2005. From 2R to 3R: evidence for a fish-specific genome duplication (FSGD). *Bioessays* 27:937–945.

Michaud, J. L., T. Rosenquist, N. R. May, and C. M. Fan. 1998. Development of neuro-endocrine lineages requires the bHLH-PAS transcription factor SIM1. *Genes Dev* 12:3264–3275.

Miljkovic-Licina, M., S. Chera, L. Ghila, and B. Galliot. 2007. Head regeneration in wild-type *Hydra* requires de novo neurogenesis. *Development* 134:1191–1201.

Miller, D. J., D. C. Hayward, J. S. ReeceHoyes, I. Scholten, J. Catmull, W. J. Gehring, P. Callaerts, J. E. Larsen, and E. E. Ball. 2000. Pax gene diversity in the basal cnidarian *Acropora millepora* (Cnidaria, Anthozoa): implications for the evolution of the Pax gene family. *Proc Natl Acad Sci USA* 97:4475–4480.

Minelli, A. 2000. Limbs and tail as evolutionarily diverging duplicates of the main body axis. *Evol Dev*2:157–165.

Mirabeau, O., and J. S. Joly. 2013. Molecular evolution of peptidergic signaling systems in bilaterians. *Proc Natl Acad Sci USA* 110:E2028–E2037.

Mire-Thibodeaux, P., and G. M. Watson. 1994. Morphodynamic hair bundles arising from sensory cell/supporting cell complexes frequency-tune nematocyst discharge in sea anemones. *J Exp Zool* 268:282–292.

Miya, T., and H. Nishida. 2003. Expression pattern and transcriptional Control of SoxB1 in embryos of the ascidian *Halocynthia roretzi*. *Zoolog Sci* 20:59–67.

Miyako, Y., K. Arikawa, and E. Eguchi. 1993. Ultrastructure of the extraocular photoreceptor in the genitalia of a butterfly, *Papilio xuthus*. *J Comp Neurol* 327:458–468.

Miyamoto, N., and H. Wada. 2013. Hemichordate neurulation and the origin of the neural tube. *Nat Commun* 4:2713.

Miyawaki, K., T. Mito, I. Sarashina, H. Zhang, Y. Shinmyo, H. Ohuchi, and S. Noji. 2004. Involvement of Wingless/Armadillo signaling in the posterior sequential segmentation in the cricket, *Gryllus bimaculatus* (Orthoptera), as revealed by RNAi analysis. *Mech Dev* 121:119–130.

Moczek, A. P. 2008. On the origins of novelty in development and evolution. *Bioessays* 30:432–447.

Modrell, M. S., D. Hockman, B. Uy, D. Buckley, T. Sauka-Spengler, M. E. Bronner, and C. V. Baker. 2014. A fate-map for cranial sensory ganglia in the sea lamprey. *Dev Biol* 385:405–416.

Modrell, M. S., M. Lyne, A. R. Carr, H. H. Zakon, D. Buckley, A. S. Campbell, M. C. Davis, G. Micklem, and C. V. Baker. 2017. Insights into electrosensory organ development, physiology and evolution from a lateral line-enriched transcriptome. *Elife* 6:e24197.

Monahan, K., and S. Lomvardas. 2015. Monoallelic expression of olfactory receptors. *Annu Rev Cell Dev Biol* 31:721–740.

Molina, M. D., N. de Croze, E. Haillot, and T. Lepage. 2013. Nodal: master and commander of the dorsal-ventral and left-right axes in the sea urchin embryo. *Curr Opin Genet Dev* 23:445–53.

Montell, C. 2012. *Drosophila* visual transduction. *Trends Neurosci* 35:356–363.

Moody, S. A., and A. S. LaMantia. 2015. Transcriptional regulation of cranial sensory placode development. *Curr Top Dev Biol* 111:301–350.

Morales-Delgado, N., B. Castro-Robles, J. L. Ferran, M. Martinez-de-la-Torre, L. Puelles, and C. Díaz. 2014. Regionalized differentiation of CRH, TRH, and GHRH peptidergic neurons in the mouse hypothalamus. *Brain Struct Funct* 219:1083–1111.

Morales-Delgado, N., P. Merchan, S. M. Bardet, J. L. Ferran, L. Puelles, and C. Díaz. 2011. Topography of somatostatin gene expression relative to molecular progenitor domains during ontogeny of the mouse hypothalamus. *Front Neuroanat* 5:10.

Morante, J., C. Desplan, and A. Celik. 2007. Generating patterned arrays of photoreceptors. *Curr Opin Genet Dev* 17:314–319.

Moret, F., L. Christiaen, C. Deyts, M. Blin, P. Vernier, and J. S. Joly. 2005. Regulatory gene expressions in the ascidian ventral sensory vesicle: evolutionary relationships with the vertebrate hypothalamus. *Dev Biol* 277:567–579.

Morita, H., and K. Hanai. 1987. Taste receptor proteins in invertebrates with special reference to glutathione receptor of *Hydra*. *Chem Senses* 12:245–250.

Morov, A. R., T. Ukizintambara, R. M. Sabirov, and K. Yasui. 2016. Acquisition of the dorsal structures in chordate amphioxus. *Open Biol* 6:160062.

Moroz, L. L. 2009. On the independent origins of complex brains and neurons. *Brain Behav Evol* 74:177–190.

Moroz, L. L., K. M. Kocot, M. R. Citarella, S. Dosung, T. P. Norekian, I. S. Povolotskaya, A. P. Grigorenko, C. Dailey, E. Berezikov, K. M. Buckley, A. Ptitsyn, D. Reshetov, K. Mukherjee, T. P. Moroz, Y. Bobkova, F. Yu, V. V. Kapitonov, J. Jurka, Y. V. Bobkov, J. J. Swore, D. O. Girardo, A. Fodor, F. Gusev, R. Sanford, R. Bruders, E. Kittler, C. E. Mills, J. P. Rast, R. Derelle, V. V. Solovyev, F. A. Kondrashov, B. J. Swalla, J. V. Sweedler, E. I. Rogaev, K. M. Halanych, and A. B. Kohn. 2014. The ctenophore genome and the evolutionary origins of neural systems. *Nature* 510:109–114.

Morris, S. C., and J. B. Caron. 2012. *Pikaia gracilens* Walcott, a stem-group chordate from the Middle Cambrian of British Columbia. *Biol Rev Camb Philos Soc* 87:480–512.

Morris, S. C., and J. B. Caron. 2014. A primitive fish from the Cambrian of North America. *Nature* 512:419–422.

Morrison, K. M., G. R. Miesegaes, E. A. Lumpkin, and S. M. Maricich. 2009. Mammalian Merkel cells are descended from the epidermal lineage. *Dev Biol* 336:76–83.

Mousley, A., G. Polese, N. J. Marks, and H. L. Eisthen. 2006. Terminal nerve-derived neuropeptide y modulates physiological responses in the olfactory epithelium of hungry axolotls (*Ambystoma mexicanum*). *J Neurosci* 26:7707–7717.

Müller, G. B., and G. P. Wagner. 1991. Novelty in evolution: restructuring the concept. *Annu Rev Ecol Syst* 22:229–256.

Müller, P., K. Seipel, N. Yanze, S. Reber-Müller, R. Streitwolf-Engel, M. Stierwald, J. Spring, and V. Schmid. 2003. Evolutionary aspects of developmentally regulated helix-loop-helix transcription factors in striated muscle of jellyfish. *Dev Biol* 255:216–229.

Murdock, D. J., X. P. Dong, J. E. Repetski, F. Marone, M. Stampanoni, and P. C. Donoghue. 2013. The origin of conodonts and of vertebrate mineralized skeletons. *Nature* 502:546–549.

Musser, J. M., K. J. Schippers, M. Nickel, G. Mizzon, A.B. Kohn, C. Pape, J. U. Hammel, Wolf. F., C. Liang, A. Hernández-Plaza, K. Achim, N. L. Schieber, W. R. Francis, S. Vargas R., S. Kling, M. Renkert, R. Feuda, I. Gaspar, P. Burkhardt, P. Bork, M. Beck, A. Kreshuk, G. Wörheide, J. Huerta-Cepas, Y. Schwab, L. L. Moroz, and D. Arendt. 2019. Profiling cellular diversity in sponges informs animal cell type and nervous system evolution. *Biorxiv*:https://doi.org/10.1101/758276.

Musser, J. M., and G. P. Wagner. 2015. Character trees from transcriptome data: origin and individuation of morphological characters and the so-called "species signal". *J Exp Zool B Mol Dev Evol* 324:588–604.

Nagatomo, K., T. Ishibashi, Y. Satou, N. Satoh, and S. Fujiwara. 2003. Retinoic acid affects gene expression and morphogenesis without upregulating the retinoic acid receptor in the ascidian *Ciona intestinalis*. *Mech Dev* 120:363–372.

Nagatomo, K., and S. Fujiwara. 2003. Expression of Raldh2, Cyp26 and Hox-1 in normal and retinoic acid-treated *Ciona intestinalis* embryos. *Gene Expr Patterns* 3:273–277.

Nakanishi, N., A. C. Camara, D. C. Yuan, D. A. Gold, and D. K. Jacobs. 2015. Gene expression data from the moon jelly, *Aurelia*, provide insights into the evolution of the combinatorial code controlling animal sense organ development. *PLoS One* 10:e0132544.

Nakanishi, N., V. Hartenstein, and D. K. Jacobs. 2009. Development of the rhopalial nervous system in *Aurelia sp.1* (Cnidaria, Scyphozoa). *Dev Genes Evol* 219:301–317.

Nakanishi, N., E. Renfer, U. Technau, and F. Rentzsch. 2012. Nervous systems of the sea anemone *Nematostella vectensis* are generated by ectoderm and endoderm and shaped by distinct mechanisms. *Development* 139:347–357.

Nakanishi, N., D. Stoupin, S. M. Degnan, and B. M. Degnan. 2015. Sensory flask cells in sponge larvae regulate metamorphosis via calcium signaling. *Integr Comp Biol* 55:1018–1027.

Nakanishi, N., D. Yuan, V. Hartenstein, and D. K. Jacobs. 2010. Evolutionary origin of rhopalia: insights from cellular-level analyses of Otx and POU expression patterns in the developing rhopalial nervous system. *Evol Dev* 12:404–415.

Nakanishi, N., D. Yuan, D. K. Jacobs, and V. Hartenstein. 2008. Early development, pattern, and reorganization of the planula nervous system in *Aurelia* (Cnidaria, Scyphozoa). *Dev Genes Evol* 218:511–524.

Nakao, T. 1964. On the fine structure of the amphioxus photoreceptor. *Tohoku J Exp Med* 82:349–363.

Nakayama, S., and M. Ogasawara. 2017. Compartmentalized expression patterns of pancreatic- and gastric-related genes in the alimentary canal of the ascidian *Ciona intestinalis*: evolutionary insights into the functional regionality of the gastrointestinal tract in Olfactores. *Cell Tissue Res* 370:113–128.

Nassel, D. R. 2002. Neuropeptides in the nervous system of *Drosophila* and other insects: multiple roles as neuromodulators and neurohormones. *Prog Neurobiol* 68:1–84.

Nezlin, L. P., and V. V. Yushin. 2004. Structure of the nervous system in the tornaria larva of *Balanoglossus proterogonius* (Hemichordata: Enteropneusta) and its phylogenetic implications. *Zoomorphology* 123:1–13.

Niehrs, C. 2010. On growth and form: a Cartesian coordinate system of Wnt and BMP signaling specifies bilaterian body axes. *Development* 137:845–857.

Niehrs, C. 2012. The complex world of WNT receptor signalling. *Nat Rev Mol Cell Biol* 13:767–779.

Nielsen, C., T. Brunet, and D. Arendt. 2018. Evolution of the bilaterian mouth and anus. *Nat Ecol Evol* 2:1358–1376.

Niimura, Y. 2009. Evolutionary dynamics of olfactory receptor genes in chordates: interaction between environments and genomic contents. *Hum Genomics* 4:107–118.

Niimura, Y., A. Matsui, and K. Touhara. 2014. Extreme expansion of the olfactory receptor gene repertoire in African elephants and evolutionary dynamics of orthologous gene groups in 13 placental mammals. *Genome Res* 24:1485–1496.

Nijhout, H. F. 1998. *Insect hormones.* Princeton: Princeton University Press.

Nikitin, M. 2015. Bioinformatic prediction of *Trichoplax adhaerens* regulatory peptides. *Gen Comp Endocrinol* 212:145–155.

Nilsson, D. E., and D. Arendt. 2008. Eye evolution: the blurry beginning. *Curr Biol* 18:R1096–R1098.

Nilsson, D. E., L. Gislén, M. M. Coates, C. Skogh, and A. Garm. 2005. Advanced optics in a jellyfish eye. *Nature* 435:201–205.

Nishida, H. 1987. Cell lineage analysis in ascidian embryos by intracellular injection of a tracer enzyme. III. Up to the tissue restricted stage. *Dev Biol* 121:526–541.

Nishida, H. 2008. Development of the appendicularian *Oikopleura dioica*: culture, genome, and cell lineages. *Dev Growth Differ* 50 Suppl 1:S239–S256.

Nishino, A., and N. Satoh. 2001. The simple tail of chordates: phylogenetic significance of appendicularians. *Genesis* 29:36–45.

Noah, T. K., A. Kazanjian, J. Whitsett, and N. F. Shroyer. 2010. SAM pointed domain ETS factor (SPDEF) regulates terminal differentiation and maturation of intestinal goblet cells. *Exp Cell Res* 316:452–465.

Nolo, R., L. A. Abbott, and H. J. Bellen. 2000. Senseless, a Zn finger transcription factor, is necessary and sufficient for sensory organ development in *Drosophila. Cell* 102:349–362.

Nomaksteinsky, M., S. Kassabov, Z. Chettouh, H. C. Stoekle, L. Bonnaud, G. Fortin, E. R. Kandel, and J. F. Brunet. 2013. Ancient origin of somatic and visceral neurons. *BMC Biol* 11:53.

Nomaksteinsky, M., E. Röttinger, H. D. Dufour, Z. Chettouh, C. J. Lowe, M. Q. Martindale, and J. F. Brunet. 2009. Centralization of the deuterostome nervous system predates chordates. *Curr Biol* 19:1264–1269.

Nordström, K., R. Wallen, J. Seymour, and D. Nilsson. 2003. A simple visual system without neurons in jellyfish larvae. *Proc Biol Sci* 270:2349–2354.

Norekian, T. P., and L. L. Moroz. 2019a. Neural system and receptor diversity in the ctenophore *Beroe abyssicola. J Comp Neurol* 527:1986–2008.

Norekian, T. P., and L. L. Moroz. 2019b. Neuromuscular organization of the ctenophore *Pleurobrachia bachei. J Comp Neurol* 527:406–436.

Nørrevang, A., and K. G. Wingstrand. 1970. On the occurrence and structure of choanocyte-like cells in some echinoderms. *Acta Zool* 51:249–270.

Norris, D. O., and J. A. Carr. 2020. *Vertebrate endocrinology.* 6th ed. London: Academic Press.

Northcutt, R. G. 1997. Evolution of gnathostome lateral line ontogenies. *Brain Behav Evol* 50:25–37.

Northcutt, R. G. 2005. The new head hypothesis revisited. *J Exp Zoolog B Mol Dev Evol* 304:274–297.

Northcutt, R. G. 2008. Forebrain evolution in bony fishes. *Brain Res Bull* 75:191–205.

Northcutt, R. G. 2012. Evolution of centralized nervous systems: two schools of evolutionary thought. *Proc Natl Acad Sci USA* 109 Suppl 1:10626–10633.

Northcutt, R. G., and C. Gans. 1983. The genesis of neural crest and epidermal placodes: a reinterpretation of vertebrate origins. *Q Rev Biol* 58:1–28.

Nozaki, M., and A. Gorbman. 1992. The question of functional homology of Hatschek's pit of amphioxus (*Branchiostoma belcheri*) and the vertebrate adenohypophysis. *Zool Sci* 9:387–395.

Nübler-Jung, K., and D. Arendt. 1996. Enteropneusts and chordate evolution. *Curr Biol* 6:352–353.

Nyhart, L. K. 1995. *Biology takes form. Animal morphology and the German universities, 1800-1900.* Chicago: University of Chicago Press.

O'Brien, E. K., and B. M. Degnan. 2000. Expression of POU, Sox, and Pax genes in the brain ganglia of the tropical abalone *Haliotis asinina. Mar Biotechnol* 2:545–557.

O'Brien, E. K., and B. M. Degnan. 2002. Developmental expression of a class IV POU gene in the gastropod *Haliotis asinina* supports a conserved role in sensory cell development in bilaterians. *Dev Genes Evol* 212:394–398.

O'Brien, E. K., and B. M. Degnan. 2003. Expression of Pax258 in the gastropod statocyst: insights into the antiquity of metazoan geosensory organs. *Evol Dev* 5:572–578.

Oakley, T. H. 2017. Furcation and fusion: the phylogenetics of evolutionary novelty. *Dev Biol* 431:69–76.

Ocampo Daza, D., and D. Larhammar. 2018. Evolution of the growth hormone, prolactin, prolactin 2 and somatolactin family. *Gen Comp Endocrinol* 264:94–112.

Oda-Ishii, I., A. Kubo, W. Kari, N. Suzuki, U. Rothbächer, and Y. Satou. 2016. A maternal system initiating the zygotic developmental program through combinatorial repression in the ascidian embryo. *PLoS Genet* 12:e1006045.

Oda, I., and H. Saiga. 2001. Hremx, the ascidian homologue of ems/emx, is expressed in the anterior and posterior-lateral epidermis but not in the central nervous system during embryogenesis. *Dev Genes Evol* 211:291–298.

Ogasawara, M., H. Wada, H. Peters, and N. Satoh. 1999. Developmental expression of Pax1/9 genes in urochordate and hemichordate gills: insight into function and evolution of the pharyngeal epithelium. *Development* 126:2539–2550.

Ogura, Y., and Y. Sasakura. 2016. Developmental control of cell-cycle compensation provides a switch for patterned mitosis at the onset of chordate neurulation. *Dev Cell* 37:148–161.

Ohtsuka, K., T. Atsumi, Y. Fukushima, and K. Shiomi. 2011. Identification of a cis-regulatory element that directs prothoracicotropic hormone gene expression in the silkworm *Bombyx mori*. *Insect Biochem Mol Biol* 41:356–361.

Ohtsuki, H. 1990. Statocyte and ocellar pigment cell in embryos and larvae of the ascidian, *Styela plicata* (Lesueur). *Dev Growth Differ* 32:85–90.

Oliveri, P., K. D. Walton, E. H. Davidson, and D. R. McClay. 2006. Repression of mesodermal fate by foxa, a key endoderm regulator of the sea urchin embryo. *Development* 133:4173–4181.

Olsson, R., K. Holmberg, and Y. Lilliemarck. 1990. Fine-structure of the brain and brain nerves of *Oikopleura dioica* (Urochordata, Appendicularia). *Zoomorphology* 110:1–7.

Onai, T., N. Adachi, and S. Kuratani. 2017. Metamerism in cephalochordates and the problem of the vertebrate head. *Int J Dev Biol* 61:621–632.

Onai, T., H. C. Lin, M. Schubert, D. Koop, P. W. Osborne, S. Alvarez, R. Alvarez, N. D. Holland, and L. Z. Holland. 2009. Retinoic acid and Wnt/beta-catenin have complementary roles in anterior/posterior patterning embryos of the basal chordate amphioxus. *Dev Biol* 332:223–233.

Onai, T., A. Takai, D. H. Setiamarga, and L. Z. Holland. 2012. Essential role of Dkk3 for head formation by inhibiting Wnt/beta-catenin and Nodal/Vg1 signaling pathways in the basal chordate amphioxus. *Evol Dev* 14:338–350.

Onai, T., J. K. Yu, I. L. Blitz, K. W. Cho, and L. Z. Holland. 2010. Opposing Nodal/Vg1 and BMP signals mediate axial patterning in embryos of the basal chordate amphioxus. *Dev Biol* 344:377–389.

Oonuma, K., and T. G. Kusakabe. 2019. Spatio-temporal regulation of Rx and mitotic patterns shape the eye-cup of the photoreceptor cells in *Ciona*. *Dev Biol* 445:245–255.

Oron-Karni, V., C. Farhy, M. Elgart, T. Marquardt, L. Remizova, O. Yaron, Q. Xie, A. Cvekl, and R. Ashery-Padan. 2008. Dual requirement for Pax6 in retinal progenitor cells. *Development* 135:4037–4047.

Orquera, D. P., M. B. Tavella, F. S. J. de Souza, S. Nasif, M. J. Low, and M. Rubinstein. 2019. The homeodomain transcription factor NKX2.1 is essential for the early specification of melanocortin neuron identity and activates Pomc expression in the developing hypothalamus. *J Neurosci* 39:4023–4035.

Osugi, T., Y. Sasakura, and H. Satake. 2017. The nervous system of the adult ascidian *Ciona intestinalis* Type A (*Ciona robusta*): insights from transgenic animal models. *PLoS One* 12:e0180227.

Osugi, T., Y. Sasakura, and H. Satake. 2020. The ventral peptidergic system of the adult ascidian *Ciona robusta* (*Ciona intestinalis* Type A) insights from a transgenic animal model. *Sci Rep* 10:1892.

Osugi, T., Y. L. Son, T. Ubuka, H. Satake, and K. Tsutsui. 2016. RFamide peptides in agnathans and basal chordates. *Gen Comp Endocrinol* 227:94–100.

Ota, K. G., S. Fujimoto, Y. Oisi, and S. Kuratani. 2011. Identification of vertebra-like elements and their possible differentiation from sclerotomes in the hagfish. *Nat Commun* 2:373.

Owen, R. 1843. *Lectures on comparative anatomy and physiology of the invertebrate animals, delivered at the Royal College of Surgeons in 1843*. London: Longman, Brown, Green & Longmans.

Owen, R. 1848. *On the archetype and homologies of the vertebrate skeleton*. London: J. van Voorst.

Packer, J. S., Q. Zhu, C. Huynh, P. Sivaramakrishnan, E. Preston, D. Dueck, D. Stefanik, K. Tan, C. Trapnell, J. Kim, R. H. Waterston, and J. I. Murray. 2019. A lineage-resolved molecular atlas of *C. elegans* embryogenesis at single-cell resolution. *Science* 365:eaax1971.

Page, L. R., and S. C. Parries. 2000. Comparative study of the apical ganglion in planktotrophic caenogastropod larvae: ultrastructure and immunoreactivity to serotonin. *J Comp Neurol* 418:383–401.

Pan, B., G. S. Geleoc, Y. Asai, G. C. Horwitz, K. Kurima, K. Ishikawa, Y. Kawashima, A. J. Griffith, and J. R. Holt. 2013. TMC1 and TMC2 are components of the mechanotransduction channel in hair cells of the mammalian inner ear. *Neuron* 79:504–515.

Pan, Y., R. I. Martinez-De Luna, C. H. Lou, S. Nekkalapudi, L. E. Kelly, A. K. Sater, and H. M. El-Hodiri. 2010. Regulation of photoreceptor gene expression by the retinal homeobox (Rx) gene product. *Dev Biol* 339:494–506.

Pang, K., and M. Q. Martindale. 2008. Developmental expression of homeobox genes in the ctenophore *Mnemiopsis leidyi*. *Dev Genes Evol* 218:307–319.

Pangršič, T., E. Reisinger, and T. Moser. 2012. Otoferlin: a multi-C2 domain protein essential for hearing. *Trends Neurosci* 35:671–680.

Pani, A. M., E. E. Mullarkey, J. Aronowicz, S. Assimacopoulos, E. A. Grove, and C. J. Lowe. 2012. Ancient deuterostome origins of vertebrate brain signalling centres. *Nature* 483:289–294.

Pankey, S., H. Sunada, T. Horikoshi, and M. Sakakibara. 2010. Cyclic nucleotide-gated channels are involved in phototransduction of dermal photoreceptors in *Lymnaea stagnalis*. *J Comp Physiol B* 180:1205–1211.

Pantzartzi, C. N., J. Pergner, I. Kozmikova, and Z. Kozmik. 2017. The opsin repertoire of the European lancelet: a window into light detection in a basal chordate. *Int J Dev Biol* 61:763–772.

Papaioannou, V. E. 2014. The T-box gene family: emerging roles in development, stem cells and cancer. *Development* 141:3819–3833.

Paps, J., P. W. Holland, and S. M. Shimeld. 2012. A genome-wide view of transcription factor gene diversity in chordate evolution: less gene loss in amphioxus? *Brief Funct Genomics* 11:177–186.

Park, D., and H. L. Eisthen. 2003. Gonadotropin releasing hormone (GnRH) modulates odorant responses in the peripheral olfactory system of axolotls. *J Neurophysiol* 90:731–738.

Park, D., H. Jia, V. Rajakumar, and H. M. Chamberlin. 2006. Pax2/5/8 proteins promote cell survival in *C. elegans*. *Development* 133:4193–4202.

Park, J. I., J. Semyonov, C. L. Chang, and S. Y. Hsu. 2005. Conservation of the heterodimeric glycoprotein hormone subunit family proteins and the LGR signaling system from nematodes to humans. *Endocrine* 26:267–276.

Parlier, D., V. Moers, C. Van Campenhout, J. Preillon, L. Leclère, A. Saulnier, M. Sirakov, H. Busengdal, S. Kricha, J. C. Marine, F. Rentzsch, and E. J. Bellefroid. 2013. The *Xenopus* doublesex-related gene Dmrt5 is required for olfactory placode neurogenesis. *Dev Biol* 373:39–52.

Parrilla, M., I. Chang, A. Degl'Innocenti, and M. Omura. 2016. Expression of homeobox genes in the mouse olfactory epithelium. *J Comp Neurol* 524:2713–2739.

Pascual-Anaya, J., N. Adachi, S. Alvarez, S. Kuratani, S. D'Aniello, and J. Garcia-Fernàndez. 2012. Broken colinearity of the amphioxus Hox cluster. *Evodevo* 3:28.

Pasini, A., A. Amiel, U. Rothbächer, A. Roure, P. Lemaire, and S. Darras. 2006. Formation of the ascidian epidermal sensory neurons: insights into the origin of the chordate peripheral nervous system. *PLoS Biol* 4:1173–1186.

Pasini, A., R. Manenti, U. Rothbächer, and P. Lemaire. 2012. Antagonizing retinoic acid and FGF/MAPK pathways control posterior body patterning in the invertebrate chordate *Ciona intestinalis*. *PLoS One* 7:e46193.

Passamaneck, Y. J., and A. Di Gregorio. 2005. *Ciona intestinalis*: chordate development made simple. *Dev Dyn* 233:1–19.

Passamaneck, Y. J., N. Furchheim, A. Hejnol, M. Q. Martindale, and C. Lüter. 2011. Ciliary photoreceptors in the cerebral eyes of a protostome larva. *Evodevo* 2:6.

Passamaneck, Y. J., A. Hejnol, and M. Q. Martindale. 2015. Mesodermal gene expression during the embryonic and larval development of the articulate brachiopod *Terebratalia transversa*. *Evodevo* 6:10.

Passamaneck, Y. J., L. Katikala, L. Perrone, M. P. Dunn, I. Oda-Ishii, and A. Di Gregorio. 2009. Direct activation of a notochord cis-regulatory module by Brachyury and FoxA in the ascidian *Ciona intestinalis*. *Development* 136:3679–3689.

Patthey, C., G. Schlosser, and S. M. Shimeld. 2014. The evolutionary history of vertebrate cranial placodes–I: cell type evolution. *Dev Biol* 389:82–97.

Paulus, H. F. 1979. Eye structure and the monophyly of the Arthropoda. In *Arthropod phylogeny*, edited by A. P. Gupta, 299–382. New York: Van Nostrand Reinhold Company.

Peel, A. D., A. D. Chipman, and M. Akam. 2005. Arthropod segmentation: beyond the *Drosophila* paradigm. *Nat Rev Genet* 6:905–916.

Peinado, G., T. Osorno, P. Gómez Mdel, and E. Nasi. 2015. Calcium activates the light-dependent conductance in melanopsin-expressing photoreceptors of amphioxus. *Proc Natl Acad Sci USA* 112:7845–7850.

Pelosi, B., S. Migliarini, G. Pacini, M. Pratelli, and M. Pasqualetti. 2014. Generation of Pet1210-Cre transgenic mouse line reveals non-serotonergic expression domains of Pet1 both in CNS and periphery. *PLoS One* 9:e104318.

Peng, G. H., and S. Chen. 2007. Crx activates opsin transcription by recruiting HAT-containing co-activators and promoting histone acetylation. *Hum Mol Genet* 16:2433–2452.

Peng, G., X. Shi, and T. Kadowaki. 2015. Evolution of TRP channels inferred by their classification in diverse animal species. *Mol Phylogenet Evol* 84:145–157.

Perez, L. N., J. Lorena, C. M. Costa, M. S. Araujo, G. N. Frota-Lima, G. E. Matos-Rodrigues, R. A. Martins, G. M. Mattox, and P. N. Schneider. 2017. Eye development in the four-eyed fish *Anableps anableps*: cranial and retinal adaptations to simultaneous aerial and aquatic vision. *Proc Biol Sci* 284:2017015.

Pergner, J., and Z. Kozmik. 2017. Amphioxus photoreceptors: insights into the evolution of vertebrate opsins, vision and circadian rhythmicity. *Int J Dev Biol* 61:665–681.

Perron, M., S. Kanekar, M. L. Vetter, and W. A. Harris. 1998. The genetic sequence of retinal development in the ciliary margin of the *Xenopus* eye. *Dev Biol* 199:185–200.

Pestarino, M. 1983. Prolactinergic neurons in a protochordate. *Cell Tissue Res* 233:471–474.

Pestarino, M. 1984. Immunocytochemical demonstration of prolactin-like activity in the neural gland of the ascidian *Styela plicata*. *Gen Comp Endocrinol* 54:444–449.

Peter, I. S., and E. H. Davidson. 2015. *Genomic control process: development and evolution*. Amsterdam: Elsevier.

Petersen, C. P., and P. W. Reddien. 2008. Smed-betacatenin-1 is required for anteroposterior blastema polarity in planarian regeneration. *Science* 319:327–330.

Petersen, C. P., and P. W. Reddien. 2009. Wnt signaling and the polarity of the primary body axis. *Cell* 139:1056–1068.

Peterson, T., and G. B. Müller. 2013. What is evolutionary novelty? Process versus character based definitions. *J Exp Zool B Mol Dev Evol* 320:345–350.

Pettigrew, J. D. 1999. Electroreception in monotremes. *J Exp Biol* 202:1447–1454.

Philippe, H., H. Brinkmann, R. R. Copley, L. L. Moroz, H. Nakano, A. J. Poustka, A. Wallberg, K. J. Peterson, and M. J. Telford. 2011. Acoelomorph flatworms are deuterostomes related to *Xenoturbella*. *Nature* 470:255–258.

Philippe, H., N. Lartillot, and H. Brinkmann. 2005. Multigene analyses of bilaterian animals corroborate the monophyly of Ecdysozoa, Lophotrochozoa, and Protostomia. *Mol Biol Evol* 22:1246–1253.

Philippe, H., A. J. Poustka, M. Chiodin, K. J. Hoff, C. Dessimoz, B. Tomiczek, P. H. Schiffer, S. Müller, D. Domman, M. Horn, H. Kuhl, B. Timmermann, N. Satoh, T. Hikosaka-Katayama, H. Nakano, M. L. Rowe, M. R. Elphick, M. Thomas-Chollier, T. Hankeln, F. Mertes, A. Wallberg, J. P. Rast, R. R. Copley, P. Martinez, and M. J. Telford. 2019. Mitigating anticipated effects of systematic errors supports sister-group relationship between Xenacoelomorpha and Ambulacraria. *Curr Biol* 29:1818–1826 e6.

Phillips, C. E., and W. O. Friesen. 1982. Ultrastructure of the water-movement-sensitive sensilla in the medicinal leech. *J Neurobiol* 13:473–486.

Piatigorsky, J., and Z. Kozmik. 2004. Cubozoan jellyfish: an Evo/Devo model for eyes and other sensory systems. *Int J Dev Biol* 48:719–729.

Picard, M. A., C. Cosseau, G. Mouahid, D. Duval, C. Grunau, E. Toulza, J. F. Allienne, and J. Boissier. 2015. The roles of Dmrt (Double sex/Male-abnormal-3 Related Transcription factor) genes in sex determination and differentiation mechanisms: ubiquity and diversity across the animal kingdom. *C R Biol* 338:451–462.

Picciani, N., J. R. Kerlin, N. Sierra, A. J. M. Swafford, M. D. Ramirez, N. G. Roberts, J. T. Cannon, M. Daly, and T. H. Oakley. 2018. Prolific origination of eyes in Cnidaria with co-option of non-visual opsins. *Curr Biol* 28:2413–2419 e4.

Pieper, M., G. W. Eagleson, W. Wosniok, and G. Schlosser. 2011. Origin and segregation of cranial placodes in *Xenopus laevis*. *Dev Biol* 360:257–275.

Pierce, M. L., M. D. Weston, B. Fritzsch, H. W. Gabel, G. Ruvkun, and G. A. Soukup. 2008. MicroRNA-183 family conservation and ciliated neurosensory organ expression. *Evol Dev* 10:106–113.

Piette, J., and P. Lemaire. 2015. Thaliaceans, the neglected pelagic relatives of ascidians: a developmental and evolutionary enigma. *Q Rev Biol* 90:117–145.

Pignoni, F., B. Hu, K. H. Zavitz, J. Xiao, P. A. Garrity, and S. L. Zipursky. 1997. The eye-specification proteins So and Eya form a complex and regulate multiple steps in *Drosophila* eye development. *Cell* 91:881–891.

Pineda, D., J. Gonzalez, P. Callaerts, K. Ikeo, W. J. Gehring, and E. Salo. 2000. Searching for the prototypic eye genetic network: sine oculis is essential for eye regeneration in planarians. *Proc Nat Acad Sci USA* 97:4525–4529.

Pineda, D., L. Rossi, R. Batistoni, A. Salvetti, M. Marsal, V. Gremigni, A. Falleni, J. Gonzalez-Linares, P. Deri, and E. Salo. 2002. The genetic network of prototypic planarian eye regeneration is Pax6 independent. *Development* 129:1423–1434.

Piotrowski, T., and C. V. Baker. 2014. The development of lateral line placodes: taking a broader view. *Dev Biol* 389:68–81.

Piraino, S., G. Zega, C. Di Benedetto, A. Leone, A. Dell'Anna, R. Pennati, D. C. Carnevali, V. Schmid, and H. Reichert. 2011. Complex neural architecture in the diploblastic larva of *Clava multicornis* (Hydrozoa, Cnidaria). *J Comp Neurol* 519:1931–1951.

Pires-daSilva, A., and R. J. Sommer. 2003. The evolution of signalling pathways in animal development. *Nat Rev Genet* 4:39–49.

Plachetzki, D. C., B. M. Degnan, and T. H. Oakley. 2007. The origins of novel protein interactions during animal opsin evolution. *PLoS One* 2:e1054.

Plachetzki, D. C., C. R. Fong, and T. H. Oakley. 2010. The evolution of phototransduction from an ancestral cyclic nucleotide gated pathway. *Proc Biol Sci* 277:1963–1969.

Plachetzki, D. C., C. R. Fong, and T. H. Oakley. 2012. Cnidocyte discharge is regulated by light and opsin-mediated phototransduction. *BMC Biol* 10:17.

Pogoda, H. M., H. S. von der, W. Herzog, C. Kramer, H. Schwarz, and M. Hammerschmidt. 2006. The proneural gene ascl1a is required for endocrine differentiation and cell survival in the zebrafish adenohypophysis. *Development* 133:1079–1089.

Poole, R. J., E. Bashllari, L. Cochella, E. B. Flowers, and O. Hobert. 2011. A genome-wide RNAi screen for factors involved in neuronal specification in *Caenorhabditis elegans*. *PLoS Genet* 7:e1002109.

Porter, M. L. 2016. Beyond the eye: molecular evolution of extraocular photoreception. *Integr Comp Biol* 56:842–852.

Porter, M. L., J. R. Blasic, M. J. Bok, E. G. Cameron, T. Pringle, T. W. Cronin, and P. R. Robinson. 2012. Shedding new light on opsin evolution. *Proc Biol Sci* 279:3–14.

Posnien, N., N. Koniszewski, and G. Bucher. 2011. Insect Tc-six4 marks a unit with similarity to vertebrate placodes. *Dev Biol* 350:208–216.

Posnien, N., N. D. Koniszewski, H. J. Hein, and G. Bucher. 2011. Candidate gene screen in the red flour beetle *Tribolium* reveals six3 as ancient regulator of anterior median head and central complex development. *PLoS Genet* 7:e1002416.

Proske, U., J. E. Gregory, and A. Iggo. 1998. Sensory receptors in monotremes. *Philos Trans R Soc Lond B Biol Sci* 353:1187–1198.

Provencio, I., G. Jiang, W. J. De Grip, W. P. Hayes, and M. D. Rollag. 1998. Melanopsin: an opsin in melanophores, brain, and eye. *Proc Natl Acad Sci USA* 95:340–345.

Pulido, C., G. Malagón, C. Ferrer, J. K. Chen, J. M. Angueyra, E. Nasi, and P. Gómez Mdel. 2012. The light-sensitive conductance of melanopsin-expressing Joséph and Hesse cells in amphioxus. *J Gen Physiol* 139:19–30.

Purnell, M. A. 2001. Scenarios, selection and the ecology of early vertebrates. In *Major Eventsin Early Vertebrate Evolution: Palaeontology, Phylogeny, Genetics, and Development. Systematics Association Special Volume Series, no. 61*, edited by P. E. Ahlberg, 188–208. London: Taylor and Francis.

Purschke, G. 2005. Sense organs in polychaetes (Annelida). *Hydrobiologia* 535:53–78.

Purschke, G. 2015. Annelida: Basal groups and pleistoannelida. In *Structure and Evolution of Invertebrate Nervous Systems*, edited by A. Schmidt-Rhaesa, S. Harzsch and G. Purschke, 254-312. Oxford: Oxford Univ Press.

Putnam, N. H., T. Butts, D. E. Ferrier, R. F. Furlong, U. Hellsten, T. Kawashima, M. Robinson-Rechavi, E. Shoguchi, A. Terry, J. K. Yu, E. L. Benito-Gutiérrez, I. Dubchak, J. Garcia-Fernàndez, J. J. Gibson-Brown, I. V. Grigoriev, A. C. Horton, P. J. de Jong, J. Jurka, V. V. Kapitonov, Y. Kohara, Y. Kuroki, E. Lindquist, S. Lucas, K. Osoegawa, L. A. Pennacchio, A. A. Salamov, Y. Satou, T. Sauka-Spengler, J. Schmutz, I. Shin, A. Toyoda, M. Bronner-Fraser, A. Fujiyama, L. Z. Holland, P. W. Holland, N. Satoh, and D. S. Rokhsar. 2008. The amphioxus genome and the evolution of the chordate karyotype. *Nature* 453:1064–1071.

Pyenson, N. D., J. A. Goldbogen, A. W. Vogl, G. Szathmary, R. L. Drake, and R. E. Shadwick. 2012. Discovery of a sensory organ that coordinates lunge feeding in rorqual whales. *Nature* 485:498–501.

Qian, J., N. Esumi, Y. Chen, Q. Wang, I. Chowers, and D. J. Zack. 2005. Identification of regulatory targets of tissue-specific transcription factors: application to retina-specific gene regulation. *Nucleic Acids Res* 33:3479–3491.

Qin, H., and J. A. Powell-Coffman. 2004. The *Caenorhabditis elegans* aryl hydrocarbon receptor, AHR-1, regulates neuronal development. *Dev Biol* 270:64–75.

Quick, K., J. Zhao, N. Eijkelkamp, J. E. Linley, F. Rugiero, J. J. Cox, R. Raouf, M. Gringhuis, J. E. Sexton, J. Abramowitz, R. Taylor, A. Forge, J. Ashmore, N. Kirkwood, C. J. Kros, G. P. Richardson, M. Freichel, V. Flockerzi, L. Birnbaumer, and J. N. Wood. 2012. TRPC3 and TRPC6 are essential for normal mechanotransduction in subsets of sensory neurons and cochlear hair cells. *Open Biol* 2:120068.

Quiring, R., U. Walldorf, U. Kloter, and W. J. Gehring. 1994. Homology of the eyeless gene of *Drosophila* to the Small eye gene in mice and Aniridia in humans. *Science* 265:785–789.

Quiroga Artigas, G., P. Lapebie, L. Leclère, N. Takeda, R. Deguchi, G. Jékely, T. Momose, and E. Houliston. 2018. A gonad-expressed opsin mediates light-induced spawning in the jellyfish Clytia. *Elife* 7:e29555.

Racioppi, C., A. K. Kamal, F. Razy-Krajka, G. Gambardella, L. Zanetti, D. di Bernardo, R. Sanges, L. A. Christiaen, and F. Ristoratore. 2014. Fibroblast growth factor signalling controls nervous system patterning and pigment cell formation in *Ciona intestinalis*. *Nat Commun* 5:4830.

Raft, S., and A. K. Groves. 2015. Segregating neural and mechanosensory fates in the developing ear: patterning, signaling, and transcriptional control. *Cell Tissue Res* 359:315–332.

Raft, S., E. J. Koundakjian, H. Quinones, C. S. Jayasena, L. V. Goodrich, J. E. Johnson, N. Segil, and A. K. Groves. 2007. Cross-regulation of Ngn1 and Math1 coordinates the production of neurons and sensory hair cells during inner ear development. *Development* 134:4405–4415.

Raible, F., K. Tessmar-Raible, E. Arboleda, T. Kaller, P. Bork, D. Arendt, and M. I. Arnone. 2006. Opsins and clusters of sensory G-protein-coupled receptors in the sea urchin genome. *Dev Biol* 300:461–475.

Ramirez, M. D., A. N. Pairett, M. S. Pankey, J. M. Serb, D. I. Speiser, A. J. Swafford, and T. H. Oakley. 2016. The last common ancestor of most bilaterian animals possessed at least nine opsins. *Genome Biol Evol* 8:3640–3652.

Ramirez, M. D., D. I. Speiser, M. S. Pankey, and T. H. Oakley. 2011. Understanding the dermal light sense in the context of integrative photoreceptor cell biology. *Vis Neurosci* 28:265–279.

Ranade, S. S., D. Yang-Zhou, S. W. Kong, E. C. McDonald, T. A. Cook, and F. Pignoni. 2008. Analysis of the Otd-dependent transcriptome supports the evolutionary conservation of CRX/OTX/OTD functions in flies and vertebrates. *Dev Biol* 315:521–534.

Randel, N., L. A. Bezares-Calderón, M. Gühmann, R. Shahidi, and G. Jékely. 2013. Expression dynamics and protein localization of rhabdomeric opsins in *Platynereis* larvae. *Integr Comp Biol* 53:7–16.

Randel, N., and G. Jékely. 2016. Phototaxis and the origin of visual eyes. *Philos Trans R Soc Lond B Biol Sci*. 371:20150042.

Rasmussen, S. L., L. Z. Holland, M. Schubert, L. Beaster-Jones, and N. D. Holland. 2007. Amphioxus AmphiDelta: evolution of delta protein structure, segmentation, and neurogenesis. *Genesis* 45:113–122.

Rawlinson, K. A., F. Lapraz, E. R. Ballister, M. Terasaki, J. Rodgers, R. J. McDowell, J. Girstmair, K. E. Criswell, M. Boldogkoi, F. Simpson, D. Goulding, C. Cormie, B. Hall, R. J. Lucas, and M. J. Telford. 2019. Extraocular, rod-like photoreceptors in a flatworm express xenopsin photopigment. *Elife* 8:e45465.

Reddy, G. V., and V. Rodrigues. 1999. Sibling cell fate in the *Drosophila* adult external sense organ lineage is specified by prospero function, which is regulated by Numb and Notch. *Development* 126:2083–2092.

Reim, I., H. H. Lee, and M. Frasch. 2003. The T-box-encoding Dorsocross genes function in amnioserosa development and the patterning of the dorsolateral germ band downstream of Dpp. *Development* 130:3187–3204.

Reinecke, M. 1981. Immunohistochemical localization of polypeptide hormones in endocrine cells of the digestive tract of *Branchiostoma lanceolatum*. *Cell Tissue Res* 219:445–456.

Renard, E., J. Vacelet, E. Gazave, P. Lapebie, C. Borchiellini, and A. V. Ereskovsky. 2009. Origin of the neuro-sensory system: new and expected insights from sponges. *Integr Zool* 4:294–308.

Rentzsch, F., J. H. Fritzenwanker, C. B. Scholz, and U. Technau. 2008. FGF signalling controls formation of the apical sensory organ in the cnidarian *Nematostella vectensis*. *Development* 135:1761–1769.

Rentzsch, F., C. Juliano, and B. Galliot. 2019. Modern genomic tools reveal the structural and cellular diversity of cnidarian nervous systems. *Curr Opin Neurobiol* 56:87–96.

Rentzsch, F., M. Layden, and M. Manuel. 2017. The cellular and molecular basis of cnidarian neurogenesis. *Wiley Interdiscip Rev Dev Biol* 6:e257.

Richards, G. S., and F. Rentzsch. 2014. Transgenic analysis of a SoxB gene reveals neural progenitor cells in the cnidarian *Nematostella vectensis*. *Development* 141:4681–4689.

Richards, G. S., and F. Rentzsch. 2015. Regulation of *Nematostella* neural progenitors by SoxB, Notch and bHLH genes. *Development* 142:3332–3342.

Richards, G. S., E. Simionato, M. Perron, M. Adamska, M. Vervoort, and B. M. Degnan. 2008. Sponge genes provide new insight into the evolutionary origin of the neurogenic circuit. *Curr Biol* 18:1156–1161.

Riddiford, N., and G. Schlosser. 2016. Dissecting the pre-placodal transcriptome to reveal presumptive direct targets of Six1 and Eya1 in cranial placodes. *Elife* 5:e17666.

Riedl, R. 1975. *Die Ordnung des Lebendigen*. Hamburg: Parey.

Rigon, F., F. Gasparini, S. M. Shimeld, S. Candiani, and L. Manni. 2018. Developmental signature, synaptic connectivity and neurotransmission are conserved between vertebrate hair cells and tunicate coronal cells. *J Comp Neurol* 526:957–971.

Rigon, F., T. Stach, F. Caicci, F. Gasparini, P. Burighel, and L. Manni. 2013. Evolutionary diversification of secondary mechanoreceptor cells in Tunicata. *BMC Evol Biol* 13:112.

Rindi, G., A. B. Leiter, A. S. Kopin, C. Bordi, and E. Solcia. 2004. The "normal" endocrine cell of the gut: changing concepts and new evidences. *Ann N Y Acad Sci* 1014:1–12.

Rivera, A. S., N. Ozturk, B. Fahey, D. C. Plachetzki, B. M. Degnan, A. Sancar, and T. H. Oakley. 2012. Blue-light-receptive cryptochrome is expressed in a sponge eye lacking neurons and opsin. *J Exp Biol* 215:1278–1286.

Rivera, A., I. Winters, A. Rued, S. Ding, D. Posfai, B. Cieniewicz, K. Cameron, L. Gentile, and A. Hill. 2013. The evolution and function of the Pax/Six regulatory network in sponges. *Evol Dev* 15:186–196.

Rizzoti, K. 2015. Genetic regulation of murine pituitary development. *J Mol Endocrinol* 54:R55–R73.

Roach, G., R. Heath Wallace, A. Cameron, R. Emrah Ozel, C. F. Hongay, R. Baral, S. Andreescu, and K. N. Wallace. 2013. Loss of ascl1a prevents secretory cell differentiation within the zebrafish intestinal epithelium resulting in a loss of distal intestinal motility. *Dev Biol* 376:171–186.

Robertson, H. M. 2015. The insect chemoreceptor superfamily is ancient in animals. *Chem Senses* 40:609–614.

Robertson, H. M. 2019. Molecular evolution of the major arthropod chemoreceptor gene families. *Annu Rev Entomol* 64:227–242.

Robertson, H. M., and J. H. Thomas. 2006. The putative chemoreceptor families of *C. elegans*. *WormBook* ed. The C. elegans Research Community, WormBook, doi: 10.1895/wormbook.1.66.1:1–12.

Roch, G. J., E. R. Busby, and N. M. Sherwood. 2011. Evolution of GnRH: diving deeper. *Gen Comp Endocrinol* 171:1–16.

Roch, G. J., and N. M. Sherwood. 2014. Glycoprotein hormones and their receptors emerged at the origin of metazoans. *Genome Biol Evol* 6:1466–1479.

Roch, G. J., J. A. Tello, and N. M. Sherwood. 2014. At the transition from invertebrates to vertebrates, a novel GnRH-like peptide emerges in amphioxus. *Mol Biol Evol* 31:765–778.

Röttinger, E., and C. J. Lowe. 2012. Evolutionary crossroads in developmental biology: hemichordates. *Development* 139:2463–2475.

Röhlich, P., S. Viragh, and B. Aros. 1970. Fine structure of photoreceptor cells in the earthworm, *Lumbricus terrestris*. *Z Zellforsch Mikrosk Anat* 104:345–357.

Romer, A. S. 1959. *The vertebrate story*. Chicago: University of Chicago Press.

Romer, A. S. 1972. The vertebrate as a dual animal—somatic and visceral. In *Evolutionary biology*, edited by T. Dobzhansky, M. K. Hecht and W. C. Steere, 121–156. New York: Appleton-Century-Crofts.

Roth, G., and M. F. Wullimann. 1996. Evolution der Nervensysteme und der Sinnesorgane. In *Neurowissenschaft: Vom Molekül zur Kognition*, edited by J. Dudel, R. Menzel and R. F. Schmidt, 1–31. Berlin: Springer.

Roth, V. L. 1988. The biological basis of homology. In *Ontogeny and systematics*, edited by C. J. Humphries, 1–26. New York: Columbia University Press.

Roth, V. L. 1991. Homology and hierarchies: problems solved and unresolved. *J Evol Biol* 4:167–194.

Rothbächer, U., V. Bertrand, C. Lamy, and P. Lemaire. 2007. A combinatorial code of maternal GATA, Ets and {beta}-catenin-TCF transcription factors specifies and patterns the early ascidian ectoderm. *Development* 134:4023–4032.

Roure, A., P. Lemaire, and S. Darras. 2014. An otx/nodal regulatory signature for posterior neural development in ascidians. *PLoS Genet* 10:e1004548.

Rouse, G. W., N. G. Wilson, J. I. Carvajal, and R. C. Vrijenhoek. 2016. New deep-sea species of *Xenoturbella* and the position of Xenacoelomorpha. *Nature* 530:94–97.

Rowe, M. L., and M. R. Elphick. 2012. The neuropeptide transcriptome of a model echinoderm, the sea urchin *Strongylocentrotus purpuratus*. *Gen Comp Endocrinol* 179:331–344.

Ruiz-Trillo, I., M. Rintort, D. T. J. Littlewood, E. A. Hernion, and J. Baguñà. 1999. Acoel flatworms: earliest extant bilaterian metazoans, not members of platyhelminthes. *Science* 283:1919–1923.

Ruiz, M. S., and R. Anadón. 1991a. Some considerations on the fine structure of rhabdomeric photoreceptors in the amphioxus, *Branchiostoma lanceolatum* (Cephalochordata). *J Hirnforsch* 32:159–164.

Ruiz, S., and R. Anadón. 1991b. The fine structure of lamellate cells in the brain of amphioxus (*Branchiostoma lanceolatum*, Cephalochordata). *Cell Tissue Res* 263:597–600.

Ruppert, E. E. 1990. Structure, ultrastructure and function of the neural gland complex of *Ascidia interrupta* (Chordata, Ascidiacea): clarification of hypotheses regarding the evolution of the vertebrate anterior pituitary. *Acta Zool* 71:135–149.

Ruppert, E. E. 1997. Cephalochordata (Acrania). In *Microscopic anatomy of invertebrates, Vol. 15: Hemichordata, Chaetognatha, and the invertebrate chordates*, edited by F. W. Harrison and E. E. Ruppert, 349–504. New York: Wiley-Liss.

Ruvinsky, I., L. M. Silver, and J. J. Gibson-Brown. 2000. Phylogenetic analysis of T-Box genes demonstrates the importance of amphioxus for understanding evolution of the vertebrate genome. *Genetics* 156:1249–1257.

Ryan, J. F., and M. Chiodin. 2015. Where is my mind? How sponges and placozoans may have lost neural cell types. *Philos Trans R Soc Lond B Biol Sci* 370:20150059.

Ryan, J. F., K. Pang, C. E. Schnitzler, A. D. Nguyen, R. T. Moreland, D. K. Simmons, B. J. Koch, W. R. Francis, P. Havlak, S. A. Smith, N. H. Putnam, S. H. Haddock, C. W. Dunn, T. G. Wolfsberg, J. C. Mullikin, M. Q. Martindale, and A. D. Baxevanis. 2013. The genome of the ctenophore *Mnemiopsis leidyi* and its implications for cell type evolution. *Science* 342:1242592.

Ryan, K., Z. Lu, and I. A. Meinertzhagen. 2016. The CNS connectome of a tadpole larva of *Ciona intestinalis* (L.) highlights sidedness in the brain of a chordate sibling. *Elife* 5:e16962.

Ryan, K., Z. Lu, and I. A. Meinertzhagen. 2018. The peripheral nervous system of the ascidian tadpole larva: types of neurons and their synaptic networks. *J Comp Neurol* 526:583–608.

Ryan, K., and I. A. Meinertzhagen. 2019. Neuronal identity: the neuron types of a simple chordate sibling, the tadpole larva of *Ciona intestinalis*. *Curr Opin Neurobiol* 56:47–60.

Rybina, O. Y., M. I. Schelkunov, E. R. Veselkina, S. V. Sarantseva, A. V. Krementsova, M. Y. Vysokikh, P. A. Melentev, M. A. Volodina, and E. G. Pasyukova. 2019. Knockdown of the neuronal gene Lim3 at the early stages of development affects mitochondrial function and lifespan in *Drosophila*. *Mech Ageing Dev* 181:29–41.

Rychel, A. L., S. E. Smith, H. T. Shimamoto, and B. J. Swalla. 2006. Evolution and development of the chordates: collagen and pharyngeal cartilage. *Mol Biol Evol* 23:541–549.

Safieddine, S., A. El-Amraoui, and C. Petit. 2012. The auditory hair cell ribbon synapse: from assembly to function. *Annu Rev Neurosci* 35:509–528.

Sagane, Y., J. Hosp, K. Zech, and E. M. Thompson. 2011. Cytoskeleton-mediated templating of complex cellulose-scaffolded extracellular structure and its association with oikosins in the urochordate *Oikopleura*. *Cell Mol Life Sci* 68:1611–1622.

Sahlin, K., and R. Olsson. 1986. The wheel organ and Hatschek's groove in the lancelet, *Branchiostoma lanceolatum* (Cephalochordata). *Acta Zool* 67:201–209.

Saina, M., H. Busengdal, C. Sinigaglia, L. Petrone, P. Oliveri, F. Rentzsch, and R. Benton. 2015. A cnidarian homologue of an insect gustatory receptor functions in developmental body patterning. *Nat Commun* 6:6243.

Sakai, T., M. Aoyama, T. Kawada, T. Kusakabe, M. Tsuda, and H. Satake. 2012. Evidence for differential regulation of GnRH signaling via heterodimerization among GnRH receptor paralogs in the protochordate, *Ciona intestinalis*. *Endocrinology* 153:1841–1849.

Sakai, T., M. Aoyama, T. Kusakabe, M. Tsuda, and H. Satake. 2010. Functional diversity of signaling pathways through G protein-coupled receptor heterodimerization with a species-specific orphan receptor subtype. *Mol Biol Evol* 27:1097–1106.

Sakai, T., A. Shiraishi, T. Kawada, S. Matsubara, M. Aoyama, and H. Satake. 2017. Invertebrate gonadotropin-releasing hormone-related peptides and their receptors: an update. *Front Endocrinol (Lausanne)* 8:217.

Sakarya, O., K. A. Armstrong, M. Adamska, M. Adamski, I. F. Wang, B. Tidor, B. M. Degnan, T. H. Oakley, and K. S. Kosik. 2007. A post-synaptic scaffold at the origin of the animal kingdom. *PLoS One* 2:e506.

Sakurai, D., M. Goda, Y. Kohmura, T. Horie, H. Iwamoto, H. Ohtsuki, and M. Tsuda. 2004. The role of pigment cells in the brain of ascidian larva. *J Comp Neurol* 475:70–82.

Salas, P., V. Vinaithirthan, E. Newman-Smith, M. J. Kourakis, and W. C. Smith. 2018. Photoreceptor specialization and the visuomotor repertoire of the primitive chordate *Ciona*. *J Exp Biol* 221:jeb177972.

Salo, E., D. Pineda, M. Marsal, J. Gonzalez, V. Gremigni, and R. Batistoni. 2002. Genetic network of the eye in Platyhelminthes: expression and functional analysis of some players during planarian regeneration. *Gene* 287:67–74.

Salvini-Plawen, L. V., and E. Mayr. 1977. *On the evolution of photoreceptors and eyes.* New York: Plenum Press.

Salzet, M., B. Salzet-Raveillon, C. Cocquerelle, M. Verger-Bocquet, S. C. Pryor, C. M. Rialas, V. Laurent, and G. B. Stefano. 1997. Leech immunocytes contain proopiomelanocortin: nitric oxide mediates hemolymph proopiomelanocortin processing. *J Immunol* 159:5400–5411.

Sansom, R. S., S. E. Gabbott, and M. A. Purnell. 2010. Non-random decay of chordate characters causes bias in fossil interpretation. *Nature* 463:797–800.

Santagata, S., C. Resh, A. Hejnol, M. Q. Martindale, and Y. J. Passamaneck. 2012. Development of the larval anterior neurogenic domains of *Terebratalia transversa* (Brachiopoda) provides insights into the diversification of larval apical organs and the spiralian nervous system. *Evodevo* 3:3.

Sasai, Y., B. Lu, H. Steinbeisser, and E. M. De Robertis. 1995. Regulation of neural induction by the Chd and Bmp-4 antagonistic patterning signals in *Xenopus. Nature* 376:333–336.

Sasakura, Y., and A. Hozumi. 2018. Formation of adult organs through metamorphosis in ascidians. *Wiley Interdiscip Rev Dev Biol* 7:doi: 10.1002/wdev.304.

Sasakura, Y., K. Mita, Y. Ogura, and T. Horie. 2012. Ascidians as excellent chordate models for studying the development of the nervous system during embryogenesis and metamorphosis. *Dev Growth Differ* 54:420–437.

Satoh, G. 2005. Characterization of novel GPCR gene coding locus in amphioxus genome: gene structure, expression, and phylogenetic analysis with implications for its involvement in chemoreception. *Genesis* 41:47–57.

Satoh, G., Y. Wang, P. Zhang, and N. Satoh. 2001. Early development of amphioxus nervous system with special reference to segmental cell organization and putative sensory cell precursors: a study based on the expression of pan-neuronal marker gene Hu/elav. *J Exp Zool* 291:354–364.

Satoh, N. 1994. *Developmental biology of ascidians.* New York: Cambridge University Press.

Satoh, N. 2009. An advanced filter-feeder hypothesis for urochordate evolution. *Zoolog Sci* 26:97–111.

Satoh, T., and D. M. Fekete. 2005. Clonal analysis of the relationships between mechanosensory cells and the neurons that innervate them in the chicken ear. *Development* 132:1687–1697.

Satou, Y., K. S. Imai, and N. Satoh. 2001. Early embryonic expression of a LIM-homeobox gene Cs-lhx3 is downstream of beta-catenin and responsible for the endoderm differentiation in *Ciona savignyi* embryos. *Development* 128:3559–3570.

Saudemont, A., N. Dray, B. Hudry, M. Le Gouar, M. Vervoort, and G. Balavoine. 2008. Complementary striped expression patterns of NK homeobox genes during segment formation in the annelid *Platynereis. Dev Biol* 317:430–443.

Saudemont, A., E. Haillot, F. Mekpoh, N. Bessodes, M. Quirin, F. Lapraz, V. Duboc, E. Röttinger, R. Range, A. Oisel, L. Besnardeau, P. Wincker, and T. Lepage. 2010. Ancestral regulatory circuits governing ectoderm patterning downstream of Nodal and BMP2/4 revealed by gene regulatory network analysis in an echinoderm. *PLoS Genet* 6:e1001259.

Sauka-Spengler, T., and M. Bronner-Fraser. 2008. Evolution of the neural crest viewed from a gene regulatory perspective. *Genesis* 46:673–682.

Scalettar, B. A. 2006. How neurosecretory vesicles release their cargo. *Neuroscientist* 12:164–176.

Schaeper, N. D., N. M. Prpic, and E. A. Wimmer. 2010. A clustered set of three Sp-family genes is ancestral in the Metazoa: evidence from sequence analysis, protein domain structure, developmental expression patterns and chromosomal location. *BMC Evol Biol* 10:88.

Schlosser, G. 2002a. Development and evolution of lateral line placodes in amphibians. I Development. *Zoology* 105:119–146.

Schlosser, G. 2002b. Development and evolution of lateral line placodes in amphibians. II. Evolutionary diversification. *Zoology* 105:177–193.

Schlosser, G. 2002c. Modularity and the units of evolution. *Theory Biosci* 121:1–80.

Schlosser, G. 2004. The role of modules in development and evolution. In *Modularity in development and evolution*, edited by G. Schlosser and G. P. Wagner, 519–582. Chicago: University of Chicago Press.

Schlosser, G. 2005. Evolutionary origins of vertebrate placodes: insights from developmental studies and from comparisons with other deuterostomes. *J Exp Zoolog B Mol Dev Evol* 304B:347–399.

Schlosser, G. 2006. Induction and specification of cranial placodes. *Dev Biol* 294:303–351.

Schlosser, G. 2007. How old genes make a new head: redeployment of Six and Eya genes during the evolution of vertebrate cranial placodes. *Integr Comp Biol* 47:343–359.

Schlosser, G. 2008. Do vertebrate neural crest and cranial placodes have a common evolutionary origin? *Bioessays* 30:659–672.

Schlosser, G. 2011. Wie alte Gene neue Sinne machen – Sinnesevolution bei Wirbeltieren. *Biospektrum* 17 630–633.

Schlosser, G. 2015. Vertebrate cranial placodes as evolutionary innovations – the ancestor's tale. *Curr Top Dev Biol* 111:235–300.

Schlosser, G. 2017. From so simple a beginning – what amphioxus can teach us about placode evolution. *Int J Dev Biol* 61:633–648.

Schlosser, G. 2018. A short history of nearly every sense – the evolutionary history of vertebrate sensory cell types. *Integr Comp Biol* 58:301–316.

Schlosser, G. 2019. Treacherous trees: trials and tribulations in tracing the trajectories of traits. In *Perspectives on evolutionary and developmental biology. Essays for Alessandro Minelli*, edited by G. Fusco, 379–388. Padova: Padova University Press.

Schlosser, G. 2021. *Development of sensory and neurosecretory cell types. Vertebrate cranial placodes*, Vol. 1. Boca Raton: CRC Press.

Schlosser, G., and K. Ahrens. 2004. Molecular anatomy of placode development in *Xenopus laevis*. *Dev Biol* 271:439–466.

Schlosser, G., T. Awtry, S. A. Brugmann, E. D. Jensen, K. Neilson, G. Ruan, A. Stammler, D. Voelker, B. Yan, C. Zhang, M. W. Klymkowsky, and S. A. Moody. 2008. Eya1 and Six1 promote neurogenesis in the cranial placodes in a SoxB1-dependent fashion. *Dev Biol* 320:199–214.

Schlosser, G., C. Patthey, and S. M. Shimeld. 2014. The evolutionary history of vertebrate cranial placodes II. Evolution of ectodermal patterning. *Dev Biol* 389:98–119.

Schmitz, F., A. Königstorfer, and T. C. Südhof. 2000. RIBEYE, a component of synaptic ribbons: a protein's journey through evolution provides insight into synaptic ribbon function. *Neuron* 28:857–872.

Schnitzler, C. E., K. Pang, M. L. Powers, A. M. Reitzel, J. F. Ryan, D. Simmons, T. Tada, M. Park, J. Gupta, S. Y. Brooks, R. W. Blakesley, S. Yokoyama, S. H. Haddock, M. Q. Martindale, and A. D. Baxevanis. 2012. Genomic organization, evolution, and expression of photoprotein and opsin genes in *Mnemiopsis leidyi*: a new view of ctenophore photocytes. *BMC Biol* 10:107.

Schubert, M., H. Escriva, J. Xavier-Neto, and V. Laudet. 2006. Amphioxus and tunicates as evolutionary model systems. *Trends Ecol Evol* 21:269–277.

Schubert, M., L. Z. Holland, M. D. Stokes, and N. D. Holland. 2001. Three amphioxus Wnt genes (AmphiWnt3, AmphiWnt5, and AmpjWnt6) associated with the tailbud: the evolution of somitogenesis in chordates. *Dev Biol* 240:262–273.

Schubert, M., N. D. Holland, H. Escriva, L. Z. Holland, and V. Laudet. 2004. Retinoic acid influences anteroposterior positioning of epidermal sensory neurons and their gene expression in a developing chordate (amphioxus). *Proc Natl Acad Sci USA* 101:10320–10325.

Schubert, M., N. D. Holland, V. Laudet, and L. Z. Holland. 2006. A retinoic acid-Hox hierarchy controls both anterior/posterior patterning and neuronal specification in the developing central nervous system of the cephalochordate amphioxus. *Dev Biol* 296:190–202.

Schüler, A., G. Schmitz, A. Reft, S. Özbek, U. Thurm, and E. Bornberg-Bauer. 2015. The rise and fall of TRP-N, an ancient family of mechanogated ion channels, in Metazoa. *Genome Biol Evol* 7:1713–1727.

Schulte, E., and R. Riehl. 1977. Elektronenmikroskopische Untersuchungen an den Oralcirren und der Haut von *Branchiostoma lanceolatum*. *Helgoland wiss Meeresunters* 29:337–357.

Schwartz, H. T., and H. R. Horvitz. 2007. The *C. elegans* protein CEH-30 protects male-specific neurons from apoptosis independently of the Bcl-2 homolog CED-9. *Genes Dev* 21:3181–3194.

Sebé-Pedrós, A., E. Chomsky, K. Pang, D. Lara-Astiaso, F. Gaiti, Z. Mukamel, I. Amit, A. Hejnol, B. M. Degnan, and A. Tanay. 2018. Early metazoan cell type diversity and the evolution of multicellular gene regulation. *Nat Ecol Evol* 2:1176–1188.

Sebé-Pedrós, A., B. M. Degnan, and I. Ruiz-Trillo. 2017. The origin of Metazoa: a unicellular perspective. *Nat Rev Genet* 18:498–512.

Sebé-Pedrós, A., B. Saudemont, E. Chomsky, F. Plessier, M. P. Mailhe, J. Renno, Y. Loe-Mie, A. Lifshitz, Z. Mukamel, S. Schmutz, S. Novault, P. R. H. Steinmetz, F. Spitz, A. Tanay, and H. Marlow. 2018. Cnidarian cell type diversity and regulation revealed by whole-organism single-cell RNA-Seq. *Cell* 173:1520–1534 e20.

Seimiya, M., H. Ishiguro, K. Miura, Y. Watanabe, and Y. Kurosawa. 1994. Homeobox-containing genes in the most primitive metazoa, the sponges. *Eur J Biochem* 221:219–225.

Seipel, K., N. Yanze, and V. Schmid. 2004. Developmental and evolutionary aspects of the basic helix-loop-helix transcription factors Atonal-like 1 and Achaete-scute homolog 2 in the jellyfish. *Dev Biol* 269:331–345.

Sellami, A., H. J. Agricola, and J. A. Veenstra. 2011. Neuroendocrine cells in Drosophila melanogaster producing GPA2/GPB5, a hormone with homology to LH, FSH and TSH. *Gen Comp Endocrinol* 170:582–588.

Sen, S., H. Reichert, and K. VijayRaghavan. 2013. Conserved roles of ems/Emx and otd/Otx genes in olfactory and visual system development in *Drosophila* and mouse. *Open Biol* 3:120177.

Sensenbaugh, T., and A. Franzén. 1987. Fine structural observations of the apical organ in the larva of *Polygordius* (Annelida; Polychaeta). *Scanning Microsc* 1:181–189.

Seo, H. C., J. Curtiss, M. Mlodzik, and A. FJosé. 1999. Six class homeobox genes in *Drosophila* belong to three distinct families and are involved in head development. *Mech Dev* 83:127–139.

Seo, H. C., R. B. Edvardsen, A. D. Maeland, M. Bjordal, M. F. Jensen, A. Hansen, M. Flaat, J. Weissenbach, H. Lehrach, P. Wincker, R. Reinhardt, and D. Chourrout. 2004. Hox cluster disintegration with persistent anteroposterior order of expression in *Oikopleura dioica*. *Nature* 431:67–71.

Serb, J. M., and T. H. Oakley. 2005. Hierarchical phylogenetics as a quantitative analytical framework for evolutionary developmental biology. *Bioessays* 27:1158–1166.

Serikaku, M. A., and J. E. O'Tousa. 1994. sine oculis is a homeobox gene required for *Drosophila* visual system development. *Genetics* 138:1137–1150.

Serizawa, S., K. Miyamichi, H. Nakatani, M. Suzuki, M. Saito, Y. Yoshihara, and H. Sakano. 2003. Negative feedback regulation ensures the one receptor-one olfactory neuron rule in mouse. *Science* 302:2088–2094.

Serrano-Saiz, E., E. Leyva-Díaz, E. De La Cruz, and O. Hobert. 2018. BRN3-type POU homeobox genes maintain the identity of mature postmitotic neurons in nematodes and mice. *Curr Biol* 28:2813–2823 e2.

Servetnick, M., and R. M. Grainger. 1991. Changes in neural and lens competence in *Xenopus* ectoderm: evidence for an autonomous developmental timer. *Development* 112:177–188.

Sexton, J. E., T. Desmonds, K. Quick, R. Taylor, J. Abramowitz, A. Forge, C. J. Kros, L. Birnbaumer, and J. N. Wood. 2016. The contribution of TRPC1, TRPC3, TRPC5 and TRPC6 to touch and hearing. *Neurosci Lett* 610:36–42.

Sharma, S., W. Wang, and A. Stolfi. 2019. Single-cell transcriptome profiling of the *Ciona* larval brain. *Dev Biol* 448:226–236.

Sharman, A. C., S. M. Shimeld, and P. W. H. Holland. 1999. An amphioxus Msx gene expressed predominantly in the dorsal neural tube. *Dev Genes Evol* 209:260–263.

Sheng, G., E. Thouvenot, D. Schmucker, D. S. Wilson, and C. Desplan. 1997. Direct regulation of rhodopsin 1 by Pax-6/eyeless in *Drosophila*: evidence for a conserved function in photoreceptors. *Genes Dev* 11:1122–1131.

Sherwood, N. M., B. A. Adams, and J. A. Tello. 2005. Endocrinology of protochordates. *Can J Zool* 83:225–255.

Shichida, Y., and T. Matsuyama. 2009. Evolution of opsins and phototransduction. *Philos Trans R Soc Lond B Biol Sci* 364:2881–2895.

Shimeld, S. M. 1997. Characterisation of amphioxus HNF-3 genes: conserved expression in the notochord and floor plate. *Dev Biol* 183:74–85.

Shimeld, S. M., M. J. Boyle, T. Brunet, G. N. Luke, and E. C. Seaver. 2010. Clustered Fox genes in lophotrochozoans and the evolution of the bilaterian Fox gene cluster. *Dev Biol* 340:234–248.

Shimeld, S. M., A. G. Purkiss, R. P. Dirks, O. A. Bateman, C. Slingsby, and N. H. Lubsen. 2005. Urochordate betagamma-crystallin and the evolutionary origin of the vertebrate eye lens. *Curr Biol* 15:1684–1689.

Shimizu, T., and M. Hibi. 2009. Formation and patterning of the forebrain and olfactory system by zinc-finger genes Fezf1 and Fezf2. *Dev Growth Differ* 51:221–231.

Shiomi, K., Y. Fujiwara, Y. Yasukochi, Z. Kajiura, M. Nakagaki, and T. Yaginuma. 2007. The Pitx homeobox gene in *Bombyx mori*: regulation of DH-PBAN neuropeptide hormone gene expression. *Mol Cell Neurosci* 34:209–218.

Shirasaki, R., and S. L. Pfaff. 2002. Transcriptional codes and the control of neuronal identity. *Annu Rev Neurosci.* 25:251–281.

Shu, D.-G., H.-L. Luo, S. Conway Morris, X.-L. Zhang, S.-X. Hu, L. Chen, J. Han, M. Zhu, Y. Li, and L.-Z. Chen. 1999. Lower Cambrian vertebrates from south China. *Nature* 402:42–46.

Shu, D. G., S. C. Morris, J. Han, Z. F. Zhang, K. Yasui, P. Janvier, L. Chen, X. L. Zhang, J. N. Liu, Y. Li, and H. Q. Liu. 2003. Head and backbone of the Early Cambrian vertebrate *Haikouichthys*. *Nature* 421:526–529.

Shu, D., S. C. Morris, Z. F. Zhang, J. N. Liu, J. Han, L. Chen, X. L. Zhang, K. Yasui, and Y. Li. 2003. A new species of yunnanozoan with implications for deuterostome evolution. *Science* 299:1380–1384.

Shu, D., C. Zhang, and L. Chen. 1996. Reinterpretation of Yunnanozoon as the earliest known hemichordate. *Nature* 380:428–430.

Shubin, N., C. Tabin, and S. Carroll. 2009. Deep homology and the origins of evolutionary novelty. *Nature* 457:818–823.

Siebert, S., J. A. Farrell, J. F. Cazet, Y. Abeykoon, A. S. Primack, C. E. Schnitzler, and C. E. Juliano. 2019. Stem cell differentiation trajectories in *Hydra* resolved at single-cell resolution. *Science* 365:eaav9314.

Sigg, M. A., T. Menchen, C. Lee, J. Johnson, M. K. Jungnickel, S. P. Choksi, G. Garcia, 3rd, H. Busengdal, G. W. Dougherty, P. Pennekamp, C. Werner, F. Rentzsch, H. M. Florman, N. Krogan, J. B. Wallingford, H. Omran, and J. F. Reiter. 2017. Evolutionary proteomics uncovers ancient associations of cilia with signaling pathways. *Dev Cell* 43:744–762 e11.

Silva, L., and A. Antunes. 2017. Vomeronasal receptors in vertebrates and the evolution of pheromone detection. *Annu Rev Anim Biosci* 5:353–370.

Silver, S. J., and I. Rebay. 2005. Signaling circuitries in development: insights from the retinal determination gene network. *Development* 132:3–13.

Simeone, A., M. R. D'Apice, V. Nigro, J. Casanova, F. Graziani, D. Acampora, and V. Avantaggiato. 1994. Orthopedia, a novel homeobox-containing gene expressed in the developing CNS of both mouse and *Drosophila*. *Neuron* 13:83–101.

Simion, P., H. Philippe, D. Baurain, M. Jager, D. J. Richter, A. Di Franco, B. Roure, N. Satoh, E. Queinnec, A. Ereskovsky, P. Lapebie, E. Corre, F. Delsuc, N. King, G. Wörheide, and M. Manuel. 2017. A large and consistent phylogenomic dataset supports sponges as the sister group to all other animals. *Curr Biol* 27:958–967.

Simionato, E., P. Kerner, N. Dray, M. Le Gouar, V. Ledent, D. Arendt, and M. Vervoort. 2008. atonal- and achaete-scute-related genes in the annelid *Platynereis dumerilii*: insights into the evolution of neural basic-Helix-Loop-Helix genes. *BMC Evol Biol* 8:170.

Simionato, E., V. Ledent, G. Richards, M. Thomas-Chollier, P. Kerner, D. Coornaert, B. M. Degnan, and M. Vervoort. 2007. Origin and diversification of the basic helix-loop-helix gene family in metazoans: insights from comparative genomics. *BMC Evol Biol* 7:33.

Simmons, D. K., K. Pang, and M. Q. Martindale. 2012. Lim homeobox genes in the ctenophore *Mnemiopsis leidyi*: the evolution of neural cell type specification. *Evodevo* 3:2.

Singh, S., and A. K. Groves. 2016. The molecular basis of craniofacial placode development. *Wiley Interdiscip Rev Dev Biol* 5:363–376.

Singla, C. L. 1974. Ocelli of hydromedusae. *Cell Tissue Res* 149:413–429.

Singla, C. L. 1975. Statocysts of hydromedusae. *Cell Tissue Res* 158:391–407.

Sinigaglia, C., H. Busengdal, L. Leclère, U. Technau, and F. Rentzsch. 2013. The bilaterian head patterning gene six3/6 controls aboral domain development in a cnidarian. *PLoS Biol* 11:e1001488.

Sinigaglia, C., H. Busengdal, A. Lerner, P. Oliveri, and F. Rentzsch. 2015. Molecular characterization of the apical organ of the anthozoan *Nematostella vectensis*. *Dev Biol* 398:120–133.

Slota, L. A., E. M. Miranda, and D. R. McClay. 2019. Spatial and temporal patterns of gene expression during neurogenesis in the sea urchin *Lytechinus variegatus*. *Evodevo* 10:2.

Smidt, M. P., and J. P. Burbach. 2009. A passport to neurotransmitter identity. *Genome Biol* 10:229.

Smith, C. L., F. Varoqueaux, M. Kittelmann, R. N. Azzam, B. Cooper, C. A. Winters, M. Eitel, D. Fasshauer, and T. S. Reese. 2014. Novel cell types, neurosecretory cells, and body plan of the early-diverging metazoan *Trichoplax adhaerens*. *Curr Biol* 24:1565–1572.

Smith, J. A., P. McGarr, and J. S. Gilleard. 2005. The *Caenorhabditis elegans* GATA factor elt-1 is essential for differentiation and maintenance of hypodermal seam cells and for normal locomotion. *J Cell Sci* 118:5709–5719.

Smith, M. D., S. J. Dawson, and D. S. Latchman. 1997. Inhibition of neuronal process outgrowth and neuronal specific gene activation by the Brn-3b transcription factor. *J Biol Chem* 272:1382–1388.

Solek, C. M., P. Oliveri, M. Loza-Coll, C. S. Schrankel, E. C. Ho, G. Wang, and J. P. Rast. 2013. An ancient role for Gata-1/2/3 and Scl transcription factor homologs in the development of immunocytes. *Dev Biol* 382:280–292.

Sorrentino, M., L. Manni, N. J. Lane, and P. Burighel. 2000. Evolution of cerebral vesicles and their sensory organs in an ascidian larva. *Acta Zool* 81:243–258.

Soukup, V. 2017. Left-right asymmetry specification in amphioxus: review and prospects. *Int J Dev Biol* 61:611–620.

Soukup, V., and Z. Kozmik. 2016. Zoology: a new mouth for amphioxus. *Curr Biol* 26:R367–R368.

Sower, S. A., M. Freamat, and S. I. Kavanaugh. 2009. The origins of the vertebrate hypothalamic-pituitary-gonadal (HPG) and hypothalamic-pituitary-thyroid (HPT) endocrine systems: new insights from lampreys. *Gen Comp Endocrinol* 161:20–29.

Spehr, M., and S. D. Munger. 2009. Olfactory receptors: G protein-coupled receptors and beyond. *J Neurochem* 109:1570–1583.

Spencer, W. C., and E. S. Deneris. 2017. Regulatory mechanisms controlling maturation of serotonin neuron identity and function. *Front Cell Neurosci* 11:215.

Spitz, F., and E. E. Furlong. 2012. Transcription factors: from enhancer binding to developmental control. *Nat Rev Genet* 13:613–626.

Squarzoni, P., F. Parveen, L. Zanetti, F. Ristoratore, and A. Spagnuolo. 2011. FGF/MAPK/ Ets signaling renders pigment cell precursors competent to respond to Wnt signal by directly controlling Ci-Tcf transcription. *Development* 138:1421–1432.

Srivastava, M., E. Begovic, J. Chapman, N. H. Putnam, U. Hellsten, T. Kawashima, A. Kuo, T. Mitros, A. Salamov, M. L. Carpenter, A. Y. Signorovitch, M. A. Moreno, K. Kamm, J. Grimwood, J. Schmutz, H. Shapiro, I. V. Grigoriev, L. W. Buss, B. Schierwater, S. L. Dellaporta, and D. S. Rokhsar. 2008. The *Trichoplax* genome and the nature of placozoans. *Nature* 454:955–960.

Srivastava, M., C. Larroux, D. R. Lu, K. Mohanty, J. Chapman, B. M. Degnan, and D. S. Rokhsar. 2010. Early evolution of the LIM homeobox gene family. *BMC Biol* 8:4.

Srivastava, M., O. Simakov, J. Chapman, B. Fahey, M. E. Gauthier, T. Mitros, G. S. Richards, C. Conaco, M. Dacre, U. Hellsten, C. Larroux, N. H. Putnam, M. Stanke, M. Adamska, A. Darling, S. M. Degnan, T. H. Oakley, D. C. Plachetzki, Y. Zhai, M. Adamski, A. Calcino, S. F. Cummins, D. M. Goodstein, C. Harris, D. J. Jackson, S. P. Leys, S. Shu, B. J. Woodcroft, M. Vervoort, K. S. Kosik, G. Manning, B. M. Degnan, and D. S. Rokhsar. 2010. The *Amphimedon queenslandica* genome and the evolution of animal complexity. *Nature* 466:720–726.

Stach, T. 2002. On the homology of the protocoel in Cephalochordata and 'lower' Deuterostomia. *Acta Zool* 83:25–31.

Stach, T. 2005. Comparison of the serotonergic nervous system among Tunicata: implications for its evolution within Chordata. *Organ Divers Evol* 5:15–24.

Stach, T., and S. Kaul. 2011. The postanal tail of the enteropneust *Saccoglossus kowalevskii* is a ciliary creeping organ without distinct similarities to the chordate tail. *Acta Zool* 92:150–160.

Stach, T., J. Winter, J. M. Bouquet, D. Chourrout, and R. Schnabel. 2008. Embryology of a planktonic tunicate reveals traces of sessility. *Proc Natl Acad Sci USA* 105:7229–7234.

Stainier, D. Y. 2002. A glimpse into the molecular entrails of endoderm formation. *Genes Dev* 16:893–907.

Stathopoulos, A., M. Van Drenth, A. Erives, M. Markstein, and M. Levine. 2002. Whole-genome analysis of dorsal-ventral patterning in the *Drosophila* embryo. *Cell* 111:687–701.

Stefano, G. B., B. Salzet-Raveillon, and M. Salzet. 1999. *Mytilus edulis* hemolymph contains pro-opiomelanocortin: LPS and morphine stimulate differential processing. *Brain Res Mol Brain Res* 63:340–350.

Steinmetz, P. R. H. 2019. A non-bilaterian perspective on the development and evolution of animal digestive systems. *Cell Tissue Res* 377:321–339.

Steinmetz, P. R. H., A. Aman, J. E. M. Kraus, and U. Technau. 2017. Gut-like ectodermal tissue in a sea anemone challenges germ layer homology. *Nat Ecol Evol* 1:1535–1542.

Steinmetz, P. R., R. P. Kostyuchenko, A. Fischer, and D. Arendt. 2011. The segmental pattern of otx, gbx, and Hox genes in the annelid *Platynereis dumerilii*. *Evol Dev* 13:72–79.

Steinmetz, P. R., R. Urbach, N. Posnien, J. Eriksson, R. P. Kostyuchenko, C. Brena, K. Guy, M. Akam, G. Bucher, and D. Arendt. 2010. Six3 demarcates the anterior-most developing brain region in bilaterian animals. *Evodevo* 1:14.

Stierwald, M., N. Yanze, R. P. Bamert, L. Kammermeier, and V. Schmid. 2004. The Sine oculis/Six class family of homeobox genes in jellyfish with and without eyes: development and eye regeneration. *Dev Biol* 274:70–81.

Stokes, M. D., and N. D. Holland. 1995. Embryos and larvae of a lancelet, *Branchiostoma floridae*, from hatching through metamorphosis: growth in the laboratory and external morphology. *Acta Zool* 76:105–120.

Stolfi, A., T. B. Gainous, J. J. Young, A. Mori, M. Levine, and L. Christiaen. 2010. Early chordate origins of the vertebrate second heart field. *Science* 329:565–568.

Stolfi, A., K. Ryan, I. A. Meinertzhagen, and L. Christiaen. 2015. Migratory neuronal progenitors arise from the neural plate borders in tunicates. *Nature* 527:371–374.

Strausfeld, N. J., and F. Hirth. 2013. Deep homology of arthropod central complex and vertebrate basal ganglia. *Science* 340:157–161.

Streit, A. 2018. Specification of sensory placode progenitors: signals and transcription factor networks. *Int J Dev Biol* 62:195–205.

Striedter, G. F., and R. G. Northcutt. 1991. Biological hierarchies and the concept of homology. *Brain Behav Evol* 38:177–189.

Sudo, S., Y. Kuwabara, J. I. Park, S. Y. Hsu, and A. J. Hsueh. 2005. Heterodimeric fly glycoprotein hormone-alpha2 (GPA2) and glycoprotein hormone-beta5 (GPB5) activate fly leucine-rich repeat-containing G protein-coupled receptor-1 (DLGR1) and stimulation of human thyrotropin receptors by chimeric fly GPA2 and human GPB5. *Endocrinology* 146:3596–3604.

Suga, H., V. Schmid, and W. J. Gehring. 2008. Evolution and functional diversity of jellyfish opsins. *Curr Biol* 18:51–55.

Suga, H., P. Tschopp, D. F. Graziussi, M. Stierwald, V. Schmid, and W. J. Gehring. 2010. Flexibly deployed Pax genes in eye development at the early evolution of animals demonstrated by studies on a hydrozoan jellyfish. *Proc Natl Acad Sci USA* 107:14263–14268.

Suh, E., J. Bohbot, and L. J. Zwiebel. 2014. Peripheral olfactory signaling in insects. *Curr Opin Insect Sci* 6:86–92.

Sundström, G., S. Dreborg, and D. Larhammar. 2010. Concomitant duplications of opioid peptide and receptor genes before the origin of jawed vertebrates. *PLoS One* 5:e10512.

Suzuki, D. G., Y. Murakami, H. Escriva, and H. Wada. 2015. A comparative examination of neural circuit and brain patterning between the lamprey and amphioxus reveals the evolutionary origin of the vertebrate visual center. *J Comp Neurol* 523:251–261.

Suzuki, T., and K. Saigo. 2000. Transcriptional regulation of atonal required for *Drosophila* larval eye development by concerted action of eyes absent, sine oculis and hedgehog signaling independent of fused kinase and cubitus interruptus. *Development* 127:1531–1540.

Swalla, B. J., C. B. Cameron, L. S. Corley, and J. R. Garey. 2000. Urochordates are monophyletic within the deuterostomes. *Syst Biol* 49:52–64.

Swalla, B. J., and A. B. Smith. 2008. Deciphering deuterostome phylogeny: molecular, morphological and palaeontological perspectives. *Philos Trans R Soc Lond B Biol Sci*. 363:1557–1568.

Tagawa, K. 2016. Hemichordate models. *Curr Opin Genet Dev* 39:71–78.

Tahayato, A., R. Sonneville, F. Pichaud, M. F. Wernet, D. Papatsenko, P. Beaufils, T. Cook, and C. Desplan. 2003. Otd/Crx, a dual regulator for the specification of ommatidia subtypes in the *Drosophila* retina. *Dev Cell* 5:391–402.

Takacs, C. M., G. Amore, P. Oliveri, A. J. Poustka, D. Wang, R. D. Burke, and K. J. Peterson. 2004. Expression of an NK2 homeodomain gene in the apical ectoderm defines a new territory in the early sea urchin embryo. *Dev Biol* 269:152–164.

Takamura, K. 1998. Nervous network in larvae of the ascidian *Ciona intestinalis*. *Dev Genes Evol* 208:1–8.

Takasu, N., and M. Yoshida. 1983. Photic effects on photosensory microvilli in the seastar *Asterias amurensis* (Echinodermata: Asteroida). *Zoomorphology* 103:135–148.

Takatoh, J., V. Prevosto, and F. Wang. 2018. Vibrissa sensory neurons: linking distinct morphology to specific physiology and function. *Neuroscience* 368:109–114.

Takatori, N., K. Hotta, Y. Mochizuki, G. Satoh, Y. Mitani, N. Satoh, Y. Satou, and H. Takahashi. 2004. T-box genes in the ascidian *Ciona intestinalis*: characterization of cDNAs and spatial expression. *Dev Dyn* 230:743–753.

Takezaki, N., F. Figueroa, Z. Zaleska-Rutczynska, and J. Klein. 2003. Molecular phylogeny of early vertebrates: monophyly of the agnathans as revealed by sequences of 35 genes. *Mol Biol Evol* 20:287–292.

Tamm, S. L. 2014. Formation of the statolith in the ctenophore *Mnemiopsis leidyi*. *Biol Bull* 227:7–18.

Tando, Y., and K. Kubokawa. 2009. A homolog of the vertebrate thyrostimulin glycoprotein hormone alpha subunit (GPA2) is expressed in Amphioxus neurons. *Zoolog Sci* 26:409–414.

Tang, P. C., and G. M. Watson. 2014. Cadherin-23 may be dynamic in hair bundles of the model sea anemone *Nematostella vectensis*. *PLoS One* 9:e86084.

Tardent, P. 1995. The cnidarian cnidocyte, a high-tech cellular weaponry. *Bioessays* 17:351–362.

Tautz, D. 2004. Segmentation. *Dev Cell* 7:301–312.

Telford, M. J., G. E. Budd, and H. Philippe. 2015. Phylogenomic insights into animal evolution. *Curr Biol*. 25:R876–R887.

Tello, J. A., and N. M. Sherwood. 2009. Amphioxus: beginning of vertebrate and end of invertebrate type GnRH receptor lineage. *Endocrinology* 150:2847–2856.

Terakado, K. 2001. Induction of gamete release by gonadotropin-releasing hormone in a protochordate, *Ciona intestinalis*. *Gen Comp Endocrinol* 124:277–284.

Terakado, K., M. Ogawa, K. Inoue, K. Yamamoto, and S. Kikuyama. 1997. Prolactin-like immunoreactivity in the granules of neural complex cells in the ascidian *Halocynthia roretzi*. *Cell Tissue Res* 289:63–71.

Terakubo, H. Q., Y. Nakajima, Y. Sasakura, T. Horie, A. Konno, H. Takahashi, K. Inaba, K. Hotta, and K. Oka. 2010. Network structure of projections extending from peripheral neurons in the tunic of ascidian larva. *Dev Dyn* 239:2278–2287.

Terazawa, K., and N. Satoh. 1997. Formation of the chordamesoderm in the amphioxus embryo: analysis with Brachyury and fork head/HNF-3 genes. *Dev Genes Evol* 207:1–11.

Terry, N. A., E. R. Walp, R. A. Lee, K. H. Kaestner, and C. L. May. 2014. Impaired enteroendocrine development in intestinal-specific Islet1 mouse mutants causes impaired glucose homeostasis. *Am J Physiol Gastrointest Liver Physiol* 307:G979–G991.

Tessmar-Raible, K. 2004. The evolution of sensory and neurosecretory cell types in bilaterian brains. *Department of Biology*, PhD thesis Philipps University Marburg.

Tessmar-Raible, K. 2007. The evolution of neurosecretory centers in bilaterian forebrains: insights from protostomes. *Semin Cell Dev Biol* 18:492–501.

Tessmar-Raible, K., F. Raible, F. Christodoulou, K. Guy, M. Rembold, H. Hausen, and D. Arendt. 2007. Conserved sensory-neurosecretory cell types in annelid and fish forebrain: insights into hypothalamus evolution. *Cell* 129:1389–1400.

Thomas, M. B., and N. C. Edwards. 1991. Cnidaria: hydrozoa. In *Microscopic anatomy of invertebrates*, edited by Westfall J. A. Harrison F. W., 91–183. New York: Wiley-Liss.

Thomas, P., and R. Beddington. 1996. Anterior primitive endoderm may be responsible for patterning the anterior neural plate in the mouse embryo. *Curr Biol* 6:1487–1496.

Thor, S., S. G. Andersson, A. Tomlinson, and J. B. Thomas. 1999. A LIM-homeodomain combinatorial code for motor-neuron pathway selection. *Nature* 397:76–80.

Thor, S., and J. B. Thomas. 1997. The *Drosophila* islet gene governs axon pathfinding and neurotransmitter identity. *Neuron* 18:397–409.

Thor, S., and J. B. Thomas. 2002. Motor neuron specification in worms, flies and mice: conserved and 'lost' mechanisms. *Curr Opin Genet Dev* 12:558–564.

Thorndyke, M. L., and D. Georges. 1988. Functional aspects of peptide neurohormones in protochordates. *Sem Ser Soc Exp Biol* 33:235–258.

Thurm, U., M. Brinkmann, R. Golz, M. Holtmann, D. Oliver, and T. Sieger. 2004. Mechanoreception and synaptic transmission of hydrozoan nematocytes. *Hydrobiologia* 530:97–105.

Tiozzo, S., L. Christiaen, C. Deyts, L. Manni, J. S. Joly, and P. Burighel. 2005. Embryonic versus blastogenetic development in the compound ascidian *Botryllus schlosseri*: insights from Pitx expression patterns. *Dev Dyn* 232:468–478.

Tjoa, L. T., and U. Welsch. 1974. Electron microscopical observations on Kölliker's and Hatschek's pit and on the wheel organ in the head region of Amphioxus (*Branchiostoma lanceolatum*). *Cell Tissue Res* 153:175–187.

Todi, S. V., J. D. Franke, D. P. Kiehart, and D. F. Eberl. 2005. Myosin VIIA defects, which underlie the Usher 1B syndrome in humans, lead to deafness in *Drosophila*. *Curr Biol* 15:862–868.

Tomarev, S. I., P. Callaerts, L. Kos, R. Zinovieva, G. Halder, W. Gehring, and J. Piatigorsky. 1997. Squid pax-6 and eye development. *Proc Nat Acad Sci USA* 94:2421–2426.

Tomer, R., A. S. Denes, K. Tessmar-Raible, and D. Arendt. 2010. Profiling by image registration reveals common origin of annelid mushroom bodies and vertebrate pallium. *Cell* 142:800–809.

Toresson, H., J. P. Martinez-Barbera, A. Bardsley, X. Caubit, and S. Krauss. 1998. Conservation of BF-1 expression in amphioxus and zebrafish suggests evolutionary ancestry of anterior cell types that contribute to the vertebrate forebrain. *Dev Genes Evol* 208:431–439.

Torrence, S. A., and R. A. Cloney. 1982. Nervous systems of ascidian larvae: caudal primary sensory neurons. *Zoomorphol* 99:103–115.

Torrence, S. A., and R. A. Cloney. 1983. Ascidian larval nervous system: primary sensory neurons in adhesive papillae. *Zoomorphol* 102:111–123.

Tosches, M. A., and D. Arendt. 2013. The bilaterian forebrain: an evolutionary chimaera. *Curr Opin Neurobiol* 23:1080–1089.

Tosches, M. A., D. Bucher, P. Vopalensky, and D. Arendt. 2014. Melatonin signaling controls circadian swimming behavior in marine zooplankton. *Cell* 159:46–57.

Tournière, O., D. Dolan, G. S. Richards, K. Sunagar, Y. Y. Columbus-Shenkar, Y. Moran, and F. Rentzsch. 2020. NvPOU4/Brain3 functions as a terminal selector gene in the nervous system of the cnidarian *Nematostella vectensis*. *Cell Rep* 30:4473–4489 e5.

Travaglini, K. J., A. N. Nabhan, L. Penland, R. Sinha, A. Gillich, R. V. Sit, S. Chang, S. D. Conle, Y. Mori, J. Seita, G. J. Berry, J. B. Shrager, R. J. Metzger, C. S. Kuo, N. Neff, I. L. Weissman, S. R. Quake, and M. A. Krasnow. 2020. A molecular cell atlas of the human lung from single cell RNA sequencing. *Nature* 587:619–625.

Tresser, J., S. Chiba, M. Veeman, D. El-Nachef, E. Newman-Smith, T. Horie, M. Tsuda, and W. C. Smith. 2010. doublesex/mab3 related-1 (dmrt1) is essential for development of anterior neural plate derivatives in *Ciona*. *Development* 137:2197–2203.

True, J. R., and E. S. Haag. 2001. Developmental system drift and flexibility in evolutionary trajectories. *Evol Dev* 3:109–119.

Tsagkogeorga, G., X. Turon, N. Galtier, E. J. Douzery, and F. Delsuc. 2010. Accelerated evolutionary rate of housekeeping genes in tunicates. *J Mol Evol* 71:153–167.

Tsuda, M., D. Sakurai, and M. Goda. 2003. Direct evidence for the role of pigment cells in the brain of ascidian larvae by laser ablation. *J Exp Biol* 206:1409–1417.

Tsutsui, H., N. Yamamoto, H. Ito, and Y. Oka. 1998. GnRH-immunoreactive neuronal system in the presumptive ancestral chordate, *Ciona intestinalis* (Ascidian). *Gen Comp Endocrinol* 112:426–432.

Tu, Q., C. T. Brown, E. H. Davidson, and P. Oliveri. 2006. Sea urchin Forkhead gene family: phylogeny and embryonic expression. *Dev Biol* 300:49–62.

Ullrich-Lüter, E. M., S. D'Aniello, and M. I. Arnone. 2013. C-opsin expressing photoreceptors in echinoderms. *Integr Comp Biol* 53:27–38.

Ullrich-Lüter, E. M., S. Dupont, E. Arboleda, H. Hausen, and M. I. Arnone. 2011. Unique system of photoreceptors in sea urchin tube feet. *Proc Natl Acad Sci USA* 108:8367–8372.

Umesono, Y., K. Watanabe, and K. Agata. 1999. Distinct structural domains in the planarian brain defined by the expression of evolutionarily conserved homeobox genes. *Dev Genes Evol* 209:31–39.

Urbach, R., and G. M. Technau. 2003. Molecular markers for identified neuroblasts in the developing brain of *Drosophila*. *Development* 130:3621–3637.

Urbach, R., and G. M. Technau. 2008. Dorsoventral patterning of the brain: a comparative approach. *Adv Exp Med Biol* 628:42–56.

Vadodaria, K. C., M. C. Marchetto, J. Mertens, and F. H. Gage. 2016. Generating human serotonergic neurons in vitro: methodological advances. *Bioessays* 38:1123–1129.

Valencia, J. E., R. Feuda, D. O. Mellott, R. D. Burke, and I. S. Peter. 2019. Ciliary photoreceptors in sea urchin larvae indicate pan-deuterostome cell type conservation. *BioRxiv*:doi: https://doi.org/10.1101/683318.

Valero-Gracia, A., L. Petrone, P. Oliveri, D. Nilsson, and M. I. Arnone. 2016. Non-directional photoreceptors in the pluteus of *Strongylocentrotus purpuratus*. *Front Ecol Evol* 4.

Van Buskirk, C., and P. W. Sternberg. 2010. Paired and LIM class homeodomain proteins coordinate differentiation of the *C. elegans* ALA neuron. *Development* 137:2065–2074.

Van de Peer, Y., S. Maere, and A. Meyer. 2009. The evolutionary significance of ancient genome duplications. *Nat Rev Genet* 10:725–732.

Van Keymeulen, A., G. Mascre, K. K. Youseff, I. Harel, C. Michaux, N. De Geest, C. Szpalski, Y. Achouri, W. Bloch, B. A. Hassan, and C. Blanpain. 2009. Epidermal progenitors give rise to Merkel cells during embryonic development and adult homeostasis. *J Cell Biol* 187:91–100.

Vandendries, E. R., D. Johnson, and R. Reinke. 1996. orthodenticle is required for photoreceptor cell development in the *Drosophila* eye. *Dev Biol* 173:243–255.

Vanfleteren, J. R., and A. Coomans. 1976. Photoreceptor evolution and phylogeny. *Z Zool Syst Evol Forsch* 14:157–169.

Veeman, M. T., E. Newman-Smith, D. El Nachef, and W. C. Smith. 2010. The ascidian mouth opening is derived from the anterior neuropore: reassessing the mouth/neural tube relationship in chordate evolution. *Dev Biol* 344:138–149.

Velarde, R. A., C. D. Sauer, K. K. Walden, S. E. Fahrbach, and H. M. Robertson. 2005. Pteropsin: a vertebrate-like non-visual opsin expressed in the honey bee brain. *Insect Biochem Mol Biol* 35:1367–1377.

Venkatachalam, K., J. Luo, and C. Montell. 2014. Evolutionarily conserved, multitasking TRP channels: lessons from worms and flies. *Handb Exp Pharmacol* 223:937–962.

Venkatachalam, K., and C. Montell. 2007. TRP channels. *Annu Rev Biochem* 76:387–417.

Venkatesh, T. V., N. D. Holland, L. Z. Holland, M. T. Su, and R. Bodmer. 1999. Sequence and developmental expression of amphioxus AmphiNk2-1: insights into the evolutionary origin of the vertebrate thyroid gland and forebrain. *Dev Genes Evol* 209:254–259.

Venuti, J. M., and W. R. Jeffery. 1989. Cell lineage and determination of cell fate in ascidian embryos. *Int J Dev Biol* 33:197–212.

Veraszto, C., M. Gühmann, H. Jia, V. B. V. Rajan, L. A. Bezares-Calderón, C. Pineiro-López, N. Randel, R. Shahidi, N. K. Michiels, S. Yokoyama, K. Tessmar-Raible, and G. Jékely. 2018. Ciliary and rhabdomeric photoreceptor-cell circuits form a spectral depth gauge in marine zooplankton. *Elife* 7:e36440.

Vergara, H. M., P. Y. Bertucci, P. Hantz, M. A. Tosches, K. Achim, P. Vopalensky, and D. Arendt. 2017. Whole-organism cellular gene-expression atlas reveals conserved cell types in the ventral nerve cord of *Platynereis dumerilii*. *Proc Natl Acad Sci USA* 114:5878–5885.

Viets, K., K. Eldred, and R. J. Johnston, Jr. 2016. Mechanisms of photoreceptor patterning in vertebrates and invertebrates. *Trends Genet* 32:638–659.

Vöcking, O., I. Kourtesis, and H. Hausen. 2015. Posterior eyespots in larval chitons have a molecular identity similar to anterior cerebral eyes in other bilaterians. *Evodevo* 6:40.

Vöcking, O., I. Kourtesis, S. C. Tumu, and H. Hausen. 2017. Co-expression of xenopsin and rhabdomeric opsin in photoreceptors bearing microvilli and cilia. *Elife* 6:e23435.

Vopalensky, P., and Z. Kozmik. 2009. Eye evolution: common use and independent recruitment of genetic components. *Philos Trans R Soc Lond B Biol Sci* 364:2819–2832.

Vopalensky, P., J. Pergner, M. Liegertova, E. Benito-Gutiérrez, D. Arendt, and Z. Kozmik. 2012. Molecular analysis of the amphioxus frontal eye unravels the evolutionary origin of the retina and pigment cells of the vertebrate eye. *Proc Natl Acad Sci USA* 109:15383–15388.

Vorbrüggen, G., R. Constien, O. Zilian, E. A. Wimmer, G. Dowe, H. Taubert, M. Noll, and H. Jäckle. 1997. Embryonic expression and characterization of a Ptx1 homolog in *Drosophila*. *Mech Dev* 68:139–147.

Voutev, R., R. Keating, E. J. Hubbard, and L. G. Vallier. 2009. Characterization of the *Caenorhabditis elegans* Islet LIM-homeodomain ortholog, lim-7. *FEBS Lett* 583:456–464.

Wada, H. 1998. Evolutionary history of free-swimming and sessile lifestyles in urochordates as deduced from 18S rDNA molecular phylogeny. *Mol Biol Evol* 15:1189–1194.

Wada, H., P. W. H. Holland, S. Sato, H. Yamamoto, and N. Satoh. 1997. Neural tube is partially dorsalized by overexpression of HrPax-37: the ascidian homologue of Pax-3 and Pax-7. *Dev Biol* 187:240–252.

Wada, H., P. W. H. Holland, and N. Satoh. 1996. Origin of patterning in neural tubes. *Nature* 384:123.

Wada, H., H. Saiga, N. Satoh, and P. W. H. Holland. 1998. Tripartite organization of the ancestral chordate brain and the antiquity of placodes: insights from ascidian Pax-2/5/8, Hox and Otx genes. *Development* 125:1113–1122.

Wada, H., and N. Satoh. 1994. Details of the evolutionary history from invertebrates to vertebrates, as deduced from the sequences of 18S rDNA. *Proc Natl Acad Sci USA* 91:1801–1804.

Wada, S., Y. Katsuyama, Y. Sato, C. Itoh, and H. Saiga. 1996. Hroth an orthodenticle-related homeobox gene of the ascidian, *Halocynthia roretzi*: its expression and putative roles in the axis formation during embryogenesis. *Mech Dev* 60:59–71.

Wada, S., and H. Saiga. 2002. HrzicN, a new Zic family gene of ascidians, plays essential roles in the neural tube and notochord development. *Development* 129:5597–5608.

Wada, S., N. Sudou, and H. Saiga. 2004. Roles of Hroth, the ascidian otx gene, in the differentiation of the brain (sensory vesicle) and anterior trunk epidermis in the larval development of *Halocynthia roretzi*. *Mech Dev* 121:463–474.

Wada, S., M. Tokuoka, E. Shoguchi, K. Kobayashi, A. Di Gregorio, A. Spagnuolo, M. Branno, Y. Kohara, D. Rokhsar, M. Levine, H. Saiga, N. Satoh, and Y. Satou. 2003. A genomewide survey of developmentally relevant genes in *Ciona intestinalis*. II. Genes for homeobox transcription factors. *Dev Genes Evol* 213:222–234.

Waddington, C. H. 1957. *The Strategy of the Genes.* London: George Allen and Unwin.

Wagner, E., and M. Levine. 2012. FGF signaling establishes the anterior border of the *Ciona* neural tube. *Development* 139:2351–2359.

Wagner, E., A. Stolfi, Y. Gi Choi, and M. Levine. 2014. Islet is a key determinant of ascidian palp morphogenesis. *Development* 141:3084–3092.

Wagner, G. P. 1989. The biological homology concept. *Annu Rev Ecol Syst* 20:51–69.

Wagner, G. P. 1995. The biological role of homologues: a building block hypothesis. *N Jb Geol Paläont Abh* 195 36–43.

Wagner, G. P. 2001. *The Character Concept in Evolutionary Biology,* edited by G. P. Wagner. San Diego: Academic Press.

Wagner, G. P. 2007. The developmental genetics of homology. *Nat Rev Genet* 8:473–479.

Wagner, G. P. 2014. *Homology, Genes, and Evolutionary Innovation.* Princeton: Princeton University Press.

Wagner, G. P., and V. J. Lynch. 2010. Evolutionary novelties. *Curr Biol* 20:R48–R52.

Wake, D. B., M. H. Wake, and C. D. Specht. 2011. Homoplasy: from detecting pattern to determining process and mechanism of evolution. *Science* 331:1032–1035.

Waki, K., K. S. Imai, and Y. Satou. 2015. Genetic pathways for differentiation of the peripheral nervous system in ascidians. *Nat Commun* 6:8719.

Walentek, P., T. Beyer, C. Hagenlocher, C. Müller, K. Feistel, A. Schweickert, R. M. Harland, and M. Blum. 2015. ATP4a is required for development and function of the *Xenopus* mucociliary epidermis - a potential model to study proton pump inhibitor-associated pneumonia. *Dev Biol* 408:292–304.

Wallis, D., M. Hamblen, Y. Zhou, K. J. Venken, A. Schumacher, H. L. Grimes, H. Y. Zoghbi, S. H. Orkin, and H. J. Bellen. 2003. The zinc finger transcription factor Gfi1, implicated in lymphomagenesis, is required for inner ear hair cell differentiation and survival. *Development* 130:221–232.

Wang, C., X. Guo, K. Dou, H. Chen, and R. Xi. 2015. Ttk69 acts as a master repressor of enteroendocrine cell specification in *Drosophila* intestinal stem cell lineages. *Development* 142:3321–3331.

Wang, F., Y. Yu, D. Ji, and H. Li. 2012. The DMRT gene family in amphioxus. *J Biomol Struct Dyn* 30:191–200.

Wang, P., S. Liu, Q. Yang, Z. Liu, and S. Zhang. 2018. Functional characterization of thyrostimulin in amphioxus suggests an ancestral origin of the TH signaling pathway. *Endocrinology* 159:3536–3548.

Wang, W. D., P. Lo, M. Frasch, and T. Lufkin. 2000. Hmx: an evolutionary conserved homeobox gene family expressed in the developing nervous system in mice and *Drosophila*. *Mech Dev* 99:123–137.

Wang, W., and T. Lufkin. 2000. The murine Otp homeobox gene plays an essential role in the specification of neuronal cell lineages in the developing hypothalamus. *Dev Biol* 227:432–449.

Wang, X., J. F. Greenberg, and H. M. Chamberlin. 2004. Evolution of regulatory elements producing a conserved gene expression pattern in *Caenorhabditis*. *Evol Dev* 6:237–245.

Wang, Y., P. J. Zhang, K. Yasui, and H. Saiga. 2002. Expression of Bblhx3, a LIM-homeobox gene, in the development of amphioxus *Branchiostoma belcheri tsingtauense*. *Mech Dev* 117:315–319.

Ward, A., J. Liu, Z. Feng, and X. Z. Xu. 2008. Light-sensitive neurons and channels mediate phototaxis in *C. elegans*. *Nat Neurosci* 11:916–922.

Ward, S., N. Thomson, J. G. White, and S. Brenner. 1975. Electron microscopical reconstruction of the anterior sensory anatomy of the nematode *Caenorhabditis elegans*. *J Comp Neurol* 160:313–337.

Watanabe, H., T. Fujisawa, and T. W. Holstein. 2009. Cnidarians and the evolutionary origin of the nervous system. *Dev Growth Differ* 51:167–183.

Watanabe, H., A. Kuhn, M. Fushiki, K. Agata, S. Ozbek, T. Fujisawa, and T. W. Holstein. 2014. Sequential actions of beta-catenin and Bmp pattern the oral nerve net in *Nematostella vectensis*. *Nat Commun* 5:5536.

Watanabe, T., and M. Yoshida. 1986. Morphological and histochemical studies on Joséph cells of amphioxus, *Branchiostoma belcheri* Gray. *Exp Biol* 46:67–73.

Watson, G. M., and D. A. Hessinger. 1992. Receptors for N-acetylated sugars may stimulate adenylate-cyclase to sensitize and tune mechanoreceptors involved in triggering nematocyst discharge. *Exp Cell Res* 198:8–16.

Watson, G. M., and R. R. Hudson. 1994. Frequency and amplitude tuning of nematocyst discharge by proline. *J Exp Zool* 268:177–185.

Wawersik, S., and R. L. Maas. 2000. Vertebrate eye development as modeled in *Drosophila*. *Hum Mol Genet* 9:917–925.

Weinreich, D. M., R. A. Watson, and L. Chao. 2005. Perspective: sign epistasis and genetic constraint on evolutionary trajectories. *Evolution* 59:1165–1174.

Welsch, L. T., and U. Welsch. 1978. Histologische und elektronenmikroskopische Untersuchungen an der präoralen Wimpergrube von Saccoglossus horsti (Hemichordata) und der Hatschekschen Grube von *Branchiostoma lanceolatum* (Acrania). Ein Beitrag zur phylogenetischen Entwicklung der Adenohypophyse. *Zool Jb Anat* 100:564–578.

Welsch, U. 1968. Die Feinstruktur der Joséphschen Zellen im Gehirn von Amphioxus. *Z Zellforsch Mikrosk Anat* 86:252–261.

Westfall, J. A., C. F. Elliott, and R. W. Carlin. 2002. Ultrastructural evidence for two-cell and three-cell neural pathways in the tentacle epidermis of the sea anemone *Aiptasia pallida*. *J Morphol* 251:83–92.

Westfall, J. A., and J. C. Kinnamon. 1978. A second sensory-motor-interneuron with neurosecretory granules in *Hydra*. *J Neurocytol* 7:365–379.

Westfall, J. A., K. L. Sayyar, and C. F. Elliott. 1998. Cellular origins of kinocilia, stereocilia, and microvilli on tentacles of sea anemones of the genus *Calliactis* (Cnidaria: Anthozoa). *Invertebr Biol* 117:186–193.

Whelan, N. V., K. M. Kocot, T. P. Moroz, K. Mukherjee, P. Williams, G. Paulay, L. L. Moroz, and K. M. Halanych. 2017. Ctenophore relationships and their placement as the sister group to all other animals. *Nat Ecol Evol* 1:1737–1746.

Whitear, M. 1989. Merkel cells in lower vertebrates. *Arch Histol Cytol* 52 Suppl:415–422.

Whitfield, J. F. 2008. The solitary (primary) cilium – a mechanosensory toggle switch in bone and cartilage cells. *Cell Signal* 20:1019–1024.

Whitfield, P. J., and R. H. Emson. 1983. Presumptive ciliated receptors associated with the fibrillar glands of the spines of the echinoderm *Amphipholis squamata*. *Cell Tissue Res* 232:609–624.

Whittle, A. C. 1976. Reticular specializations in photoreceptors: a review. *Zool Scr* 5:191–206.

Wichmann, C., and T. Moser. 2015. Relating structure and function of inner hair cell ribbon synapses. *Cell Tissue Res* 361:95–114.

Wicht, H., and T. C. Lacalli. 2005. The nervous system of amphioxus: structure, development, and evolutionary significance. *Can J Zool* 83:122–150.

Wilder. 1909. *History of the human body*. New York: Holt and Co.

Wilkens, V., and G. Purschke. 2009. Pigmented eyes, photoreceptor-like sense organs and central nervous system in the polychaete *Scoloplos armiger* (Orbiniidae, Annelida) and their phylogenetic importance. *J Morphol* 270:1296–310.

Williams, E. A., and G. Jékely. 2019. Neuronal cell types in the annelid *Platynereis dumerilii*. *Curr Opin Neurobiol* 56:106–116.

Williams, E. A., C. Veraszto, S. Jasek, M. Conzelmann, R. Shahidi, P. Bauknecht, O. Mirabeau, and G. Jékely. 2017. Synaptic and peptidergic connectome of a neurosecretory center in the annelid brain. *Elife* 6:e26349.

Williams, N. A., and P. W. H. Holland. 1996. Old head on young shoulders. *Nature* 383:490.

Wirmer, A., S. Bradler, and R. Heinrich. 2012. Homology of insect corpora allata and vertebrate adenohypophysis? *Arthropod Struct Dev* 41:409–417.

Wollesen, T., C. McDougall, and D. Arendt. 2019. Remnants of ancestral larval eyes in an eyeless mollusk? Molecular characterization of photoreceptors in the scaphopod *Antalis entalis*. *Evodevo* 10:25.

Wollesen, T., C. McDougall, B. M. Degnan, and A. Wanninger. 2014. POU genes are expressed during the formation of individual ganglia of the cephalopod central nervous system. *Evodevo* 5:41.

Wollesen, T., S. V. Rodriguez Monje, C. Todt, B. M. Degnan, and A. Wanninger. 2015. Ancestral role of Pax2/5/8 in molluscan brain and multimodal sensory system development. *BMC Evol Biol* 15:231.

Wolpert, L., C. Tickle, and A. Martinez-Arias. 2019. *Principles of development*. 6th ed. Oxford: Oxford University Press.

Wong, E., J. Molter, V. Anggono, S. M. Degnan, and B. M. Degnan. 2019. Co-expression of synaptic genes in the sponge *Amphimedon queenslandica* uncovers ancient neural submodules. *Sci Rep* 9:15781.

Wong, K. S., and C. Arenas-Mena. 2016. Expression of GATA and POU transcription factors during the development of the planktotrophic trochophore of the polychaete serpulid *Hydroides elegans*. *Evol Dev* 18:254–266.

Woo, S. H., E. A. Lumpkin, and A. Patapoutian. 2015. Merkel cells and neurons keep in touch. *Trends Cell Biol* 25:74–81.

Wood, N. J., T. Mattiello, M. L. Rowe, L. Ward, M. Perillo, M. I. Arnone, M. R. Elphick, and P. Oliveri. 2018. Neuropeptidergic systems in pluteus larvae of the sea urchin *Strongylocentrotus purpuratus*: neurochemical complexity in a "simple" nervous system. *Front Endocrinol* 9:628.

Wright, K. A. 1980. Nematode sense organs. In *Nematodes as biological models*, edited by B. M. Zuckerman, 237–295. New York: Academic Press.

Wu, Z., N. Grillet, B. Zhao, C. Cunningham, S. Harkins-Perry, B. Coste, S. Ranade, N. Zebarjadi, M. Beurg, R. Fettiplace, A. Patapoutian, and U. Mueller. 2017. Mechanosensory hair cells express two molecularly distinct mechanotransduction channels. *Nat Neurosci* 20:24–33.

Wyeth, R. C., and R. P. Croll. 2011. Peripheral sensory cells in the cephalic sensory organs of *Lymnaea stagnalis*. *J Comp Neurol* 519:1894–1913.

Xiang, Y., Q. Yuan, N. Vogt, L. L. Looger, L. Y. Jan, and Y. N. Jan. 2010. Light-avoidance-mediating photoreceptors tile the *Drosophila* larval body wall. *Nature* 468:921–926.

Xie, B., M. Charlton-Perkins, E. McDonald, B. Gebelein, and T. Cook. 2007. Senseless functions as a molecular switch for color photoreceptor differentiation in *Drosophila*. *Development* 134:4243–4253.

Xu, P. X. 2013. The EYA-SO/SIX complex in development and disease. *Pediatr Nephrol* 28:843–854.

Xu, P. X., W. Zheng, C. Laclef, P. Maire, R. L. Maas, H. Peters, and X. Xu. 2002. Eya1 is required for the morphogenesis of mammalian thymus, parathyroid and thyroid. *Development* 129:3033–3044.

Xu, Z., H. Jiang, P. Zhong, Z. Yan, S. Chen, and J. Feng. 2016. Direct conversion of human fibroblasts to induced serotonergic neurons. *Mol Psychiatry* 21:62–70.

Xue, D., Y. Tu, and M. Chalfie. 1993. Cooperative interactions between the *Caenorhabditis elegans* homeoproteins UNC-86 and MEC-3. *Science* 261:1324–1328.

Yamasu, T., and M. Yoshida. 1976. Fine-structure of complex ocelli of a cubomedusan, *Tamoya bursaria* Haeckel. *Cell Tissue Res* 170:325–339.

Yan, Z., W. Zhang, Y. He, D. Gorczyca, Y. Xiang, L. E. Cheng, S. Meltzer, L. Y. Jan, and Y. N. Jan. 2013. *Drosophila* NOMPC is a mechanotransduction channel subunit for gentle-touch sensation. *Nature* 493:221–225.

Yang, Q., N. A. Bermingham, M. J. Finegold, and H. Y. Zoghbi. 2001. Requirement of Math1 for secretory cell lineage commitment in the mouse intestine. *Science* 294:2155–2158.

Yankura, K. A., C. S. Koechlein, A. F. Cryan, A. Cheatle, and V. F. Hinman. 2013. Gene regulatory network for neurogenesis in a sea star embryo connects broad neural speci-fication and localized patterning. *Proc Natl Acad Sci USA* 110:8591–8596.

Yankura, K. A., M. L. Martik, C. K. Jennings, and V. F. Hinman. 2010. Uncoupling of com-plex regulatory patterning during evolution of larval development in echinoderms. *BMC Biol* 8:143.

Yasui, K., and T. Kaji. 2008. The lancelet and ammocoete mouths. *Zoolog Sci* 25:1012–1019.

Yasui, K., S. C. Zhang, M. Uemura, S. Aizawa, and T. Ueki. 1998. Expression of a twist-related gene, Bbtwist, during the development of a lancelet species and its relation to cephalochordate anterior structures. *Dev Biol* 195:49–59.

Yasui, K., S. Zhang, M. Uemura, and H. Saiga. 2000. Left-right asymmetric expression of BbPtx, a Ptx-related gene, in a lancelet species and the developmental left-sidedness in deuterostomes. *Development* 127:187–195.

Yasuoka, Y., M. Kobayashi, D. Kurokawa, K. Akasaka, H. Saiga, and M. Taira. 2009. Evolutionary origins of blastoporal expression and organizer activity of the vertebrate gastrula organizer gene lhx1 and its ancient metazoan paralog lhx3. *Development* 136:2005–2014.

Ye, D. Z., and K. H. Kaestner. 2009. Foxa1 and Foxa2 control the differentiation of goblet and enteroendocrine L- and D-cells in mice. *Gastroenterology* 137:2052–2062.

Yokoyama, T. D., K. Hotta, and K. Oka. 2014. Comprehensive morphological analysis of indi-vidual peripheral neuron dendritic arbors in ascidian larvae using the photoconvertible protein Kaede. *Dev Dyn* 243:1362–1373.

York, J. R., and D. W. McCauley. 2020. The origin and evolution of vertebrate neural crest cells. *Open Biol* 10:190285.

Yoshida, K., and H. Saiga. 2008. Left-right asymmetric expression of Pitx is regulated by the asymmetric Nodal signaling through an intronic enhancer in *Ciona intestinalis*. *Dev Genes Evol* 218:353-360.

Yoshida, K., M. Ueno, T. Niwano, and H. Saiga. 2012. Transcription regulatory mechanism of Pitx in the papilla-forming region in the ascidian, *Halocynthia roretzi*, implies con-served involvement of Otx as the upstream gene in the adhesive organ development of chordates. *Dev Growth Differ* 54:649–659.

Yoshida, M., N. Takasu, and S. Tamotsu. 1984. Photoreception in echinoderms. In *Photoreception and vision in invertebrates. NATO ASI Series (Series A: Life Sciences), vol 74*, edited by M. A. Ali, 743–771. Boston: Springer.

Yoshida, R., T. Kusakabe, M. Kamatani, M. Daitoh, and M. Tsuda. 2002. Central nervous system-specific expression of G protein alpha subunits in the ascidian *Ciona intestina-lis*. *Zoolog Sci* 19:1079–1088.

Yu, J. K., L. Z. Holland, M. Jamrich, I. L. Blitz, and N. D. Hollan. 2002. AmphiFoxE4, an amphioxus winged helix/forkhead gene encoding a protein closely related to vertebrate thyroid transcription factor-2: expression during pharyngeal development. *Evol Dev* 4:9–15.

Yu, J. K., N. D. Holland, and L. Z. Holland. 2002. An amphioxus winged helix/forkhead gene, AmphiFoxD: insights into vertebrate neural crest evolution. *Dev Dyn* 225:289–297.

Yu, J. K., D. Meulemans, S. J. McKeown, and M. Bronner-Fraser. 2008. Insights from the amphioxus genome on the origin of vertebrate neural crest. *Genome Res* 18:1127–1132.

Yu, J. K., Y. Satou, N. D. Holland, I. Shin, Y. Kohara, N. Satoh, M. Bronner-Fraser, and L. Z. Holland. 2007. Axial patterning in cephalochordates and the evolution of the organizer. *Nature* 445:613–617.

Yuan, T., J. R. York, and D. W. McCauley. 2020. Neural crest and placode roles in formation and patterning of cranial sensory ganglia in lamprey. *Genesis* 58:e23356.

Yue, F., Z. Zhou, L. Wang, M. Wang, and L. Song. 2014. A conserved zinc finger transcription factor GATA involving in the hemocyte production of scallop *Chlamys farreri*. *Fish Shellfish Immunol* 39:125–135.

Zaraisky, A. G., V. Ecochard, O. V. Kazanskaya, S. A. Lukyanov, I. V. Fesenko, and A. M. Duprat. 1995. The homeobox-containing gene XANF-1 may control development of the Spemann organizer. *Development* 121:3839–3847.

Zaret, K. S., and J. S. Carroll. 2011. Pioneer transcription factors: establishing competence for gene expression. *Genes Dev* 25:2227–2241.

Zembrzycki, A., G. Griesel, A. Stoykova, and A. Mansouri. 2007. Genetic interplay between the transcription factors Sp8 and Emx2 in the patterning of the forebrain. *Neural Dev* 2:8.

Zeng, F., J. Wunderer, W. Salvenmoser, M. W. Hess, P. Ladurner, and U. Rothbächer. 2019. Papillae revisited and the nature of the adhesive secreting collocytes. *Dev Biol* 448:183–198.

Zhang, F., A. Bhattacharya, J. C. Nelson, N. Abe, P. Gordon, C. Lloret-Fernàndez, M. Maicas, N. Flames, R. S. Mann, D. A. Colon-Ramos, and O. Hobert. 2014. The LIM and POU homeobox genes ttx-3 and unc-86 act as terminal selectors in distinct cholinergic and serotonergic neuron types. *Development* 141:422–435.

Zhang, T., S. Ranade, C. Q. Cai, C. Clouser, and F. Pignoni. 2006. Direct control of neurogenesis by selector factors in the fly eye: regulation of atonal by Ey and So. *Development* 133:4881–4889.

Zhang, Y., and B. Mao. 2009. Developmental expression of an amphioxus (*Branchiostoma belcheri*) gene encoding a GATA transcription factor. *Zool Record* 30:137–143.

Zhao, C., and S. W. Emmons. 1995. A transcription factor controlling development of peripheral sense organs in *C. elegans*. *Nature* 373:74–78.

Zhao, D., S. Chen, and X. Liu. 2019. Lateral neural borders as precursors of peripheral nervous systems: a comparative view across bilaterians. *Dev Growth Differ* 61:58–72.

Zheng, C., F. Q. Jin, B. L. Trippe, J. Wu, and M. Chalfie. 2018. Inhibition of cell fate repressors secures the differentiation of the touch receptor neurons of *Caenorhabditis elegans*. *Development* 145:dev168096.

Zhu, X., A. S. Gleiberman, and M. G. Rosenfeld. 2007. Molecular physiology of pituitary development: signaling and transcriptional networks. *Physiol Rev* 87:933–963.

Zieger, E., S. Candiani, G. Garbarino, J. C. Croce, and M. Schubert. 2018. Roles of retinoic acid signaling in shaping the neuronal architecture of the developing amphioxus nervous system. *Mol Neurobiol* 55:5210–5229.

Zieger, E., G. Garbarino, N. S. M. Robert, J. K. Yu, J. C. Croce, S. Candiani, and M. Schubert. 2018. Retinoic acid signaling and neurogenic niche regulation in the developing peripheral nervous system of the cephalochordate amphioxus. *Cell Mol Life Sci* 75:2407–2429.

Zou, D., C. Erickson, E. H. Kim, D. Jin, B. Fritzsch, and P. X. Xu. 2008. Eya1 gene dosage
    critically affects the development of sensory epithelia in the mammalian inner ear.
    Hum Mol Genet 17:3340–3356.

Zou, D., D. Silvius, J. Davenport, R. Grifone, P. Maire, and P. X. Xu. 2006. Patterning of
    the third pharyngeal pouch into thymus/parathyroid by Six and Eya1. *Dev Biol*
    293:499–512.

Zylstra, U. 1971. Distribution and ultrastructure of epidermal sensory cells in the freshwater
    snails *Lymnaea stagnalis* and *Biomphalaria pfeifferi*. *Netherlands J Zool* 22:283–298.

Index (Volume II)

Note: Locators in *italics* represent figures and **bold** indicate tables in the text.

Printed and bound by CPI Group (UK) Ltd, Croydon, CR0 4YY

12/10/2024

01773184-0001